MAKING SENSE OF NATURE

We listen to a cacophony of voices instructing us how to think and feel about nature, including our own bodies. The news media, wildlife documentaries, science magazines and environmental NGOs are among those clamouring for our attention. But are we empowered by all this knowledge or is our dependence on various epistemic communities allowing our thoughts, sentiments and activities to be unduly governed by others?

Making sense of nature shows that what we call 'nature' is made sense of for us in ways that make it central to social order, social change and social dissent. By utilising insights and extended examples from anthropology, cultural studies, human geography, philosophy, politics, sociology, science studies, this interdisciplinary text asks whether we can better make sense of nature for ourselves, and thus participate more meaningfully in momentous decisions about the future of life – human and non-human – on the planet. This book shows how 'nature' can be made sense of without presuming its naturalness. The challenge is not so much to rid ourselves of the idea of nature and its 'collateral concepts' (such as genes) but instead, we need to be more alert to how, why and with what effects ideas about 'nature' get fashioned and deployed in specific situations. Among other things, the book deals with science and scientists, the mass media and journalists, ecotourism, literature and cinema, environmentalists, advertising and big business.

This innovative text contains numerous case studies and examples from daily life to put theory and subject matter into context, as well as study tasks, a glossary and suggested further reading. The case studies cover a range of topics, from forestry in Canada and Guinea, to bestiality in Washington State, to how human genetics is reported in Western newspapers, to participatory science experiments in the United Kingdom. *Making sense of nature* seeks to enlighten readers from a wide range of fields across the social sciences, humanities and physical sciences.

Noel Castree is Professor of Geography at Manchester University, UK. He's the co-editor of *Social nature: theory, practice and politics* (2001) and author of *Nature: the adventures of a concept* (2004).

'Noel Castree has written another bestseller. This book effortlessly synthesises from across the social and natural sciences a huge body of knowledge and reflection on the idea of "nature". Castree engages with all the important debates and authors and makes them accessible for his reader, offering not closed answers but lively provocations to further thought. This is geographical scholarship at its very best.'

Mike Hulme, Professor of Climate and Culture, King's College London

'Challenging our sources of knowledge, from the media to activist and scientific writing, *Making sense of nature* gives us powerful tools to think with. I'd put this book in the hands of anyone who wants to jar their thinking, undermine their own assumptions, and make up their own mind about people and the natural world.'

Professor Paul Robbins, Director, Nelson Institute for Environmental Studies, University of Wisconsin-Madison, USA

MAKING SENSE OF NATURE

Representation, politics and democracy

Noel Castree

LONDON AND NEW YORK

First published 2014 by Routledge
2 Park Square, Milton Park, Abingdon, Oxon OX14 4RN

Simultaneously published in the USA and Canada
by Routledge
711 Third Avenue, New York, NY 10017

Routledge is an imprint of the Taylor & Francis Group, an informa business

© 2014 Noel Castree

The right of Noel Castree to be identified as author of this work has been asserted in accordance with sections 77 and 78 of the Copyright, Designs and Patents Act 1988.

All rights reserved. No part of this book may be reprinted or reproduced or utilised in any form or by any electronic, mechanical, or other means, now known or hereafter invented, including photocopying and recording, or in any information storage or retrieval system, without permission in writing from the publishers.

Trademark notice: Product or corporate names may be trademarks or registered trademarks, and are used only for identification and explanation without intent to infringe.

British Library Cataloguing in Publication Data
A catalogue record for this book is available from the British Library

Library of Congress Cataloging in Publication Data
Making sense of nature/Noel Castree.
Includes bibliographical references and index.
1. Philosophy of nature. 2. Epistemics. I. Title.
BD581.C298 2013
113–dc23
2012050839

ISBN: 978-0-415-54548-8 (hbk)
ISBN: 978-0-415-54550-1 (pbk)
ISBN: 978-0-203-50346-1 (ebk)

Typeset in Garamond and Futura
by Sunrise Setting Ltd, Paignton, UK

Printed and bound in Great Britain by
TJ International Ltd, Padstow, Cornwall

And as for seeing things as ... they really are ... what is a fine summer's day as ... it really is? Is the meteorological office to decide? Or the poet? Or the farmer?

Edward Thomas (1909: 13)

There can be no adequate attempt ... to explore what nature is that is not centrally concerned with what it has been said to be ...

Kate Soper (1995: 21)

When the most basic concepts – the concepts, as it is said, from which we begin – are suddenly seen to be not concepts but problems ... there is no sense in listening to their sonorous summons ...

Raymond Williams (1977: 11)

How does it happen that a relatively small number of [people] ... have control over the behaviour and destinies of the vast majority of people of the world? [Is] it . . . because the majority have not known how, or have refused, to accept responsibility? ... [This is a] failure ... that has delayed the development of democracy.

John Dewey (1973: 291)

CONTENTS

LIST OF ILLUSTRATIONS

FIGURES

MAPS

PLATES

TABLE

BOXES

ACRONYMS

AAAS	American Association for the Advancement of Science
ACLU	American Civil Liberties Union
AIDS	acquired immunodeficiency syndrome
ALD	adrenoleukodystrophy
BC	British Columbia
BEST	Berkeley Earth Surface Temperature project
BSE	bovine spongiform encephalopathy
CRU	Climate Research Unit
DNA	deoxyribonucleic acid
EA	Environment Agency
ECO	Earth Communications Office
ENGO	environmental non-governmental organisation
EPO	European Patent Office
FAO	Food and Agriculture Organisation
FIDO	Fraser Island Defenders Organization
FWA	French West Africa
GHG	greenhouse gas
GM	genetically modified
HGDP	Human Genome Diversity Project
HGP	Human Genome Project
HIV	human immunodeficiency virus
IAC	Inter-Academy Council
ICT	information and communications technologies
ID	intelligent design
IPCC	Intergovernmental Panel on Climate Change
MoA	Ministry of Agriculture
MTL	mean trophic level
NASA	National Aeronautics and Space Administration
NGO	non-governmental organisation
NIPCC	Non-Governmental International Panel on Climate Change
OMNH	Oslo Museum of Natural History
PCR	polymerase chain reaction
PETA	People for the Ethical Treatment of Animals
PKU	phenylketonuria

PXE	pseudoxanthoma elasticum
QPWS	Queensland Parks & Wildlife Service
RFRG	Ryedale Flood Research Group
RNA	ribonucleic acid
RRc	Roundup Ready canola
RSPB	Royal Society for the Protection of Birds
SCST	Select Committee on Science & Technology
SSK	sociology of scientific knowledge
STS	science and technology studies
TRIPS	Trade Related Aspects of Intellectual Property Rights
UEA	University of East Anglia
UNEP	United Nations Environment Programme
USPTO	United States Patent and Trademark Office
WCWC	Western Canadian Wilderness Committee
WIPO	World Intellectual Property Organization
WWF	World Wildlife Fund

PERMISSIONS

The author and publisher would very much like to thank the following individuals and organisations for granting permission to reproduce various images used in this book: Nature Publishing Group for allowing the reuse of Figure 1.2, taken from Pauly *et al.* (2002); Dr David Kirby for allowing the reuse of Figure 3.1 (Kirby, 2008b); Professor Roger Pielke Jr. for allowing the reuse of Figure 8.1 (Pielke, 2007); the Royal Society for the Protection of Birds for allowing the reproduction of Plate 2.1, originally an advert in *The Guardian*; Skoda Auto (Czech Republic) for allowing the reproduction of Plate 2.2, also originally an advert in *The Guardian*; Dr Irus Braverman of Buffalo University for providing Plate 2.3; Adrian Dorst from Tofino, BC, for permission to reproduce Plate 4.2; Nature Publishing Group for permission to reproduce the cover of *Nature* 409 (15 February 2001) for Plate 4.4; Innovation Norway – Tourism, London, for permission to reuse the image in Plate 4.5, originally an advert in *The Guardian*; Prometheus Books, Amherst, NY, for permission to reproduce the poster of *Zoo* as Plate 5.1; and Creators Syndicate, Hermosa Beach, CA, for reuse of the cartoon that is Plate 8.1.

ACKNOWLEDGEMENTS

This book was conceived in 2006 when preparing seminar presentations for visits to the universities of Bergen, Umeå and Wollongong. Inger Birkeland, Nora Räthzel and Lesley Head were its unwitting catalysts, but its origins lie way back in my days at the University of British Columbia (where I learnt much about nature from Bruce Braun and David Demeritt, among others). Though I should know better, the book took far, far longer to write than I anticipated when setting the initial deadlines. Researching and authoring each chapter became a major logistical challenge. Time has been an exceedingly scarce resource despite the fact that I have probably been working harder (though clearly not smarter!) than at any previous time in my academic career. Indeed, of late I've had cause to wonder if I'll ever get the chance to write another book before I retire (assuming I make it that far!). Thankfully, the atrocious British 'summer' of 2012 (which kept me chained to my PC) and study leave, kindly granted by my school in the autumn, together greatly speeded completion of the manuscript.

My thanks go to Andrew Mould at Routledge, and not for the first time. He contracted *Making sense of nature* and remained understanding as deadlines came and went. Andrew's assistant Faye Leerink prodded me along in ways that were ultimately a help not a hindrance. She also supplied the book's cover after my own fruitless search for an image. Together Andrew and Faye commissioned three reviews of the book draft in summer 2012. I had no reason to expect such detailed and constructively critical feedback from this anonymous trio of reviewers. Their observations and suggestions were instrumental in allowing me to figure out what I was really trying to achieve. I offer my sincere thanks to all three of them. At least one of them reviewed the prospectus for the book in 2008, and I should also record my thanks for formative advice received from him/her and two others at this time.

I'm also very grateful to the students of GEOG 30700, the undergraduate course (also open to master's degree students) in which this book's central ideas and basic structure were trialled in a very rudimentary form. Despite my best efforts, this course was less than perfect – though the two cohorts who took it were very generous in their feedback. The module gave me reason and opportunity to synthesise arguments in a way I'd never quite achieved before. It's testament to how teaching can powerfully inform how one uses research (rather than merely being a vehicle for 'disseminating' that research 'off the shelf').

My Manchester geography colleagues continue to make my working environment a congenial one, despite us experiencing some very challenging times since 2010. I pay particular tribute to Gavin Bridge (now in Durham), Neil Coe (now in Singapore), Jonathan Darling, Peter Dicken, Martin Dodge, James Evans, Martin Hess, Mark Jayne, Maria Kaika, Chris Perkins, James Rothwell, Fiona Smyth, Erik Swyngedouw, Kevin Ward and Jamie Woodward. I've also benefited from guiding the doctoral research of Jason Beery, Lisa Ficklin, Tomas Frederiksen, Miranda Morgan and, most recently, Laura Pottinger, Craig Thomas and Daniel Banoub. Simon Guy's professional support and personal friendship has been central to my enjoyment of life in a multidisciplinary School of Environment and Development.

Further afield, the university's Society and Environment Research Group (SERG) has also been a valuable resource – and I make special mention of Dan Brockington, Rosaleen Duffy (now in Kent) and John O'Neill here. I'm glad also to have crossed paths over the years with campus colleagues Pete Wade, Karen Sykes, Elaine Graham (now at Chester University) and Vladimir Jankovich. More recently, Michelle Bastian inspired new thoughts and passions before her move to Edinburgh. Nick Scarle has, once again, been the alpha and omega when it comes to graphics. I must also acknowledge John Moore for ensuring that the essential mind–body connection remains intact, so too Sam Hickey (who, among other things, got me to hospital when my body was unexpectedly ailing). Away from the university, Pat Devine has created a valued forum for unfashionable thinking, despite my frequent absence from the gatherings he's arranged. My thanks too to Angela Yates and Karen Greening for their sterling work on the manuscript during production.

Without my wonderful family, writing *Making sense of nature* wouldn't have been nearly as rewarding as it's turned out to be. Thankyou, thankyou to Marie-Noel, Thomas, little Felicity and my mum. Finally, I'd like to record my profound gratitude to Linda and Zig Buczak for 30 years of unstinting support of me and mine. Linda's passing in April 2013 came far too soon and leaves a hole that can never be filled. I dedicate this book to her memory.

PREFACE: NATURE IS HERE, THERE AND EVERYWHERE

Do you have a love of nature? Perhaps you're worried about the destruction of the Amazon rainforest or the over-fishing of our oceans. Perhaps you've swum on the Great Barrier Reef or hiked in the Himalayas. Conversely, maybe you think too much nature is a bad thing. After all, who wants hurricanes, earthquakes, tsunamis or the Ebola virus? Maybe you appreciate the benefits afforded by controlling nature, human and non-human. In this case, you might be excited by the potentialities contained in new nature-altering technologies like gene-splicing, which permits the species barrier to be breached. Whatever your attitude, if you're interested in nature – what it is, what it does and what we do to it – then this book should interest you.

However, my approach to the topic will probably differ from other things you've read on the subject. This book explores how what is called 'nature' is made sense of for you by a myriad of others who daily seek to shape your thoughts, feelings and actions. I use the scare quotes because I'm not alone in insisting that nature isn't natural, even though continued use of the word – and its collateral terms like 'race' – suggests the contrary. In the chapters to come, I'm less interested in what nature *is* and more in *what it's considered to be*, as well as what the effects of this are. My eye will be trained on all those who speak for nature, or who make significant reference to it in what they do. This is why I say nature is made sense of *for* you, not *by* you. I believe that the vast majority of people's beliefs and sentiments about everything from climate change to their own genes are derivative. They result from an 'epistemic dependence' that makes us all potentially subject to the claims and aims of others. This is a situation in which a relative minority of the global workforce is employed to create information, knowledge, arguments, symbols, etc. to which the majority of people are exposed – and usually reliant upon.

I belong to this workforce. If you're reading this book as a university student or member of the public then you're relying on me to enlighten you – even though you may lack all the tools necessary to critically evaluate the claims that I make. This puts me at a distinct advantage. I'm one of countless 'epistemic workers' who together comprise a large and diverse set of 'epistemic communities'. These communities utilise largely one-way communication channels to shape societal ideas, tastes and practices.

They inhabit institutions like universities, think tanks, non-governmental organisations, private research laboratories, newspapers and advertising agencies. Their members often have impressive skills, expertise or credentials (like my PhD and professorial title). They devote enormous amounts of time and energy to representing different aspects of reality to the rest of society, utilising a range of genres and media, from 'serious' books like this one to political cartoons and television wildlife documentaries. They clamour for people's attention, together enriching their understanding of life – or perhaps manipulating, confusing and overwhelming them (by design or unwittingly). Because of the latter, people at times feel the need to ignore much of what is thrown their way.

References to nature, I will argue, are a significant preoccupation of a surprisingly large and diverse set of epistemic communities. It's not only, or mainly, the likes of geneticists, atmospheric scientists or environmental activists who incite us to think or care about natural phenomena. Many others insinuate references to things like elephants, water resources and stem cells into their discourses and images. They do it even when nature is not their primary concern. For instance, for years beer adverts in North America have featured snow-capped mountains and blue skies to make consumers think of pure ingredients and ice-cold refreshment. Though many of these references to nature are of little consequence in the short term, others have important medium- and long-term effects – or so I will claim. They're a key part of the ongoing process of governing the thoughts, emotions and behaviour of hundreds of millions of people worldwide. While many of these references occur outside the domains of 'government' and 'politics' proper, my argument will be that they are political nonetheless, often profoundly so. This means that – even when not designed to – these references contribute to the quality of collective deliberation and decision-making, known in much of the world as 'democracy'. And any democracy is only as good as the 'semiosphere' its citizens inhabit. This is the world of information, knowledge, debate, signs and symbols that we daily navigate through. It's a world most of us play no role whatsoever in creating. Like the weather, it appears as something we have to live with, for better or worse.

I'll explain why I think that what we call 'nature' is not natural in the first chapter. For now, let me quickly illustrate just how pervasive references to nature are in our everyday lives, and why some of these references might be highly important for all of us (at least some of the time). Consider the following.

HYBRID ORGANISMS

In November 2009, the British television station Channel 4 broadcast a science documentary entitled *Is it better to be mixed race?* It was part of a series of six programmes whose theme was *Race: science's last taboo* (see http://raceandscience.channel4.com/). The documentary examined the suggestion that 'mixed race' individuals enjoy certain biological advantages over those who descend from genetically similar populations. For instance, it assessed evidence that they might be perceived by others to be better looking facially. It reproduced the common belief that there are significant genetic differences among *homo sapiens*, even as it partly challenged that belief. A year later, and several newspapers ran stories about *cross*-species (rather than *intra*-species) hybridisation. Their focus was on an American biotechnology company called AquaBounty that had sought government permission to sell transgenic salmon (grown in fish farms) to humans. If granted a licence, the firm would have been the world's first to sell a genetically modified organism directly to consumers. Some environmentalists and organic food campaigners argued that these artificial, fast-growing salmon posed an environmental threat if they escaped and reproduced. For others, there was also the 'yuk factor' of encouraging people to eat a biologically engineered hybrid reared on manufactured feed in small marine enclosures.

ECOLOGICAL COMPENSATION

A few months before the AquaBounty story was reported, several news television programmes in the United Kingdom reported on the new Futurescapes programme announced by the Royal Society for the Protection of Birds (RSPB) (see http://www.rspb.org.uk/futurescapes/). This long-term programme is devoted to protecting wildlife outside the nature reserves already managed by the RSPB. For example, a new port able to (un)load large container ships is currently being constructed on the lower River Thames, 50 kilometres east of central London. The port's developers have been legally obliged to purchase a large area of land upstream, which will be flooded to create compensatory habitat for that lost during the construction process downstream. Thousands of water birds will benefit, most of them migratory species. Meanwhile, thousands of water voles, grass snakes and crested newts are being relocated to avoid the planned inundation. The RSPB will manage this artificially 're-wilded' estuarine site.

NATURAL ADVERTS

In 2011, the fruit juice firm Innocent ran an ad-campaign in magazines, on billboards and using flyers. Its tag line was 'Nature, bottled'. The two words were written in large, dark green letters against a blue-sky background with a see-through plastic green-capped bottle of squeezed orange juice next to them. Clearly, the intended message was that Innocent juice is pure and unprocessed – in short, good to drink in both taste and nutritional terms. Meanwhile, in the same year the fashion companies Louis Vuitton and Edun ran an ad-campaign in which nature was foregrounded, but in a very different way. The one- (sometimes two-) page adverts were placed in magazines mostly read by affluent consumers (e.g. *The Economist*). They comprised 'spontaneous' (but clearly staged) photographs of world-famous stars known to have a personal interest in addressing problems in the so-called developing world. For instance, in one the U2 lead singer Bono and his fashion designer wife Ali Hewson are seen stepping off a small plane in a savannah landscape in eastern Africa. The landscape is empty, and hills can be seen on the far horizon. Wading through yellow grasses reaching up to their knees, Bono and Ali carry expensive designer bags on their shoulders (to accompany their expensive hair cuts and fashionable clothes). Beneath the photo is the tag line 'Every journey begins in Africa'.

This is clearly a reference to the fact that *homo sapiens* are thought to have first evolved in that part of the world. Bono and Ali are not only making a symbolic journey back to humanity's common roots in a time where planes and fashion did not exist. They're also urging consumers to spend money on 'ethical' commodities, like the bags both advertise on their shoulders. The overt message is that some of the money spent will flow back from wealthy consumers' pockets to (in this case) cotton growers in rural Uganda. The implicit message is that those of us who have no first-hand knowledge of poverty or living directly off the land need reminding that there are costs, borne by people and the non-human world, when we urbanites fail to pay the 'proper' price for the many things we consume. We are encouraged to show our concern by closing income gaps and bridging geographical divides.

CRIMES AGAINST NATURE

In September 2012, a Nigerian court sentenced an actor to a 3-month prison sentence. His crime was to have sex with another man. The prosecution lawyer and sentencing judge made reference to a British colonial law that deems sodomy a transgression 'against the order of nature'. Indeed, imprisonment aside, in some parts of Nigeria sodomy is punishable by whipping or stoning. The Nigerian government also recently tried to pass a new law that makes shows of same-sex affection a crime, while banning same-sex marriages. Gay rights activists were understandably outraged. In a satirical riposte to the lawmakers, a popular comedian (John Okafor) called gay actors a 'virus' and said publicly, 'If there is any way in this world that people can make them stop it or kill it, please do it' (see: http://www.guardian.co.uk/world/2012/sep/21/nigeria-court-jails-actor-gay-offence).

SCIENCE FICTION AND THE LAWS OF NATURE

Such is the technical wizardry now employed in the film and television industries that patently fictional creatures or events can be represented in highly realistic ways. Even so, in early 2010 an American physics professor, Stanley Perkowitz, complained that too much science fiction was insufficiently tethered to the known facts of biophysical science. For instance, he complained that the giant killer bugs menacing the universe in the Hollywood blockbuster *Starship Troopers* were physically impossible. If you scaled up a similarly designed real insect to the size of the fictional bugs, Perkowitz argued, it would collapse under its own weight. He suggested that sci-fi depictions of biophysical processes and phenomena would be taken more seriously by audiences if they possessed greater scientific verisimilitude. Relatedly, the US National Academy of Science announced that it wanted its 'Science and Entertainment Exchange' to be more frequently consulted by sci-fi film and programme makers.

WILD BEHAVIOUR

Among the more notable events in recent British history were the government rescue of several large banks in 2008–9 and the urban riots of summer 2011. Subsequent to the latter, a feature writer employed by the British broadsheet *The Guardian* noticed a linguistic pattern emerging in the criticisms made of bankers and rioters. Jon Henley suggested that the word 'feral' was being repeatedly selected by otherwise different politicians and political commentators. According to the *Oxford English dictionary*, feral means (1) Existing in a wild or untamed state, (2) Having returned to an untamed state from domestication, and (3) Suggestive of a wild animal; savage. Among others using the term, Henley cited the London Mayor Boris Johnson and the former Deputy Prime Minister John Prescott. In a news interview, Johnson described the (mostly young, working-class) rioters as a "feral criminal underclass", while Prescott talked of "feral bankers" on Twitter. In both cases, the word feral was being used normatively and metaphorically. People were being compared to animals and being criticised for failing to observe the rules and norms that supposedly distinguish humans from the rest of nature (see http://www.guardian.co.uk/theguardian/2011/sep/06/use-of-feral-suddenly-everywhere?INTCMP=SRCH).

HOW TO BUILD A HUMAN

The turn of the millennium saw biologists publish the first ever 'map' of the human genome. But how do the 25,000 or so identified genes work together to create a healthy (or in some cases diseased) person? This has been the focus of the international Human Epigenome Project (http://www.epigenome.org/). The project has focused on how chemicals attach to different parts of DNA strands so as to differentiate human cells and permit their growth into body parts from birth to adulthood. In late 2009, the first fruits of the project were published in the world-famous science journal

Nature. A team led by Joseph Ecker at the Salk Institute in California issued a press release describing their research into the development of healthy skin cells. Theirs was an example of 'basic science', inquiring into the functioning of processes we currently know little about. It will form part of the foundation for future research into what causes 'malfunctions' in the epigenome. This research will, in turn, reshape preventative medicine.

* * * *

These seven vignettes are just some of a great many I could've presented. On the face of it they're very different – indeed seemingly unrelated. Yet all involve literal or metaphorical references to natural phenomena. What can we learn from them? First, references to nature are extraordinarily diverse. In the cases earlier, everything from orange juice to salmon to human genes is encompassed. Second, a remarkably wide array of people are referring to nature when addressing the rest of us: scientists, politicians, celebrities, film directors, judges, comedians and private firms (to name but a few). Third, they address us in every imaginable medium and genre of communication – from press releases to movies to websites, from realism to fantasy, from entertainment to edification. Fourth, their references to nature together span all three of our 'faculties', namely cognition, moral reasoning and aesthetic experience. They speak to head and heart, reason and emotion. Finally, the seven cases suggest how important the issues often are when nature or its collateral terms are invoked. For instance, a proper understanding of the human epigenome might allow diseases like cancer to be prevented, rather than cured after the fact. Likewise, to label some sections of a society as 'feral' is one way of highlighting a fundamental problem with their behaviour that the rest of us are enjoined to care about. Similarly, if you're not 'mixed race', how would you feel about a scientist telling you that you might be at a genetic disadvantage? If you're gay, do you think it right for homosexual intercourse to be classified as 'unnatural'?

Perhaps all of the above reflects the simple fact that nature is promiscuous in a real (or 'ontological') sense. It seems to be a large and diverse phenomenon that is literally everywhere – *in* us, as well as all *around* us. In all seven cases, nature is assumed to be a thing unto itself. It is, variously, something to be investigated, protected, properly understood, tamed, restored or modified. But is this really so? Is nature natural? Or is the *idea of naturalness* a convention that speaks volumes about the diverse values and goals of those who represent it to us day-in and day-out? I will answer this last question in the affirmative. In so doing, I hope this book will give you some important tools with which to understand your own relationship with what we call nature. Whatever your current attitude towards things like tigers or ice sheets, and however much (or little!) the topic interests you, I'm hopeful you'll learn something new and important by reading this book.

As the chapters to come demonstrate amply, *Making sense of nature* builds on over 30 years of academic and more popular writing about what is sometimes called 'the social construction of nature'. This is an ambiguous term that, these days, has a range of meanings – hence the scare quotes. However, in a broad sense it conveys the idea that what we consider to be 'natural' always bears the (usually) hidden or forgotten trace of particular assumptions, agendas and desires specific to a social group or a wider community, culture or society. I use the word 'social' to designate a wide array of shared imaginings, ideas, beliefs, norms, propositions and practices that different people employ in their everyday existence. The contents of 'the social' tend to alter slowly, usually because the words and deeds of a plethora of influential groups and institutions nudge us all into changing. But I don't naïvely presume there's a thing called 'society' that is a coherent, closed system and that's somehow separate from the biophysical world. Indeed, without a non-social domain perceived to be 'out there' in the first place, the epistemic communities to which I referred previously would be rendered largely silent. They would have nothing to tether their references to. In short, there's certainly more to the world than we humans are aware of, can imagine or can control. But the fact that we group so much of that world together and call it 'nature' is, I maintain, anything but natural.

There have been a number of books about nature's 'social construction' over the past 20 years. This one differs from those in a few key respects. First, I cover the full range of things signified by the term nature and its 'collateral concepts'. For instance, I don't just focus on the natural environment. Second, I'm interested in the full range of epistemic communities that make sense of nature. For example, whereas some authors will devote a book to news reporting of recent developments in human genetics, I cast my net much wider. Finally, I link representations of nature in *all* their forms to big issues that seem largely separate from such representations – at least at first sight. These are issues of people's identities and actions: how they are produced, governed and changed over time. It's not just what the likes of climate scientists or human biologists say to us that makes nature relevant to understanding who we (think we) are, how we treat each other and what sort of world we wish to inhabit going forward.

By insisting that references to nature and its collateral terms matter greatly in our lives, I'm being a little unfashionable. Many researchers whose publications I read these days argue that we need a new vocabulary to make sense of the world. They argue that reality does not comprise two great domains that interact (the social and the natural). Instead, it's fabricated out of all sorts of interrelations between human and non-human entities. These connections, they suggest, are so intimate that our current dichotomies (or dualisms) fail to do them justice. My own response is two-fold. First, a great many people continue to talk *as if* 'nature' is real. Second, in effect *this makes nature real*, not least because of the effects on people and non-humans of nature-talk. I defend my approach at greater length in this book's Endnote sections for those who are interested in knowing more.

I hope that degree students and their teachers in a wide range of social science and humanities subjects will find this text both relevant and stimulating. But it would please me no end if some of my readers were based in the various biophysical sciences (field, computational and laboratory) – even if they ultimately agree to disagree with my claims and contentions. I have drawn on the work of anthropologists, sociologists, philosophers, cultural studies scholars, historians, literary critics, media analysts, linguistic theorists and many others who operate beyond the perimeters of human geography, my academic home turf. The topic of 'nature' does not respect disciplinary boundaries, in either the academic world or the wider world. I have tried to honour this fact in the conception and writing of the present volume. *Making sense of nature* is thus very different in its content and aims from either *Social nature* (Castree and Braun, 2001) or *Nature* (Castree, 2005). Both of these books were strongly rooted in debates in my own discipline. By contrast, this text is not.[1]

In fact, I have no particular desire to respect the niceties and nuances of academic debate about nature, whatever the field. If I did, I would simply repeat – and no doubt get bogged down in – what have, at times, become painfully esoteric discussions rendered in a language so rarefied that only the self-selected few can understand (and thus care) about them. In the following chapters, I prefer to draw liberally upon published research from a range of fields in the service of my major arguments. At times, this will involve using others' ideas and findings in ways they did not necessarily intend or anticipate. I hope they will forgive me this licence, and tolerate my not recounting the fine details of intra- and inter-disciplinary discussions about 'nature'.[2] I also hope I'll be excused for the various simplifications I make and the many argumentative shortcuts I take. At times, in my desire to speak to those new to the ideas contained in this book, I will make general claims that specialists may regard as old hat, trite or in need of more careful elaboration.

I've written this book in what you might call an 'advanced introductory' style. This is to say that students at, or beyond, an upper bachelor's degree level ought to find this book accessible, if not always easily so. At times, *Making sense of nature* will be a very challenging read for them, but always – I hope – readable. Part 1 presents all of my major arguments and claims, while the remaining chapters amplify, deepen and illustrate them by considering a range of extended examples and cases. A principal function of universities is not only to create new knowledge (concepts, arguments, evidence, etc.) but also to ensure that this knowledge travels beyond its originators so as to participate in the drama that is human existence on the planet. Though the chapters are generally long ones, I've broken the arguments down into distinct chapter sections; so too the case material. If I can thereby engage those wholly new to this book's contents and claims, but still hold the interest of readers already steeped in some or all of the relevant literatures, then I will have realised my own ambitions.[3]

These ambitions are anchored, largely, in what otherwise different readers of this book have in common. I want to address them (*you*) not so much as professors or students (depending) but, less specifically, as members of a world in which we are *all* consumers of information, knowledge, experience and belief that we take largely on trust (happily or otherwise). Throughout the book, I'm willing to take the risk that my frequent invocation of the collective first person – 'we', 'us' – is presumptuous to the point of being thoroughly ill advised. It at least reminds us that we are members of a 'public', or rather publics in the plural, that need to be proactive in the face of a deluge of messages about our own 'nature' and that of the non-human world.

Student readers (and, I hope, their tutors) will benefit from the various study questions and exercises peppered throughout the chapters. Ideally, these should be undertaken at each stage before reading on. Numerous boxes offer an additional aid to learning, as do the several diagrams, tables and photographs. Throughout the book, terms appearing in **bold** font are defined in the glossary. Finally, a further reading section itemises my major sources for each chapter after Part 1 and the 'How to use this book' appendix may help tutors build their own module around this text, should they wish.

ENDNOTES

1 However, there is one strong point of connection with the book *Nature*: the present volume elaborates at length ideas sketched roughly in Chapter 1 of its more argumentatively circumscribed predecessor.

2 What I mean is that this book does *not* do what many survey texts do – such as Jan Golinski's (2005) *Making natural knowledge* or my colleague Peter Wade's (2002) *Race, nature and culture*. I do not here present, systematically and critically, the findings and arguments of various interlocutors within defined disciplines and fields. *Making sense of nature* is thus a 'textbook' with a twist. Unlike many literature-based books, it does not summarise ideas and findings within a single academic domain, and nor does it organise the research surveyed according to familiar categories. Instead, I package and arrange ideas, arguments and case material from a range of fields within a plenary framework that, in my view, adds value to the published work on which this book is otherwise dependent. In this endeavour, one of my models is Denis Wood's superb book *Rethinking the power of maps* (2010), based on his earlier *The power of maps*. My own book is not about maps, of course, but I take inspiration from Wood's approach to a diverse and large body of theory and empirical research. Though I cannot match his pungent, often witty prose style, and while *Making sense of nature* is (I believe) pitched at a rather more accessible level than *Rethinking the power of maps*, I hope some of the strengths of Wood's approach to presenting material are evident in this book. I should declare that when I was halfway through writing this book (December 2011) I encountered Stephanie Rutherford's (2011) *Governing the wild*. I also, belatedly, read David Delaney's (2003) superb book *Law and nature* at the same time. There are some strong generic similarities between these monographs and my own book, but – equally – some significant differences in approach.

3 In a review of his book *Geographies of nature* (2007), I chastise geographer Steve Hinchliffe for leaving novice readers adrift in his attempt to edify more expert readers (Castree, 2009). I can only hope he doesn't think the compliment worth repaying should he ever review *Making sense of nature*!

PART I

MAKING SENSE OF SENSE MAKING

1 HOW WE MAKE SENSE OF (WHAT WE CALL) NATURE

AIMS AND OBJECTIVES

What this book is *not* about

This is a book with seemingly exorbitant ambitions. Its subject matter is all of the following (and a *lot* more besides): beluga whales, natterjack toads, Siberian tigers, hermit crabs, orangutans, meerkats, skip-jack tuna, coral reefs, oak trees, human DNA, sea horses, El Niño, volcanic eruptions, photosynthesis, Mount Everest, cystic fibrosis, tectonic plates, the Marianas Trench, human hearts, quarks, mangrove swamps, ice caps, North Pacific fur seals, wolves, transgenic organisms, Hadley cells, botanical gardens, the Sahara Desert, rainforest ecosystems, haemoglobin, pebble beaches, wild orchids, manatees, intestinal bacteria, meiosis, the Humboldt squid, blueberries, iguanas, bumblebees, conjoint twins, oil reserves, dinosaur fossils, beavers, elephants, the jet stream, buzzards, igneous rock, gravity, ocean currents, algae, weeds and hot springs. I could go on, but you get the idea. My aim in the pages that follow is to make sense of nature – that 'buzzing, blooming confusion', as the philosopher William James once so beautifully described it.

It's trite but true to say that nature matters to us – enormously so. Without it we could not live, and we would (quite literally) have no past, present or future. Nature provides us with the materials required to satisfy our basic needs – the need for shelter, warmth, food and clothing. More than this, it's a source of fascination, a focus of moral concern, and an object of aesthetic appreciation. We seek to unlock nature's secrets through research; we passionately debate the ethics of using, altering, destroying or recreating it; and we take pleasure in some of its creations, be they pristine or modified by us. We also, at times, seek to keep out nature, or at least get out of its way: wildfires, tidal surges and heatwaves are just some of the natural hazards that can harm or even kill us. Where possible, we try to control nature in order to derive benefits: think of China's Three Gorges Dam, domestic pets, artificial life forms or commercial pesticides. Nature is as much a curse as a blessing, at once bane and boon, essential and unwelcome. Its central yet contradictory role in our lives has, for well over a century, found expression in debates between environmentalists (both 'deep' and 'shallow') and those for whom the 'domination of nature' is a good and necessary thing.

But, of course, nature is not only 'out there'. We consider ourselves to be *part of* nature, in two senses. Not only are we affected by and able to alter wider ecosystems; in addition we are, physiologically speaking, natural entities ourselves – in literary critic Paul Outka's (2011: 31) memorable formulation, 'a part of the earth that [has] learned to talk'. *Homo sapiens* is one of several hominid species that evolved, like all other organisms, through a very long-term process of 'natural selection'. Notwithstanding our important differences from other primates and the wider world of living creatures, we're all members of an extended biological family. This family derives from some unknown parent who emerged from the primeval slime many millions of years ago. More recently, we can trace our species origins to eastern and southern Africa.

Our physical and psychological nature is clearly important in defining what is possible for us. Large-brained bipeds with complex nervous and circulatory systems, we all require oxygen and water but can digest any number of different foodstuffs. Our linguistic skills are equalled by no other mammal, and we possess an uncommon capacity to remake our surroundings to suit our purposes. Like all other life forms, we age and eventually die – in some cases because of genetic ailments such as progeria. In short, we may not be reducible to our natural qualities, but neither can we ignore them. As the famous ethologist Konrad Lorenz once said, 'an enormous animal inheritance remains in man [*sic*] to this day' (1952: 152). We are, as some geneticists are wont to remind us, '98 per cent chimpanzee' and therefore 'the fifth ape' inhabiting planet Earth.[1]

Making sense of nature is, then, about a manifestly important subject. However, it may seem that I've bitten off far more than I (and you) can possibly chew. After all, according to one meaning of the term, nature is pretty much *everything*. It is, in the no-nonsense words of science writer Colin Tudge, 'very big, and very various indeed' (2006: 29). How can one possibly write a book about everything? The answer, of course, is that one cannot. Any attempt to describe and explain nature in a comprehensive and detailed way would have to be collaborative. The expertise of numerous specialists (e.g. zoologists, ecologists and geologists) would necessarily have to be combined. This would take years. It would yield a multivolume compendium so forbiddingly large that no one could possibly read it from start to finish without forgetting large parts of what they'd learnt along the way.

This said, Charles Darwin famously presented an overarching theory of life in a single volume: his germinal book *The origin of species* (1859). He did so by presenting factual material about diverse species selectively, in the service of what he famously called 'one long argument'. So too have two of Darwin's modern-day popularisers, the distinguished biologists Richard Dawkins and Steve Jones. So, is *Making sense of nature* an attempt to summarise (or extend) the state-of-the-art in current 'big thinking' about life on Earth? No is the short answer. And, even if I wanted to write such a book, others are far, far better equipped to do the job than I am. As a social scientist, with a broad-based training in human geography and cognate fields, I lack a deep grounding in the various 'sciences of nature', such as palaeontology, animal biology, human neurology, particle physics or organic chemistry. In truth, I don't even know much about physical geography any more (beyond the basics), having lost touch with its expanding and diversifying research frontier once my first degree ended some 20 years ago.

Making sense of nature is thus neither a compendium-cum-encyclopaedia, nor does it present an overarching theory intended to explain the material world's governing principles. Is it, then, a paean to nature? Is it a further contribution to a now established genre of writing initiated by the likes of William Wordsworth, Ralph Waldo Emerson and Henry David Thoreau well over a century ago? Is it a treatise about how we value – or should value – nature in an ethical or aesthetic sense? Or is it, perhaps, an intervention in the current debate over whether or not nature is being irreversibly destroyed, and what to do about it? This would, perhaps, be more in keeping with the talents and interests of someone who (like me) is a stranger to the intricacies of biophysical science but fascinated by public and policy discussions about 'the human impact', 'resource scarcity', 'sustainability', 'designer babies', 'artificial life' and 'genetic diseases'. But, again, the answer to all these questions is 'no'. My aim is not to debate either the moral–aesthetic value, or the current state-of-health, of the physical world (be it human or non-human nature).

What this book *is* about

It seems to me that what scientific treatises about the biophysical world have in common with books about environmental ethics and impassioned discussions of things like the shrinking Amazon rainforest or the perils/possibilities of human biotechnology is this: they all presume that nature is a distinct material domain. I did as much myself in this chapter's opening three paragraphs for the simple reason that it's *conventional* to do so. It seems natural to think in these terms, does it not? In other words, there's a widespread tendency to believe that there's a natural world existing regardless of, and separately from, our attempts to understand it. Our ordinary language is very telling in this regard. We talk *about* nature, focus *on* nature, look *at* nature, do things *to* nature, have respect *for* nature, are seen to be part *of* nature, are influenced *by* nature or subject *to* nature.[2] Events like the massive 2010 oil spill in the Gulf of Mexico and the tsunami that hit Japan in the following year seem to confirm our belief that nature is somehow a thing unto itself. It appears possessed of both distinctive attributes and autonomous causal powers.[3] Relatedly, many argue that humans remain fundamentally governed by natural drives and instincts, notwithstanding their capacity to moderate both.[4]

But what if nature were not natural at all? What if, instead of it being an object or domain we make sense of in various ways, our sense-making practices reveal something wholly 'unnatural' to us? What if it's a world whose naturalness is not given but merely appears to be so? What if the so-called Book of Nature is legible to us because we wrote, rather than simply read, the contents (plot, character, scene)? I ask these questions because in this book *I aim to make sense not of nature but of the various ways in which what we call 'nature' has been made sense of.* Nature doesn't exist 'out there' (or 'in here', within us) waiting to be understood. Rather, the very category of nature is part of the way we make sense of the world for ourselves. In other words, I'm interested not so much in what nature *is,* as in how those various things convention teaches us to call 'natural' are represented by us and to us, and with what implications.

In truth, it is more 'to us' than 'by us'. As I asserted in the Preface, most of what we know and feel about nature derives from the claims made by myriad others, for instance wildlife film-makers, journalists, chemists, environmental activists and professional ecologists. Some of these others are individuals who attain special prominence, for instance the science-documentarians and authors David Attenborough, David Suzuki and Brian Cox. But these individuals are usually members of larger epistemic communities who specialise in producing particular kinds of representations of nature. We consume and internalise these representations, placing us in a situation of 'epistemic dependence' that is, in my view, writ large in the modern world. If you like, this is a book about what philosophers call **epistemology** (i.e. knowledge and its effects). I am interested in *social* epistemology, to be specific, because knowledge and belief are not created by sovereign (Cartesian) individuals and nor do they possess absolute foundations located outside the realms of social discourse or social practice. Equally, we might say that *Making sense of nature* is about **ontology**, but not in the classic philosophical sense of specifying what's fundamentally 'real' and listing (in typically rarefied language) that 'mind-independent' reality's signature characteristics. My interest, instead, is in what various representations of nature communicate to us. They shape what they ostensibly depict – including us – and are thus worthy of close scrutiny. I thus regard 'nature' as a particularly powerful fiction: it's something made, and no less influential for being an artefact.

So, I am interested not in beluga whales, natterjack toads, Siberian tigers and hermit crabs but in *our shared ideas, feelings and beliefs about the phenomena so named* and all the other things I listed at the outset. We all too often proceed as if these ideas, feelings and beliefs refer to a 'nature' that's self-evidently 'there', waiting to be catalogued, understood, appreciated, managed, enjoyed, explored, protected, improved or otherwise altered. We often presume there to be no difference between word and world, image and actuality, thought and matter. Or we seek to 'correct' the difference so that a 'proper fit' is achieved cognitively, morally or aesthetically. As critic

Jennifer Price has observed, somewhat sardonically, we tend to think that 'Nature is where Reality lives' (1999: 210) – as exemplified by the telling phrase 'the nature of reality', one used frequently by scientists and philosophers.[5] Indeed, it's surely significant that one of the world's premier science periodicals is called, simply and unapologetically, *Nature*.

Yet, while our various representations are assuredly *about* something we call 'nature', they are not the *same* as the things to which they refer and nor are they *reducible* to them. To suppose otherwise is to commit what philosopher Roy Bhaskar once termed the 'ontic fallacy' (1993: 430). This is the belief that most of our information and knowledge is a mirror image of the physical world – except when intended to be fictional or speculative. The challenge, then, is to understand what one writer calls 'the conceptuality of real objects' (Fuller, 2005: 1). Words are also things – not simply statements about things. But, equally, the things referred to are only comprehensible in terms of the words we choose.

Denaturalising nature

Over the past 30-plus years, claims like these have become axiomatic for a generation of researchers and teachers in a wide range of social science and humanities subjects. For decades, the subject of 'nature' was left to those on the other side of university campuses: the chemists, astronomers, neuroscientists and zoologists, for instance. But since the mid-1970s, a growing number of anthropologists, sociologists, historians, philosophers, cultural critics and others have had something to say about the matter. The initial impetus for this was arguably two-fold.

First, 40 years ago there was widespread concern about both global 'over-population' and increasing scarcity of biophysical resources like oil. Emblematic of this was the well-known book *The limits to growth* (Meadows *et al.*, 1972). There was even concern about abrupt environmental change, well before the current worries about greenhouse warming. For instance, the distinguished American scientist Steve Schneider co-authored *The Genesis strategy: climate and global survival* as far back as 1976. As a reaction to this, writers on the left of the political spectrum made the following argument: attempts to control population numbers or limit resource use, they maintained, were founded on the dubious claim that 'natural limits' to economic growth could be specified without reference to the varied, power-saturated and malleable social organisation of economies and polities (see, for instance, Harvey, 1974). They thus challenged the view that something called 'nature' imposes non-negotiable restrictions that we ignore at our peril.

Second, in the human sciences the florescence of what was perceived as a resurgent biological determinism (exemplified in the writings of Arthur Jensen and J. Philippe Rushton) led to a set of strong counter-arguments. For instance, consider *The use and abuse of biology* by anthropologist

Marshall Sahlins (1976). It highlighted the moral–political dangers of suggesting that certain differences among humans are natural, and thus both intrinsic and unchangeable. Relatedly, a few years later the evolutionary geneticist Richard Lewontin co-authored *Not in our genes* (Rose *et al.*, 1984), which argued that what some took to be 'genetic behavioural traits' were, in fact, the result of a combination of people's biological 'hardware' interacting in complex, non-deterministic ways with a wider (non-natural) social environment.[6] For Sahlins and Lewontin, it was too simple to say that some people are 'naturally' athletic, clever, beautiful, thin, extrovert, etc.

Making sense of nature draws extensively on the multi-disciplinary body of work that has developed subsequent to these two early attempts to **denaturalise** what was being presented as natural. I interpret and organise its insights in ways that are, I believe, both productive and somewhat novel. In the rest of this chapter, and the two to follow, I want to describe in some detail the approach to 'nature' that I intend to take throughout this book. These are the most theoretical and programmatic portions of the book; however, I will illustrate each key point with a short example or two, and will also pose study questions intended to get you thinking. Later in the book, I will amplify the points made with reference to more extended cases, again accompanied by reflective questions directed your way.

As will become clear to readers, I'm assuming the role – sometimes maligned in an age of specialists – of a 'generalist'. My tack is to 'bring together [insights from] widely separated fields ... into a common larger area, visible only from the air' (Mumford, 1967: 16). My hope is that the proverbial whole is more than the sum of the borrowed parts. Additionally, I want to address readers in a specific way – not so much as 'students' or 'academics', but simply as inhabitants of the early twenty-first-century world. We're all members of highly complex capitalist societies marked by very elaborate divisions of mental and practical labour. As the Preface intimated, what you and I have in common, notwithstanding our differences, is this: we're obliged to understand both ourselves and the world at large via a plethora of individuals, communities, institutions and organisations that aim to shape our thoughts, attitudes, values, feelings, tastes and actions. I'll come to this shared epistemic dependence in Chapters 2 and 3 (and return to it in the last part of the book). First, and predictably enough, I begin by defining what we mean when we use the term 'nature'.

THE IDEA OF NATURE

'Nature' is a **keyword** rather than a buzzword. Keywords, as Raymond Williams (1976) argued in his famous book of this name, have three characteristics. First, they are 'ordinary', which is to say used widely and frequently by all manner of people in all manner of contexts. Second, they are enduring rather than ephemeral – they do not come-and-go in a way that buzzwords like 'globalisation' or 'post-modernism' do. Finally, keywords possess what cultural critic Tony Bennett and colleagues, in their update of Williams's book, call 'social force' (Bennett *et al.*, 2005: xxii). In other words, because their various meanings become normalised, they're able to govern (i.e. steer or direct) not only our thinking but also a wide range of practices resulting therefrom. A simple measure of the importance of 'nature' as an idea is to imagine us dispensing with the term and its meanings altogether. The 'hole' in our language would be enormous. We'd be rendered both inarticulate and incapable in large areas of our thought and action. In short, if we didn't already have the term in our present-day vocabulary, we'd probably have to invent it.

To understand what nature means for us, we need to ask three questions. In each case, we're not looking for the 'proper' meaning. Instead, we're searching for those meanings that have come, over a long period of time, to *seem* like proper ones. This immunises us from the conceit that our particular way of describing the world is a universal feature of all societies worldwide. The first question is the obvious one to pose: 'what is nature?'

What is nature?

This is an apparently difficult question to answer, at least if we remain at the level of the myriad different phenomena to which the word refers. A literal answer to the question would have us extending this chapter's opening paragraph into an almost endless inventory of organic and inorganic phenomena found on Earth (and in the stars). A more parsimonious (and sensible) approach is thus to look for the shared meanings that we attribute to these phenomena when we classify them as 'natural'. I make no claims to originality in identifying them below. First, however, why don't you have a go?

Study Task: Look around you right now or think about what you've seen since getting out of bed this morning. Then list six things you consider to be 'unnatural'. Ask yourself what it is about them that prevents you from considering them to be natural. Aside from 'unnatural', what other words come to mind to describe them (e.g. artificial)? This task should allow you to define nature from the 'outside in'.

How did you do? It seems to me that 'nature' has four principal meanings, all of which are quite venerable. First, it denotes the *non-human world*, especially those parts untouched or barely affected by humans ('the natural environment'). Second, it signifies the *entire physical world*, including humans as biological entities and products of evolutionary history.[7] Third,

NATURE			
The non-human world of living and inanimate phenomena, be they 'pristine' or modified.	The physical world in its entirety, including human beings as both products of natural history and present-day biological organisms.	The defining features or distinguishing quality of living and inanimate phenomena, including human beings.	The power, force or organising principle animating living phenomena and operating in or on inanimate phenomena.
'EXTERNAL NATURE'	**'UNIVERSAL NATURE'**	**'INTRINSIC NATURE'**	**'SUPER-ORDINATE NATURE'**

Figure 1.1 The principal meanings of the word nature in contemporary Anglophone societies

it means the *essential quality or defining property of something* (e.g. it is natural for birds to fly, fish to swim, and people to walk on two legs). Finally, it refers to the *power or force governing some or all living things* (such as gravity, the conservation of energy, the instructions contained in human DNA, or the Coriolis effect). As a shorthand, we can (respectively) call these meanings **external nature**, **universal nature**, **intrinsic nature** and **super-ordinate nature** (see Figure 1.1).[8] Their differences notwithstanding, a common semantic denominator is that nature is defined by the absence of human agency or by what remains (or endures) once human agents have altered natural processes and phenomena.

Clearly, depending on the context of usage, the idea of nature can function as a noun, a verb, an adverb or an adjective. It can also be characterised as object or subject, passive or active. For instance, we can talk of a 'nature reserve', 'natural beauty', a 'naturally destructive' hurricane or the 'natural order' being disrupted. 'Nature' is both an everyday word used in quotidian discourse, and also part of the lexicon of scientists and other credentialised 'experts'. Equally clearly, several meanings of the term nature can be in play simultaneously. Consider mathematician Marcus du Sautoy's (2008) best-selling book *Symmetry: a journey into the patterns of nature*. Here, 'nature' refers both to an object – the physical universe – and what du Sautoy regards as one of its signature characteristics – its consistency and coherence. The same is true of developmental biologist Mark Blumberg's (2008) *Freaks of nature: what anomalies tell us about development and evolution*. Blumberg argues that apparently 'freakish' living organisms like two-headed humans are in fact governed by the same growth processes as 'normal' ones. They thus possess similar inner qualities (genetic and functional), if not the same outward physical form (phenotype).

Not all the meanings of 'nature' can be used without contradiction. For instance, understood literally, the idea that nature is universal would, in philosopher John Gray's words, have us insisting that 'The internet is as natural as a spider's web' (2002: 16). And yet the idea of nature as external obliges us to say that people and their creations are not in any obvious sense 'natural' at all. Indeed, for some, our humanity consists in 'rising above' our nature, or in some sense superceding it. Hence, there's a tension between the ideas of nature as universal and as not: 'if "natural" describes everything in the universe, including ... people, it follows that it's impossible for anything to be unnatural' (Poole, 2007: 68).

In addition to combining the word nature's several meanings without (hopefully) tying ourselves in knots, we are also apt to do so with subtlety. For example, we might say that the English Lake District and Siberian tundra are both natural landscapes in the first sense of the term 'nature'. But we'd want to acknowledge that the former bears the imprint of human activity far more obviously than does the latter. This illustrates our belief that there are degrees of naturalness (in the third and fourth senses of term), and a grey area between what is deemed natural and 'unnatural'.[9] We can thus, and routinely do, distinguish 'anthropogenic' from (variously) 'pristine', 'untouched' and 'wild' nature.

Where is nature?

So far, so good. But to fully understand what nature means for us, we also need to ask where it is. Addressing this question reveals our propensity in Western societies to **spatialise** (or delimit geographically) those things we consider to be natural.

Study Task: When you think about nature, which sorts of places, countries or landscapes come to mind? Are any of the four meanings of nature identified previously less applicable than others when you think about where nature is? If so, then why?

We typically spatialise 'nature' understood in the first and third senses of the word. As an object (or 'other') possessed of distinctive qualities, we think of it in terms of specific locations. Some of these are iconic: think of Alaska, safari parks, the Sahara Desert, volcanoes, the countryside, coral reefs, the high Andes, the polar ice-caps, the deep ocean, atolls, Australia's Great Barrier Reef, Niagara Falls or the Galapagos Islands. We also think of nature as located in specific countries, such as Costa Rica. Today, the Amazon rainforest is perhaps the most recognisable 'natural space' of all. It's nothing less than a 'verdant poster child' (Slater, 2003: 21) for all that is deemed to be worth protecting against 'the human impact'.

Where what we call nature is threatened or vanishing, we as often create *ex situ* sites in which to concentrate it as we do ring-fence natural spaces and species *in situ* – think, for instance, of zoos, botanical gardens or the Millennium Seed Bank in London (metaphorical arks all). So, we routinely consider nature to be *somewhere else*: it's something we travel to, visit or dwell in prior to returning to our 'unnatural' towns and cities. Thus, when science historian Donna Haraway insisted that 'nature is not a place to which one can go, nor a treasure to fence in or bank . . .' (1992: 296), she was reminding us of this fact. Nature is not really 'over there', it's just that we think it is.

But we also spatialise nature in other ways, and this involves geographical scale. On the one side, we imagine nature (in the second sense of the term) to be almost imperceptibly small in its constituent parts (e.g. atoms, quarks and 'dark matter'). It's thus all around us, and indeed part of us, yet hidden from view. For instance, if we consider human biology and living creatures more widely, a significant shift in our collective thinking has occurred since James Watson and Francis Crick discovered the structure of deoxyribonucleic acid (DNA) in 1953. Imperceptible to the naked eye, the 'double helix' has, over the past half century, come to represent the 'building block' of all life in virtually all our imaginations. On the other hand, we also conceive nature to be global, ambient and all encompassing (i.e. far bigger than, and inclusive of, us). Notable here were the Apollo space missions of the late 1960s and early 1970s, which provided an entirely new view of the world. The striking photographs NASA astronauts and technicians produced allowed us to look at the Earth from afar as a blue sphere located in the vastness of outer space.

Notwithstanding the fact that there are just shy of 7 billion people on the planet, perhaps rising to 9.5 billion by 2050, many of us remain acutely aware of our smallness relative to the Earth's biophysical systems. One reason is because the Apollo images have been invested with specific sorts of meanings connected to the widespread concern about 'the human impact' on nature that found expression in the first Earth Day, held on 22 April 1970. The 'blue planet' imaginary that's today part of our cultural repertoire emphasises the finiteness of our home, our isolation in a vast universe, and our shared occupation of (and responsibility for) Earth, regardless of nationality or location (see Plate 1.1). Iconic visualisations of the planet subsequent to the Apollo photographs have been interpreted, at least in part, in terms of this imaginary – such as the satellite images of a 'hole' in the polar ozone layer that circulated far and wide through the 1980s and 1990s.

When is nature?

If we have a propensity to spatialise what we call 'nature', we also **temporalise** it too. Indeed, each is the analogue of the other. If I ask the seemingly strange question 'when is nature?', it soon becomes clear that we think these days that it's ever more a thing of the past. This is surely a key reason why Bill McKibben's *The end of nature* (1989) and Francis Fukuyama's (2002) *Our post-human future* became bestsellers. It also explains the bitter-sweet appeal of *Last chance to see*, a BBC television series about threatened terrestrial and marine species fronted by writer-actor Stephen Fry and wildlife photographer Mark Carwardine. The emotional charge of both books and the TV documentary resided in their reproduction of a widely

Plate 1.1 The blue planet

Since the Apollo 8 mission to the moon in 1968, we've become very accustomed to seeing the Earth from space. But how, exactly, do we 'see'? Colour images like the one reproduced here in black-and-white typically show large areas of blue ocean, white cloud and green (or sandy) coloured land. Some show the Earth floating in dark space; others show the sun's rays illuminating a portion of its surface. The overall effect is to present the Earth as, first and foremost, a biophysical unit – whole and indivisible. The 'human impact' on the environment that's been of so much concern, on-and-off, since the late 1960s, is subsumed in an image that suggests the enormity but finiteness of the planet. As environmentalists Barbara Ward and René Dubos put it in the title of their germinal 1972 book, there is *Only one Earth* (Ward and Dubos, 1972).

shared assumption. It's the assumption that because there are, today, more people, more industry, more consumption, more pollution, more travel and more 'invasive' technologies than ever before, then there's therefore less 'nature' – it seems to be a zero-sum game in which the natural world is the clear loser. In environmentalist Mark Lynas's (2011: 12) dramatic words, 'Nature no longer runs the earth. We do. It's our choice what happens from here.'

Many thus reason that we need to act decisively before it is 'too late'. We ought, they argue, to restore, protect, hoard and preserve the nature we've not yet destroyed or degraded (see Figure 1.2). We should do this, they suggest, by downscaling our own presence. What is required are fewer babies, less trade, less travel, less 'materialism' and so on. We should observe the 'precautionary principle' and take no further undue risks with our own biology or the planet's ecology. Alternatively, it's argued that we should use the

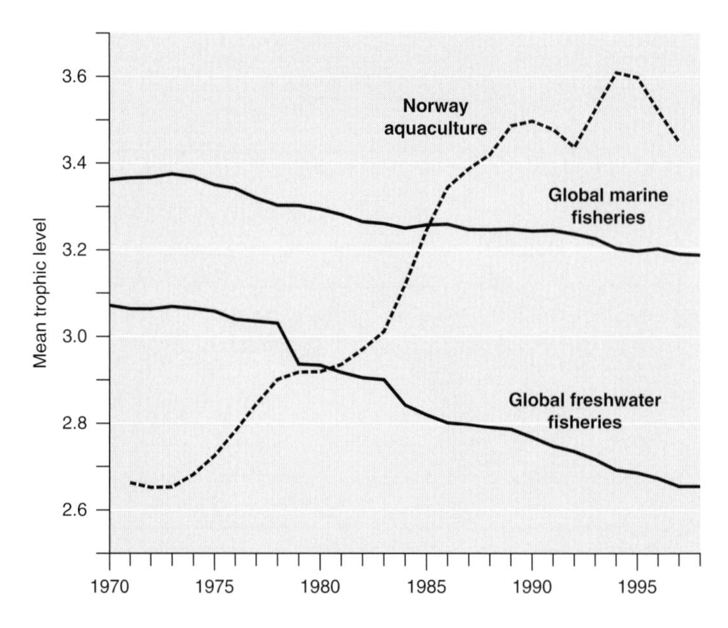

Figure 1.2 Mean trophic levels of global freshwater and marine fisheries over time: an 'aquapalypse'?

This diagram shows a continuous decline in the mean trophic level (MTL) of fish stocks worldwide since 1970. Alone and with other researchers, the French-Swiss marine biologist Daniel Pauly has persistently argued that an apparent *increase* in non-aquaculture fish harvests globally since the late nineteenth century has come at the expense of MTLs. The latter are the ecological levels at which harvested fish exist. Pauly and colleagues' claim is that the fishing industry has 'moved down the food chain' because once higher species (like cod) are harvested to virtual extinction. Writing in the current affairs magazine *The new republic* in 2009, Pauly talked dramatically of an 'aquacalypse' and 'the end of fish' (Pauly, 2009). This is just the sort of discourse about natural resource over-use that Western citizens are well familiar with. Other scientists have taken issue with Pauly's research. Redrawn based on Pauly *et al.*, (2002).

advanced technologies we have at our disposal in more 'ecofriendly' ways than heretofore. We can then recreate lost species using these technologies and bring a dead past into a living present. We can tackle 'natural problems' within our own species, such as various genetic diseases. And we can address 'wicked' environmental problems like climate change (e.g. through geoengineering solar radiation fluxes). But critics argue that we'll only end up creating unanticipated new and a larger problems with all our 'meddling'.

Indeed, when Nobel Prize winning scientist Paul Crutzen coined the term 'the Anthropocene' in 2000 he invited us to temporalise nature on a grand scale. His neologism encourages us to regard the biophysical changes to the Earth that have occurred during *homo sapiens*' brief residence on the planet as akin to those occurring during previous geological epochs. The same can be said of the dramatic name chosen for the British-based Christian environmental charity Operation Noah, with its intimations of 'environmental crisis'. The human influence, especially over the past 100 years, is thus presented as qualitatively equivalent to the profound, unplanned alterations of life on Earth previously only achieved over millennia. McKibben calls it, 'the great simplification', the 'most disconcerting, strange, jarring, out-of-the-ordinary stretch of time since we climbed down from the trees' (2006: 36).

We have, it is said, exceeded our ecological niche and consigned ever more life forms to history. We are eliminating 'natural time' – for instance, the cycles of birth, growth and death in insects – with 'social time'. This is a calendar suited to our own frenetic wishes and desires. We are, it's also said, approaching an epochal 'tipping point'; we even risk 'messing up' our own nature if new, more invasive biomedical technologies get the go-ahead. This, as readers will recognise, is a dominant narrative of our age. We regard it as a 'fact' because so much evidence (apparently) tells us it's true, such as the recent worldwide Census of Marine Life (COML). Published in summer 2010 to coincide with the International Year of Biodiversity, it was a decade in the making and as might be expected, it makes for grim reading. Or, to take a second example, look at the time-lapse images of environmental change on NASA's Earth Observatory website: in a few seconds, years of environmental degradation pass before the viewer's disbelieving eyes (go to http://earthobservatory.nasa.gov/).

* * * *

Having identified the principal meanings of the term 'nature', we're bound to conclude that it's an unusually tricky word to understand. For Candace Slater, it's 'a noun with a necessary multiplicity of modifiers, if not a singular in desperate need of pluralisation' (Cronon *et al.*, 1996a: 451). Indeed, it is arguably 'the most complex word in [our] ... language', as Williams (1976: 219) famously opined.

To use classic semiological terms, we might describe it as a **signifier** (token or sound) with several different **signifieds** (or meanings), which, in turn, get attached (alone and in combination) to a very wide array of **referents** (or material phenomena). In its range of applications, it takes us from the ordinary to the extraordinary, from the banal to the sublime, from the familiar to the strange. Of course, in any given situation we do not – and could not possibly – invoke the whole complex of meanings and referents. Instead, we deploy the term 'nature' in a wide range of different contexts in specific and circumscribed ways. It is thus remarkably promiscuous, but the point is that this single word signifies a plurality of different things in a range of ways, depending (as illustrated in Figure 1.3).[10] As Williams famously observed, references to 'nature' are 'usually selective, according to the speaker's general purpose' (Williams, 1980: 70). Thus, 'To [understand] ... the nature that is all around us', as William Cronon insists, 'we must think long and hard about the nature we carry inside our heads' (Cronon, 1996a: 22). See Box 1.1.

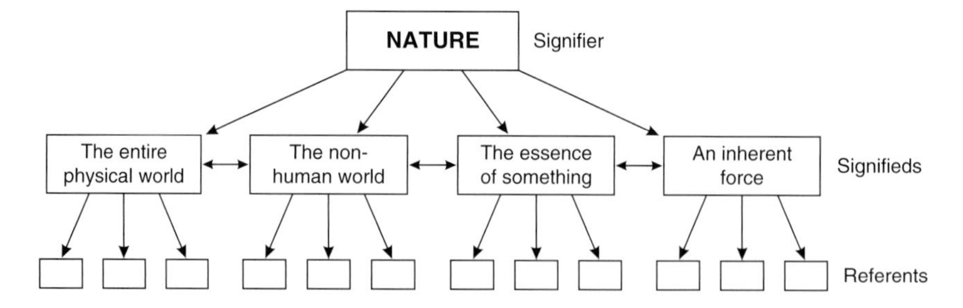

Figure 1.3 The concept of nature: its meanings and referents

The two-way horizontal arrows indicate that more than one of the four meanings can be in play in any given reference to 'nature'. The meanings can be attached to the totality of what we call 'nature' and, equally, to any of the myriad of different phenomena we consider to be natural in kind or degree.

BOX 1.1

A KEYWORD IN ACTION: THE SELECTIVE AND COMPLEX USE OF 'NATURE'

Because its meanings and referents are plural, the term nature performs diverse and complicated linguistic 'work' in the Anglophone cultures to which most readers of this book will belong. Two examples will suffice to make the point. In his collected poems, entitled *No nature*, the American 'green' poet Gary Snyder says this in the preface: 'The greatest respect that we can pay to nature is to acknowledge that it eludes us' (1992: v). Snyder here uses the term nature to denote the non-human world ('external nature'), but the 'we' he refers to does not include all of humanity. Rather, it's those of us living in highly technologised, high-consuming capitalist societies. His key point, as an environmentalist, is that 'nature' is an ultimately unknowable other. We should not kid ourselves that we can ever understand or control it completely. His use of the word 'respect' indicates clearly the moral and political aims of his poetry. He derives these aims from what he takes to be nature's ineffable qualities. For him, when properly understood, nature can tell us how to behave towards it.

A very different example of nature's invocation comes from a white New Zealander reflecting on so-called 'mixed race' individuals when interviewed by two university researchers. 'You'll find it's very common amongst people of mixed blood,' the respondent said. 'With anyone who has two cultural backgrounds', he goes on, 'one part is usually stronger than the other ... [He] has difficulty balancing themselves (*sic*). And often it comes out in the form of racial intolerance against what is actually part of themselves' (cited in Wetherell and Potter, 1992: 206). This comment is offered as a personal, though also 'factual', observation. While the term 'nature' is not used overtly, its meaning is nonetheless present in the man's choice of the words 'blood' and 'race' (which become 'collateral concepts' for 'nature' – see the following section). For him, humanity is separated into 'natural kinds', that of European colonists and indigenous Maori in the New Zealand case. People are seen as natural organisms (one part of 'universal nature'), but 'naturally differentiated' into groups (their 'intrinsic nature' varies). These groups are mixed through interracial intercourse, producing hybrid individuals. But in what does the hybridity really exist? Is it purely biological in the man's view? Here we see just how complex the interview excerpt is when scrutinised closely. The man flips-flops between talking about genetic mixing and cultural mixing. The 'intolerance' referred to seems to be the result either of an individual's desire to claim 'racial purity' in some biological sense, or of their desire to claim a clear 'cultural identity'. The two are both conflated and substituted.

NATURE'S COLLATERAL CONCEPTS

So, nature is a complicated keyword with various meanings popping up in all manner of different situations. As critical theorist Theodor Adorno insisted on many occasions, to examine a key concept like nature is, in effect, to examine the mental and practical conventions of a whole society. But it would be a mistake to limit our attention to this term alone. Williams rightly referred to 'particular *formations* of meaning' (1976: 13, emphasis added) when explaining to readers how they might interpret a signifier like 'nature'. His point was that in any given society at one historical moment there are likely to be families of keywords whose meanings bleed into, and borrow from, one another. In his book *Historical ontology*, the philosopher Ian Hacking (2002: 35) explains this by distinguishing 'concepts' (or meanings) from 'words' (the tokens we use to signify concepts). Following Hacking, we can say that the concept of nature is not exclusively associated with the word 'nature'; instead, the meanings are routinely signified by a range of other words that are (or have become) part of our collective vocabulary. In this sense, 'nature' is something of a 'ghost that is rarely visible under its own name' (Olwig, 1996a: 87). Its meanings often appear as **collateral concepts** (Earle *et al.*, 1996: xvi). That is to say, they are signified by different keywords that refer us to similar or additional referents.

Study Task: Can you identify some commonly used words that communicate one or more of the four meanings of nature? In each case, which meanings are communicated? If, initially, you find this task challenging it will help you to read the second example in Box 1.1.

In the early twenty-first-century, nature's collateral terms include the following, among others: 'environment', 'wilderness', 'gene', 'genius', 'biology', 'race', 'sex', 'biodiversity', 'animal', 'life', 'intelligence', 'human', 'instinct', 'blood', 'reality', 'climate change', 'mind' and 'ecosystem' (see Figure 1.4). Some of these terms are relatively old, others relatively new. Some appear semantically simple and straightforward (though, in actuality, they're not), others more evidently complex. The precise ways in which they partake of the meanings of 'nature' varies, according to both the word and the context of reference. Most of them feature in discussions of both human and non-human nature, but some are used more exclusively. Most of them are also freighted with meanings that go *beyond* those connoted by the term 'nature'. These collateral words are thus only *partly*, rather than exclusively, synonyms for the latter. It depends entirely on the circumstances of their invocation and usage. Let me offer two brief examples, one relating to human biology, the other to the non-human world.

'Race' is almost as old a term as 'nature' itself, and its meanings are almost as diverse and complex. Some are still apt to use the word in a universal sense ('the human race'), thereby denoting our biological differences from other species – differences that were the focus of the Human Genome Project (HGP), a 13-year international attempt to describe the

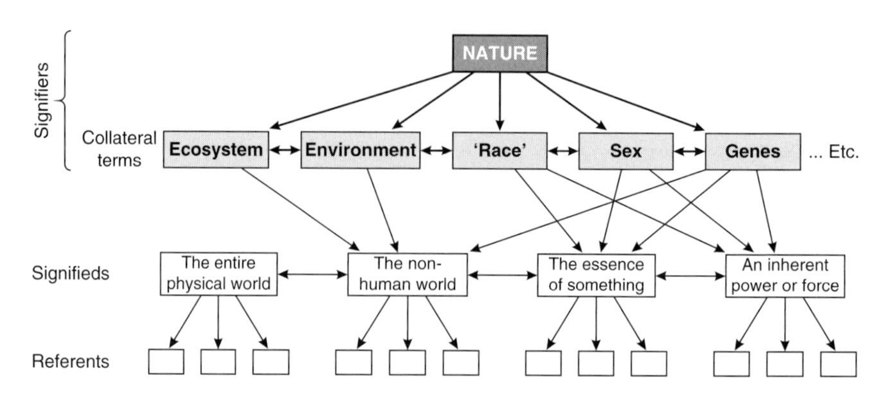

Figure 1.4 Nature and its collateral concepts

The principal meanings of the word nature (i.e. what it signifies) are routinely attached to all manner of material referents by way of other words (i.e. collateral terms). The words and meanings become conjoined in often complicated ways; however, these collateral terms may also signify meanings *beyond* the four signified by 'nature'.

genetic make-up of *homo sapiens* that began in 1990. But many people continue to apply 'race' in a more specific way in order to distinguish different groups of people according to certain criteria. This process of dividing up one species – 'humanity' – into more-or-less distinct constituencies goes back to at least the sixteenth century, the first period of sustained European contact with the 'new world' and its peoples (Hannaford, 1996). Since that time, what the term 'race' means and refers to have varied greatly. One common belief – certainly in the nineteenth century and even today – is that races are intra-specific variations in human biology rather than (say) in ethnic identity or cultural custom. These variations can be regarded as phenotypical (relating to externally visible things like skin colour) or genotypical (relating to the 'internal' character of people's minds or bodies), or both. For instance, one might presume many 'blacks' to be 'naturally' more athletic than Caucasians.

This, at least, was the claim made by science writer Jon Entine (2000) in his book *Taboo: why black athletes dominate sports and why we're afraid to talk about it*. Entine argued that athletes of West African origin tend to have a higher percentage of energy-efficient 'fast twitch' muscles when compared with other members of the human race.[11] Relatedly, the Human Genome Diversity Project (HGDP), set up in 1991 as an independent complement to the HGP, presumed there to be non-trivial genetic variations among humans. Though it studiously avoided the term 'races', the project's decision to sample among different 'populations', for instance, geographically isolated groups in different continents, risked naturalising biological difference in much the same way as Entine's book. When I say '**naturalising**', I mean that observed group-level difference was being presented as something biologically *intrinsic*: as a product of variations in the natural history of humans since the time of 'mitochondrial Eve' some 200,000 years ago. This, as I noted previously, contrasts with the idea that 'races' are – supposing we're prepared even to entertain the term – defined in non-biological terms as contrasting ethno-cultural collectives. It also contrasts with the suggestion that 'race' is a meaningless and malevolent category that we should dispense with. This argument was advanced by cultural critic Paul Gilroy, among others, in his book *Against race* (2000). For Gilroy, there's nothing natural about 'race' – and he's certainly not alone in accepting W. E. B. Du Bois's famous, counter-intuitive insight that 'Race is a . . . cultural fact.'[12]

A rather different collateral term, through which the concept of 'nature' operates, is of much more recent provenance than the keyword 'race'. 'Biodiversity' – a contraction of the term 'biological diversity' – is undoubtedly an established part of our global lingua franca. It refers to the diversity of genes, species and ecosystems within any given region of the Earth – and by implication the whole planet. Yet 30 years ago, it was a new term used only by a few biologists and environmental policymakers. It's said to have been coined by American science administrator Walter G. Rosen in 1985 when deciding on a topical focus for an expert meeting funded by the

US National Research Council. The word first appeared in a publication in 1988 when Harvard University entomologist E. O. Wilson used it as the title of the proceedings of that meeting.[13] By November of that year, the United Nations Environment Programme (UNEP) convened the Ad Hoc Working Group of Experts on Biological Diversity. Its work fed directly into the first UN 'Earth Summit' held in Rio in 1992, where national governments signed up to a global Convention on Biological Diversity geared to conservation and enhancement. That Convention has legal force and, at the time of writing, most countries worldwide are signatories.

This is not the place to explore how, in the space of a few short years, a little-known word became commonplace in a wide variety of debates and arenas worldwide. What interests me is how the meaning of 'biodiversity', notwithstanding the term's relative youth and apparent novelty, is closely linked to the meaning of the much older term 'nature'. Biodiversity is a holistic concept: it encourages us to regard life's variations, not so much as a set of discrete differences between 'natural kinds', but as a set of relationally produced differences emergent from species interaction. Ralph Waldo Emerson once called this 'the unity in variety ... that meets us everywhere' (1836/2005: 29). We're enjoined to focus on 'external nature' as a complex system – be it 'wild' or 'anthropogenic' biodiversity we're considering. Diversity is taken to be an intrinsic aspect of living nature, especially evident in those so-called 'biodiversity hotspots' found mostly in the tropics and subtropics. Thus, the idea of biodiversity trades on the venerable idea of nature as an object (or other) existing 'out there', *and* on the idea that it possesses certain innate qualities that are definitive of its character – so much so that they require active protection. Modern humanity is positioned as a threat to and custodian of biodiversity, at once a destroyer *and* potential saviour.[14]

THE DUALISMS OF WESTERN THOUGHT 1

We've come a fair way in understanding what nature means for us – but not yet far enough. We've seen that the idea of 'nature' is a tool we use to make sense of the world and aspects of ourselves. We've identified its principal meanings and its collateral concepts. And we've established that they appear in a wide range of debates and discussions. To enrich the analysis above, we now need to situate nature and its collateral terms in a much wider semantic field, whose boundaries and contents I'll summarise using the term **discourse**. I utilise this word in a deliberately loose and non-technical sense. According to my *Oxford dictionary*, discourse is communication between two or more people, either face-to-face or at a distance. It's both a thing and an action. We're all creatures of discourse and use a huge repertoire of words and meanings to affect each other and (through our actions) the world we inhabit; however, for this use to be more than babble, we have to achieve semantic consistency over time.

Discourse and discourses

As many analysts have shown, discourse exists at two metaphorical levels that, in practice, intertwine. First, it encompasses all of the major categories and distinctions that serve to make 'reality' intelligible for any given society or group of societies. This 'semantic grid' defines the parameters of what is sayable and unsayable, thinkable and unthinkable, intelligible and non-sense. Linguistic theorist Ferdinand de Saussure once famously called this *langue*. For instance, in Western society it makes no sense to say that ships walk, plants fly or grass thinks. For de Saussure, the signified–signifier–referent connection is entirely *arbitrary*, even though it seems *necessary* to those whose thoughts are conditioned by any given *langue*. Second, within this 'grid' (discourse in the singular), we are free to create a myriad of discourses (in the plural) – but only up to a point. As cultural critic Sara Mills puts it, 'Discourse as a whole consists of regulated discourses' (1997). By this, I take her to mean that any specific act of spoken, written or non-textual communication must – if it's to make sense – conform to the prevalent 'rules' and 'norms' of language.[15] Thus, for all their differences in content and style, the discourses used by Anglophone climatologists and organic farmers (say) must observe the same semantic conventions. Otherwise, these discourses would be unintelligible to non-specialists – literally a foreign language.

Though it operates on and through individuals, a discourse is social in the sense that it's a widely shared mindset and vocabulary – a sort of mental architecture with many inhabitants occupying the same building but moving between different rooms in different ways, and with often different goals in mind. Michael Frayn offer us an alternative metaphor: 'The established usages of language', he argues, 'are part of the standing timber in the great forest that surrounds us' (2007: 356). The verb 'to discourse' is thus crucially related to discourse in both the mass and count noun senses: acts of communication between people are governed by inherited linguistic conventions at both the 'meta' and more specific levels.

Discourses have histories and geographies, and thus are not homogenous through time or across space (Williams rather brilliantly traced some of the continuities and changes in *Keywords* (1976)). In the medium to long term, their parameters and content are not set in stone – though at any given moment in any given society they possess a seeming 'fixity'. Within this apparent fixity, any given discourse, in the singular sense, can accommodate the development of myriad particular discourses (in the count noun sense, akin to what de Saussure called *parole* or 'language in use'). These plural discourses are '*recurrent patterns* of speaking or writing' (Cox, 2010: 63). They posses a certain coherence and distinctiveness relative to other discourses, or they may be colloquial everyday discourses, or very formal and specialised ones. The latter are differentiated according to their aims, their domain(s) of topical interest/concern, and their communicative style (or 'rhetoric'). Typically, they are created by a relative minority of people. I'll call them **epistemic workers**. Through face-to-face and virtual interaction, these workers form **epistemic communities** with more or less distinctive languages. As the semiologist Robert Scholes once said, 'Both the sonnet and the medical prescription can be regarded as forms of discourse that are bound by rules which cover not only their verbal procedures but their social production and exchange' (Scholes, 1982: 144). The boundaries between them and lay audiences are often permeable, such that 'translation' routinely occurs – for instance, the specialised vocabulary of professional meteorologists is daily simplified into short weather reports for public consumption.

Study Task: Select two of the following professionals who produce distinctive discourses about nature: human biologists, physicists, geologists, atmospheric scientists, wildlife conservationists, fisheries managers or environmental economists. Their discourses reflect not only what elements of nature they are interested in, but also what it is about those elements that interests them and how those elements are investigated. Do some Internet or library-based research. For each group, list up to ten specialist terms that illustrate the specificity of their discourses. Despite their differences, what assumptions or terms are shared between the two discourses? These assumptions or terms are likely to be general ones pertaining to epistemology and ontology.

Clearly, my definition of discourse in the first of the two senses is rather at odds with those who, like world-famous Harvard psychologist Steven Pinker, argue that language possesses universal properties hardwired into our highly evolved human brains. For instance, in *The stuff of thought: language as a window into human nature* (one of several of his bestsellers), Pinker says that 'the mind categorizes matter into discrete things (like *a sausage*) and continuous stuff (like *meat*) . . . ' and thus acts as a 'language engine' (2007: 5, 117). Note that Pinker talks about *the* mind, in the universal singular, and (in his book's subtitle) language as a window on to another (putative) universal singular ('human nature'). In the same spirit, the cognitive philosopher Jerry Fodor suggests that people across the world possess a finite inventory of 'innate concepts', which they apply routinely, regardless of culture or location. Pinker and Fodor may well be correct (so too their famous forebear Noam Chomsky), but the fact remains that an awful lot of discursive diversity has existed in the world and still does. This much is evident from linguistics, comparative anthropology and the history of human thought. Whether the diversity is *sui generis* – a product of 'culture' in its various forms as many ethnologists would argue – or a direct result of the latitude built in to the 'universal grammar' discerned by Pinker and Fodor is a moot point.

Tim Ingold's ethnographic work on hunter-gatherer societies and Michel Foucault's studies of what he sometimes called 'epistemes' and 'discursive formations' exemplify discursive diversity in space and time respectively. In the book *Redefining nature*, Ingold argued that present-day

> hunter-gatherers do *not*, as a rule, approach their environment as an external world of nature that has to be 'grasped' conceptually . . . [Unlike Westerners]. . . [t]hey do not see themselves as mindful subjects having to contend with a . . . world of physical objects; indeed, the separation of mind and nature has no place in their thought and practice.
>
> (Ingold, 1986: 120)

Complementary to this act of cultural comparison are Foucault's highly influential attempts to describe historic mutations in ways of thinking within European societies. In both his 'grand' works, such as *The order of things*, and his more circumscribed ones, such as *Discipline and punish*, Foucault charted the rise of distinctive new templates for categorising and characterising the world (1970, 1979). In turn, these templates produced new objects of analysis and action. For instance, he argued that the rapid changes to European medicine in the late eighteenth century were not simply reflective of 'better science' (Foucault, 1973). For him, they were as much a reflection of new metaphors chosen to characterise the body. In Foucault's view, discourses were both mutable and material: as they waxed and waned over time, the world not only *looked* different but, in a real sense, *was* different because of the practical efficacy of ruling discourses.

This is why Butler prefers the term 'discursive practices' to 'discourse' because these practices can, given time, 'produce the effects [they] . . . name' (1993: 2). Whatever stability and normalcy these practices possess, she argues, arises from their repeated and routinised performance: they allow us to *do*, as much as to *know* (think of legal discourse, for example, which is eminently performative). As I will explain in Chapter 2, this is, in part, because the practices are realised in and through various material infrastructures. Their 'validity' and potency is in no way a function of some 'extra-discursive' realm that they either represent objectively or operate upon 'properly'. Instead, Butler argues, the very idea of an extra-discursive realm is *itself* a contingent part of modern Western discourse. It is a way we've come to secure the content, purchase and legitimacy of many of our utterances.

Contemporary Western discourse and its antonyms

To return to the distinction between *langue* (discourse as a mass noun) and specific discourses (e.g. those employed in medicine, ecological restoration or nuclear physics), Butler's suggestion that the representation–reality distinction is constructed (not given) speaks to a much larger set of categorical contrasts that Westerners rather take for granted (see Figure 1.5). These dualisms, as several previous commentators have noted, comprise a 'semantic rule book' that at some level governs specific acts of verbal and visual representation. They are the epistemic means by which we assure ourselves that there is a measure of 'ontological hygiene' abroad in the world. These binaries create boundaries and differences that are presumed to inhere in reality, in part 'by nature', in part by design.[16]

Have a close look at Figure 1.5. Do most of the dualisms listed appear familiar to the point of being 'common sense'? If so, it's because we're so accustomed to using and encountering these various antonyms

'HARD' AND 'SOFT' ANTONYMS		
Nature	⟷	Culture and nurture
Environment	⟷	Society
Sex	⟷	Gender
Wilderness	⟷	Cultivated, cleared and settled land
Race	⟷	Social identity
Biology	⟷	Social conventions and practices
Genes	⟷	Cultural norms, habits & rituals
Organism	⟷	Social group
Real	⟷	Fake
Instinct	⟷	Free will
Rural	⟷	Urban
Countryside	⟷	City
Natural world	⟷	Built environment
Emotion	⟷	Reason/rationality
Woman	⟷	Man
Mother	⟷	Father
Savagery	⟷	Civility and civilization
Object	⟷	Subject
Reality	⟷	Representation
Authentic	⟷	Artificial
Raw	⟷	Cooked
Wild	⟷	Tame, domesticated or disciplined
Traditional	⟷	Modern
Physical	⟷	Mental
Body	⟷	Mind
Matter	⟷	Spirit, ideology and belief
Innate	⟷	Manufactured
Animal	⟷	Human
Fact	⟷	Fiction, conjecture and speculation
Truth	⟷	Falsity
Ontology	⟷	Epistemology
Observed	⟷	Observer
Disorder	⟷	Order

Figure 1.5 Fundamental dualisms of Western thought since the European Enlightenment

Depending on the precise context of their use in acts of communication, each of the terms on one side of the figure can imply its opposite term in a 'hard' or 'soft' way. Some of the terms, again depending on the context of use, have ambivalent meanings, slipping and sliding between and across both sides of the figure. For instance, as indicated earlier in this chapter, 'race' can be used in non-biological ways to designate cultural differences between groups of people.

in a wide range of different contexts. Clearly, 'nature' and its collateral concepts assume their meanings within this larger family of dichotomies. For instance, there's long been a (dubious) equation in Western discussions of 'aboriginal' (or First Nations) peoples between certain meanings of the terms 'nature', 'savagery', 'wild' and 'emotion' presented as antonyms to 'culture', 'civilisation' and 'reason'. I say antonyms because the terms in Figure 1.5 are frequently understood to be *opposites*, or (at the very least) to designate qualitatively different aspects of the world. As geographer Gunnar Olsson long ago noted, it appears as if 'Everything is identical to itself and nothing is identical to anything else. Nothing is itself and not itself at the same time' (Olsson, 1980: 62b). For instance, the question 'do glaciers listen?', which makes sense in some aboriginal Canadian cultures (Cruikshank, 2005), simply does not compute in Western discourse for obvious reasons. Likewise, the celebrated environmentalist Aldo Leopold's (1949) injunction to 'think like a mountain' would strike most of us as nonsensical (or else counter-intuitive). They both rest on an apparent category mistake.

What's more, the oppositions in Figure 1.5 often appear to be *hierarchical*, with one side coded positively and the other not. Yet this appearance is, on closer inspection, something of a mirage. One of the paradoxes of binary thinking is that in order for any one antonym to make sense it must necessarily imply the meaning of the very thing it is the supposed opposite of – what post-structuralists are wont to call its 'constitutive outside'. In Candace Slater's words, each antonym casts a 'conceptual shadow' over its putative 'other', so functioning as an absent presence (2003: 10). This emphasises the point that, upon close inspection, words are *not* like metaphorical mirrors (or windows) whose meanings map neatly on to a world already divided into discrete chunks. But there are three other important complications too.

THE DUALISMS OF WESTERN THOUGHT 2

The first complication is that 'nature', its collateral terms and the wider set of dualisms listed in Figure 1.5 can all have cognitive, moral and aesthetic dimensions. In simple terms, **cognition** pertains to empirical questions (that is, to describing and explaining the world) or to the use of logical reasoning. Meanwhile, **morality** (or ethics) involves making value judgements. Finally, **aesthetics** concerns what we regard as beautiful or ugly, moving or offensive, and so on.

Study Task: If you read this book's Preface, refer back to the seven vignettes included there. Otherwise, read these for the first time now. Then select two examples from within each vignette that relate to cognition, morality or aesthetics, as appropriate. Note that many references to nature involve two or all three domains *simultaneously*. For instance, in the 'Crime against nature' vignette, feelings of disgust about gay sex are combined by its critics with the moral argument that it 'goes against the natural order'. However, often cognitive, moral and aesthetic references to nature are made separately, as you should discover.

As the task should have demonstrated, it is almost always possible (but not obligatory) to refer to natural phenomena in all three registers. For instance, if I were an Earth scientist I could describe and explain the nature of Arizona's Grand Canyon with reference to geological and hydrological processes operating over millennia. If I were an environmentalist, I might moralise about this self-same landscape, arguing that its striking geomorphological uniqueness demands that it be actively protected from 'human interference' in perpetuity. I might, furthermore, draw an ethical lesson from the Canyon's great age: that it reminds us how short human occupancy of the planet has been and yet how quickly we're destroying a natural world that provides us with material and spiritual sustenance. This ethical argument would, in turn, have an aesthetic basis: those fortunate enough to visit the Canyon enjoy and are often moved by the sheer beauty and scale of what they see. Here, then, our reference to a feature of the 'natural world' operates in the domains of 'reason', valuation and emotion – it involves both head and heart.

This simple example reveals that the antonyms of Western thought not only divide the world into ontological pieces, but are often **normative** too. They are frequently enrolled in arguments about how the world *ought to be* for moral, aesthetic or pragmatic/practical reasons. They help us delimit things that become objects of ethical and emotional concern (or not, depending). In this sense, they are rarely apolitical or innocent. Environmental scientist Barry Commoner's well-known 'four laws of ecology' (formulated in 1971) are a case in point (Commoner, 1971). Law number three, 'nature knows best', runs the 'is' and the 'ought' together seamlessly in one pithy, but grand, formulation. It has an aphoristic quality that makes it seem self-evidently correct to many upon first hearing.

The second complication is that the terms listed in Figure 1.5 each have more than just *one* set of established meanings and connotations. In a famous essay entitled 'Is female to male as nature is to culture'?, the anthropologist Sherry Ortner argued that, in several societies worldwide, women are unthinkingly linked to 'nature' and 'emotion' more than to 'culture', and that this is a way in which men maintain control over the minds and bodies of females (Ortner, 1974; see also Lloyd, 1993; Roach, 2003). While Ortner's observation about the symbolic and semantic aspects of patriarchy was (and remains) richly suggestive, it also gave the questionable impression that the meaning of 'keywords' remains historically invariant and one-dimensional. Williams, whose book *The country and the city* was published the year before Ortner's essay, argued otherwise (Williams, 1973). His historical analysis of what 'urban' and 'rural' had come to signify in England since around 1600 showed that, while the dualism has survived the centuries, its constitutive terms meant contradictory things *simultaneously*:

> On the country has gathered the idea of a natural way of life: of peace, innocence and simple virtue. On the city has gathered the idea of an achieved centre: of learning, communication, light. [But] ... powerful hostile associations have *also* developed: on the city as a place of noise, worldliness and ambition; on the country as a place of backwardness, ignorance, limitation.
>
> (Williams, 1973: 9, emphasis added)

A similar argument could be made about the idea of 'wilderness', which has been especially influential in North America. The raft of cognitive, moral and aesthetic meanings that have attached themselves to this term, and its antonyms, reveal inconsistency and variety in what the word is made to signify by different commentators and organisations. The variety is certainly not endless. But my wider point is that the dualisms that underpin so much of our thought and action are not employed by us in simple, mechanical or wholly consistent ways. There's a certain amplitude of meaning and use, definite – though often interfering – patterns of signification in the plural. As Michael Thompson and colleagues mention in their book *Cultural theory*, 'ideas are plastic; they can be squeezed into different configurations but, at the same time, there are some limits' (Thompson *et al.*, 1990: 25). There are, we might say, 'repertoires of meaning' attached to semantic binaries.

A third and final complication that characterises the discursive architecture I've been discussing relates to what we can call **ambivalent categories**. These categories, in both their semantics and (some or all) of their material referents, do not map neatly on to (or into) Figure 1.5's antonyms. Instead, they signify in more than one direction (e.g. they can have opposed cognitive meanings or moral implications). As such, we typically find them either troubling or immensely useful, depending on our circumstances and goals. They reveal both the *power* of dualisms in organising Western thought, and yet can be resources for *challenging that power*. Consider what terms like 'hybrid', 'transgender', 'homosexual', 'alien', 'monster', 'miscegenation', 'bisexual', 'cyborg', 'transsexual', 'clone', 'zombie' and 'androgynous' signify and refer to. Each is a boundary-crossing category that points us to things that appear to exist outside of, or in between, the antonyms of Western thought. These things strike us, ontologically speaking, as being unusual, odd, impure, syncretic and (perhaps) even threatening or alluring. They're about unconventional mixings and transgressions, and so give us pause for thought.

Mary Shelley's book *Frankenstein* (published in 1818) is emblematic in this respect. Often misinterpreted purely as a morality tale about the dangers of unregulated science and technology (its subtitle is *The modern Prometheus*), her story of Victor Frankenstein's 'monstrous' creation can also be understood as a reflection on the limitations and exclusions built in to prevailing social conventions. One of the book's most subversive ideas is that Frankenstein's progeny is not *born* a monster but *ascribed* this status by those around him. He becomes monstrous because of the way supposedly 'civilised' humans treat him rather than because of his singular biology. He's denied the possibility of ever being deemed by people to be 'rational' and 'cultured' because his origins and appearance are considered far too 'abnormal' by a family he seeks to befriend. Tragically, the murder he eventually commits is interpreted as positive proof of his intrinsic monstrosity rather than a result of his social stigmatisation and the psychological harm it inflicts on him. Shelley's monster thus acts as a critical mirror: he asks his makers, 'Can you love and treat kindly a being as ugly as me, as uncertain in his status as a "person" as me?' (Morton, 2010: 112). Almost two centuries later, and the term 'Frankenfoods' has been routinely used by critics of genetically modified crops. As with misreadings of Shelley's book, this moniker implies that a Maginot Line should be maintained between 'nature' and 'society'. Any mixing of the two, in this case by utilising recombinant DNA technology, is said by some to be wrong or undesirable for a range of practical, moral or aesthetic reasons.

Of course, boundary-transgressing entities, be they human or otherwise, do not always attract criticism or revulsion. Many people look favourably upon genetically modified organisms, and see them as being no different in kind to the sort of crossbred plants and animals that societies have utilised for centuries – they are no more (or less) 'unnatural'. The same could be said of artificial life forms, which will no doubt be commonplace by the end of this century. Gay and bisexual individuals present a rather different case. Many have defended their right to depart from the man–woman model of 'normal sex', in part by challenging the conventional idea that sexual preference is (or ought to be) dictated by one's hormones or genital anatomy.[17] Similarly, some members of the transgender and transsexual communities have contested the normative demands imposed by the expectation that 'male' and 'female' are discrete gender identities never to be blended or confused.

Then there's the extraordinarily complex notion of what it is to be 'human'. I placed this term on the right side of Figure 1.5 but, in truth, it is often used in ways that are far less clear-cut. Since most of us consider our biology to be a key part of what makes us what we are, and because this biology is widely acknowledged to be the product of a slow and astoundingly complex process of species engagement with a wider environment, our 'humanity' cannot simply reside in our capacity to, as it were, 'escape' the left-hand side of Figure 1.5 and inhabit the other side.

Study Task: Make a list of ten things you consider to be distinctive about humans relative to other living species. Then, try to map them on to one or other side of Figure 1.5, using the relevant categories listed there.

Did all ten things you have listed belong to just one side of Figure 1.5? Chances are they did not. This is why Neil Evernden calls us 'natural aliens': we appear cusped between the natural and non-natural (Evernden, 2006). Tim Ingold also expresses it well: '"human" is a word that points to the existential dilemma of a creature that can know itself and the world of which it is a part only through the renunciation of its very being in that world' (Ingold, 2011: 113–14). Our life condition appears to be 'both/and' rather than 'either/or', obliging us to use the contradictory ideas of nature as 'external' *and* 'universal' when discussing ourselves.[18]

I say contradictory because suggestions that biology 'governs' what some take to be aspects of culture, socialisation and upbringing can cause real alarm and very heated discussion. For example, there was something of an outcry when the evolutionary biologists Craig Palmer and Randy Thornhill published *A natural history of rape: biological bases of sexual coercion* in 2000. According to some, by portraying men raping women as an evolutionary adaptation, Palmer and Thornhill were diverting attention away from the important non-biological reasons why rape occurs. Some even suggested they were arguing the immoral and harmful actions are 'natural' – a 'fact' of human existence. In light of this and other controversial cases, it's no surprise at all that many philosophers, poets, novelists, sculptors, painters and scientists still feel compelled to debate the question 'what makes us human?' at some length (for example, see Fernandez-Armesto, 2004; Pasternak, 2007).

METONYMY: THE 'NATURE EFFECT'

I want to round out this description of how 'nature' takes on meaning for us by discussing a dualism I've not mentioned so far, but which has been implicit throughout: that distinguishing the particular and the general, the concrete and the abstract. Again, we might regard this as a rather Western dichotomy, even though it may appear to be universal. We're very accustomed to regarding any given thing not only as something in its own right, but also as a representative of (or else departure from) something larger or broader. This could be a law, a system, a norm, a class, a force, or a process that transcends any one moment of time or single portion of space. Whatever the exact details, it involves what literary critics call **metonymy**: reference to a given thing *also* involves implicit reference to something else, which the thing comes to stand for.[19] Arguably, the class of phenomena that any one thing is said to represent is less a function of a thing's physical properties and more a function of a society's classification systems.

Timothy Mitchell's (1988) book *Colonising Egypt* is one of several fine studies that make this point in compelling detail. A contribution to cultural history, it aims to make plain the particular and peculiar worldview of late nineteenth-century colonists from France and Britain when encountering non-Western peoples. Mitchell recounts a visit by four Egyptians to the World Exhibition held in Paris in 1889. In the late nineteenth century, such exhibitions were major events and a means for different countries to showcase their distinctive contributions (cultural, technological, etc.) to the wealth of nations. At the Paris Expo, the Egyptian visitors encountered a life-size model of a winding street in old Cairo. The street was realistic in almost every detail, thus a piece of 'real Cairo' was evoked by way of a full-scale reproduction. Later, when the Egyptians visited Stockholm they found themselves stared at in much the same way as the model street was stared at by the Europeans visiting the Expo. They were objects of curiosity because they were so visibly different in appearance to west Europeans.

Mitchell's argument was two-fold. First, he suggested that even before the twentieth century, Europeans were accustomed to supposing that specifics 'stood for' larger realities. The street model, he maintained, not only referred viewers to the 'real Cairo' by virtue of its apparent realism; less obviously, it conjured up something much bigger, namely 'the Orient'. As Edward Said (1978) showed in his celebrated study of nineteenth-century British and French colonial thinking, the Orient was not so much a large geographical region as a very simplistic European idea that was taken for granted as a general depiction of that region. It cast 'the Orient' as Europe's Other: an extended space of disorder, irrationality, darkness, libidinal energy and uncleanliness that was at once attractive and repellent. This is a case of what semiologist Roland Barthes once called **connotative reference** (in contrast to literal, 'denotative' or 'indexical' reference). In light of this, Mitchell's second point was far more radical and unsettling than his first. If a European left the Paris Expo and visited the 'real Cairo' they would not, he insisted, be any nearer to the Orient – except in a purely locational sense. Instead, 'Outside the exhibition ... one encountered ... only further models and representations of the real' (Mitchell, 1988: 12). The 'reality effect' of the Parisian model, Mitchell suggested, is that it created the illusion of a distinction between the copy and the real thing, when in fact *the latter was interpreted by visitors in the very same imaginative terms as the former*. This, then, is a peculiar form of metonym in which the process of signification is, despite appearances, purely circular.

We can, it seems to me, talk in analogous terms about a **nature effect**. Though we're often not conscious of it, when we describe specific things, processes, sites or regions as natural – genes, giant pandas, volcanoes, melting sea ice, Antarctica, the Himalayas, or polar bears, for instance – we immediately connect them to something conceived as being much larger or wider than themselves. Some people do this intentionally as a matter of professional practice, such as physical scientists or environmental activists. The reason, it seems to me, is because the conventional definitions of nature actively *invite* us to do so: each is predicated on the assumption that 'nature' is not any (or only) *one* thing in particular, but the *totality* of things, their myriad interrelations and their qualities.

Study Task: Take a minute to think about polar bears and giant pandas. Both are iconic species. The former inhabits the Arctic region and is the largest land carnivore on the Earth. Meanwhile, there may be as few as 1,500 wild pandas, with over 200 in captivity. When you think of polar bears and giant pandas, what are they metonyms for?

I earlier mentioned the 'blue planet' imaginary, a particular framing of 'global nature' or the 'web of life' that is now common sense. This imaginary, it seems to me, is a pervasive context in which specific references to particular elements of nature are these days implicitly situated.[20] Things like polar bears and giant pandas are metonyms for loss of nature at the hands of humans on an epic scale. Similarly, in 2010, there was a lot of media reporting about the death of bees in Britain, United States, Canada and worldwide – the so-called 'colony collapse disorder'. Notwithstanding their capacity to sting, bees have long been a favourite insect in Britain, my home country. Attractive and industrious, some of them produce that delicious confection my children love to have spread on their morning toast (honey). Their apparent demise thus carries an emotional charge for many, but it arguably signifies something else too: our continuing ability to 'disrupt' 'natural systems' and the 'balance of nature' worldwide. This much is obvious from Alison Benjamin and Brian McCallum's book (2008) *A world without bees*, whose title announces explicitly the link between the general and the specific, the global and the local. Such is the level of concern about the wider effects of bee decline that several British research funding bodies launched, in mid-2010, the Insect Pollinators Initiative. The largest programme to date of its kind, it was designed to look at the multiple reasons thought to be behind this devastation in the bee population. See Box 1.2 for a second example of the 'nature effect'.

BOX 1.2

THE 'NATURE EFFECT' IN THE SCIENCE OF ECOLOGY

If bee decline is today a metonym for ecosystem destruction worldwide (i.e. a shrinkage of non-anthropogenic 'external nature'), it is also a metonym for a failure to respect the particular qualities of the biophysical world (i.e. an alteration to 'intrinsic nature') and its governing logics (i.e. a disruption of 'super-ordinate nature'). The latter two forms of metonymy are scarcely new. The science of ecology makes this plain. For decades, ecologists believed that the 'natural' state of most ecosystems was to be stable and ordered. For instance, the early twentieth-century researchers Frederic Clements and Arthur Tansley suggested that plant communities, given time, adapt to the prevailing climatic, pedological and other environmental conditions found in any given locality or region. These conditions were presumed to be consistent over a period of hundreds if not thousands of years, resulting in a 'climax ecology'. Similarly, biologists studying wild animals long believed that predators and their prey existed in a homeostatic relationship: the numbers of both, it was suggested, tended to hover around a mean because an 'excess' of one would necessarily be eliminated by a 'scarcity' of the other. After 1950, these ideas slowly fell out of favour and a new paradigm emerged, namely 'non-equilibrium ecology'. It was showcased in an influential book by American biologist Daniel Botkin, counter-intuitively titled *Discordant harmonies* (Botkin, 1990; see more recently, Kricher, 2009). This 'new ecology' suggests that ecosystems and species (contrary to older beliefs) are profoundly shaped by disturbances, routinely cross operational thresholds and normally exhibit variable behaviour in time and space. Despite this, it has two key similarities with its predecessor. First, it presumes that 'nature' possesses a signature quality waiting to be discovered – in this case, its 'disorderliness'. Second, it is also assumed that any particular ecosystem or species can tell us something about this general characteristic of life on Earth.

For instance, here is the 'new' ecologist Stephen Budiansky writing about the giant beech trees of southern Chile a few years after Botkin's book was published:

> The first botanists who entered these remote and towering woods believed they had discovered a classic example of a forest climax community ... It took a long time for [them] ... to admit it, but there was something terribly wrong with this explanation. Nothofagus seeds are virtually absent from the forest floor. Other trees are present in all sizes and ages; the false beeches appear only in the narrow age-band of oldest trees that form the canopy. What keeps these other species from ousting the false beeches ... ? ... [I]t is nothing short of repeated disaster that is responsible for the illusion of stability ... Nothofagus are adapted not to stability but chaos. The mountains are regularly shaken by powerful earthquakes ... when the hillsides are laid bare by these upheavals, Nothofagus seedlings grow in abundance. By the time that shade tolerant competitor species have invaded ... the false beeches have gained a substantial head

start ... Rather than an intrusion that upsets nature's timeless balance ... disturbance is a necessity for much of the life on this planet.

(Budiansky, 1996; 69–70)

Here again we see metonymy at work, with reference to a specific biophysical entity (a single tree species) being made to speak to much grander themes concerning nature as a whole: its natural lack of stability and regularity in certain situations. The most obvious – least subtle – metonyms take the form of symbols and icons whose role is, quite deliberately, to evoke a larger or grander meaning on the basis of a single word, phrase or image. See Box 2.3 for more on icons.

Of course, many would argue that this example and that in Box 1.2, relating to bees and trees, respectively, are metonymic in an *ontological* rather than simply epistemological sense. In other words, they might say that British bees can be made to stand for environmental degradation and Chilean beeches for nature's intrinsic disorderliness because they really, actually *do* stand for these things. I do not want to deny these claims, but neither do I want to affirm them uncritically. My specific (and more limited) aim is to show that the 'nature effect' attends many (perhaps most) specific references to parts or aspects of the biophysical world. This is obvious when general categories, like 'old growth forest' or 'oil reserves', are used to refer to specific cases thereof. But think too of Dolly the sheep, the recently deceased Lonesome George, the already mentioned Amazon Basin, blood-stained ice and baby fur seal carcasses, the famous images of a human foetus floating in amniotic fluid, or the shrinking perimeter of the Antarctic ice sheet. In each case, particulars are evocative vehicles for denoting wider 'realities'. Similarly, the famous environmentalist injunction to 'act locally, think globally' resonates with many of us because it appears to be based on the common-sense idea that small parts together make a larger integrated whole. What these examples suggest is that even when the idea of 'universal nature' is not being communicated explicitly, it is often being communicated implicitly by reference to specific and visible 'natural' phenomena. As historical geographer Kenneth Olwig puts it, 'Despite the fact that nature is one of the most abstract ... concepts we have, [it] ... nevertheless signifies all that is concrete ...' (Olwig, 1996b: 380).

SUMMARY

It should by now be abundantly clear that I intend to take a thoroughly 'non-natural' approach to those things that we consider to be natural, in kind or degree. This is hardly original on my part, even if readers new to the ideas presented so far may find them thought-provoking (as I hope they do). In this chapter, I've argued that 'nature' is not a given, waiting to be analysed, experienced or interacted with. I've suggested, instead, that it's a way of categorising and labelling the world, one that's best understood in the context of its collateral concepts and the wider discursive 'rule book' characteristic of Western societies.

'Nature' should, by rights, belong to the family of what philosopher Walter Gallie (1956) famously called 'essentially contested concepts', such is its range and variety of meaning, reference and application. We should always be on the lookout for the diverse and complex uses to which the term and its filial concepts are put in a wide range of sites and situations. And yet, strangely, it's easy to forget this fact and presume that nature is a given – something 'mind independent', or at least independent of any thoughts particular minds might have about it. In the next two chapters, I focus less on what nature is made to mean by us and for us, and focus much more on meaning-making as an organised and differentiated social process, though one that most of us only participate in as recipients and consumers. By the end of these two chapters, readers should be able to understand why the ideas presented in this chapter matter – and why the overall approach I am taking to 'nature' and its collateral ideas is more than simply intellectual interest.

Before I proceed, let me make it clear that, in seeking to 'denaturalise nature', I am not advocating for the demise of this signifier or its collateral terms. On the contrary, I think 'nature' is here to stay – at least for a while longer. Literary scholar Timothy Morton's (2010) plea to dispense with the language of 'nature' in order to make space for what he calls 'the ecological thought' strikes me as utopian, if not premature. As Morton himself rightly acknowledges, 'since we've been addicted to Nature for so long, giving up will be painful' (2010: 95). It seems to me that we're not yet at the point where we're prepared to experience the pain in order to apprehend the world in new ways. Until we are, we should focus more on managing the addiction effectively rather than seeking a cure.

ENDNOTES

1 Though many of us have little difficulty acknowledging our 'animal nature', it's worth recalling that a generation ago a great many people had learnt to define their 'humanity' in strictly non-biological terms – hence the shocked reaction (even publication bans, in some countries) to Desmond Morris's (1967) bestseller *The naked ape*. This reaction recalls the even larger negative reaction in some quarters to Darwin's (1859) *The origin of species*, which scandalised Victorian society by proposing that *homo sapiens* were accidental creations of nature not God.

2 This way of talking and thinking underpins the long-standing debates in both academia and public life about our purported 'domination of nature' and the 'human impact' on the natural environment. Many believe that the society–nature dualism in Western thought is at the root of present-day environmental degradation and our casual disregard for the non-human world.

3 The so-called Deepwater Horizon incident occurred when an exploratory drilling rig caught fire and the well cap on the ocean floor then failed to stem the flow of oil. The leakage lasted many weeks and constituted the largest offshore oil spill in US

history. The 2011 Pacific Ocean tsunami affected coastal areas of Japan, causing death, destruction and damage on an epic scale.

4 This 'tethering' of otherwise diverse human customs, preferences and behaviours to biological universals is a favourite theme of evolutionary psychologists and some neuroscientists. Their work has been packaged and popularised by the likes of Sam Harris (2011) and Peter Corning (2011).

5 In some ways, both the social construction and the social use of the idea of 'reality' are at least as complex, interesting and consequential as the constitution and use of the notion of 'nature' and its various collateral concepts – my focus in this book. The same might be said of the idea of 'truth', which is intimately connected to the concept of nature in many contexts of use.

6 This argument is nicely encapsulated in the title of Lewontin's single-authored volume, published 16 years later, *The triple helix: gene, organism and environment* (Lewontin, 2000). More recently, it's articulated by neuroscientist Lise Eliot (2009), in her book *Pink brain, blue brain* – a refutation of the idea that sex differences in human mental functionings are, at base, natural and thus largely immune from social conditioning and engineering. It's a testament to how enduring the idea that 'nature' is, that the debate about what it means to be human remains so tightly framed by a 'nature–nurture' dichotomy – witness the American philosopher Jesse Prinz's (2012) book, tellingly entitled *Beyond human nature: how culture and experience shape our lives*.

7 References to 'universal nature' can also go beyond the earthly sphere, of course, and describe our solar system, the galaxy, the cosmos, the so-called Big Bang and so on.

8 In my previous book, *Nature*, I identified *three* principal meanings, while others (e.g. Habgood, 2002) identify several more. There is no consensus on the matter in the published academic literature. Even so, I'm confident – after some years of reflection on the matter – that the quartet of meanings I itemise here are the ones most commonly used when 'nature' is invoked in discussion. All four do not, of course, usually come into play at the same time in any given instance of nature being referred to.

9 One recent example of how 'degrees of naturalness' matter relates to environmental conservation and restoration: there are debates about how 'informed choices' can be made by conservation professionals about which 'nature' from the past needs protecting or recreating in the present. See, for example, Nigel Dudley's book *Authenticity in nature* (Dudley, 2011).

10 The philosopher and historian of ideas Arthur O. Lovejoy (1873–1962), in several of his philological writings, identified more than sixty meanings of 'nature', variously differentiated by degree (mostly) and kind (more rarely).

11 Entine argued that, if we accept that many Africans and those with African descendents are susceptible to sickle-cell anaemia (which medical professionals do), then why not also entertain the idea that their success in competitive sports has a biological basis? Unsurprisingly, Entine's book proved to be controversial. Some suggested that it made the science of physiology an accomplice in creating racial differences. Ben Carrington, for instance, argues that the idea that blacks are 'naturally athletic' is very much a Euro-North American one that can be dated to the late nineteenth century when, for the first time, modern nation states had to cope with the pressures of large-scale immigration and the politics of bringing different cultural and linguistic groups together in the same territorial space (Carrington, 2010). As Carrington explains, the idea is – in its origins – racist because it creates a putative biological difference between people and then seeks to control the life chances of one group at the expense of the other. In his book *Darwin's athletes*, John Hoberman (1997) offered a powerful critique of the sort of 'biological essentialism' found in Entine's book – more's the pity that Entine ignored it.

12 So-called 'race critical theory', which has been very influential in the Western social sciences and humanities since the 1960s, takes du Bois's insight as axiomatic. 'Race' is seen not as a set of natural differences but as a set of efficacious *social* categories *constructed in terms of supposedly significant biological differences*. This said, Lawrence Hirschfield (1996) has sought to provide an ingenious alternative explanation for

racism. He has argued that while humanity is not divisible into 'races' on biological grounds, humans have a mental disposition to so divide, meaning that racism is, in effect, natural while 'races' are not. Arguably, this distinction between 'race' (=culture) and racism (=nature) risks implying that there's nothing to be done about the latter! In Chapter 4, I explore the use of the idea of 'race' in molecular genetics.

13 And, like many other analysts, Wilson went on to invest the neologism with systematic content in the co-edited book *Biodiversity II* (Reaka-Kudla *et al.*, 1997) and the single-authored book *The diversity of life* (1992).

14 Two fine books that elucidate the way the idea of 'biodiversity' has been both invented and utilised are *The idea of biodiversity* (Takacs, 1996) and *Saving nature's legacy: origins of the idea of biological diversity* (Farnham, 2007). By contrast, in their book *What is biodiversity?* biologists James Maclaurin and Kim Sterelny try, unsurprisingly given their 'scientific' backgrounds, to fix the concept's meaning and referents with reference to genetic, specific and ecosystemic variation (Maclaurin and Sterelny, 2008).

15 Over the past 30 years, the Anglophone social sciences and humanities have been intensely preoccupied with 'discourse' and 'representation' (I examine the latter in Chapter 2). This interest has arisen out of a belief that language, symbolism and denotation are not secondary to 'material concerns' (like earning a living) but key elements in the structuring of the world. The key sources of inspiration for this 'linguistic, cultural and representation turn' was a convergence of traditional linguistics, semiotics, sociolinguistics and philosophies of language – all of which synergised in the field of 'cultural studies' from the late 1970s. This turn has now, it seems, run out of steam in the Anglophone academy.

16 A number of critics have suggested that the Western imaginary is 'mechanical' in a metaphorical sense, this having very practical consequences for how we engage with 'nature'. William Leiss (1974) famously went further and argued that by representing the physical world as a separate realm of 'nature', modern Westerners were able to 'dominate' it (or treat it as something to *be* dominated). I am not, however, making that argument here but rather a broader one about dualisms as providing an overarching mindset for a *range* of more specific imaginaries of the real.

17 When introducing nature's collateral concepts earlier in the chapter, I made the point that these concepts are not necessarily always simply synonyms for 'nature'. The related terms 'sex' and 'sexuality' illustrate this point well. They can refer to (1) biologically reproductive activities; (2) non-reproductive sexual acts (e.g. fellatio or anal penetration); (3) the specific physical–emotional forms of attraction underpinning (1) and (2); and (4) subject positions and identities of individuals based on (1)–(3). Here 'nature' does not necessarily feature in (2) through (4), depending on how one chooses to understand 'sex' and 'sexuality' in these cases.

18 The idea of 'human exceptionalism', much used in bioethical and legal discourse, aims for the purity of a categorical (and putatively ontological) difference between humans and other creatures. It struggles, in other words, against the evident fact of humanity's membership of the wider 'natural world'. I examine one concrete instance of the 'work' this idea performs in Chapter 5 when I consider a case of bestiality in Washington State in the United States.

19 I'm not going to quibble here about whether 'metonymy' or 'synecdoche' is the correct term. It's sometimes said that the former involves an *external* relation (e.g. pen stands for writer), while the latter stands for an *internal* relation (e.g. sail for ship). Metonymic references can be purely metaphorical, of course, but my focus is an those that are taken or intended to be literal – where one thing is used to *stand for* another.

20 Metonyms and metaphors often work together to convey meaning. A metaphor, of course, is a representation of one thing in terms of another. What I'm calling the 'nature effect', which is metonymic, is frequently achieved through metaphors – such as the Earth as a 'spaceship' or 'ark' and of life as 'system' or 'web'. Brendon Larsen's (2011) new book on the use of metaphors in discussions of the natural environment is illuminating and offers many examples of metonymy in action.

2 REPRESENTING NATURE

If the discussion in Chapter 1 seemed rather grand or abstract, let me now bring it a little closer to Earth. I said earlier that our particular beliefs about, and experiences of, nature – even including our own flesh and blood – are highly mediated. We rely on myriad others to form our own ideas, hopes, opinions, values and worries about everything from the major volcanic eruption in Iceland in 2010 to the implications of synthetic biology to the strange wonders of black holes. These others both produce and disseminate various representations of nature, which we, whoever 'we' happen to be in any given situation, are invited to consume, critically or otherwise. What, though, is **representation** – both as a process (*re*-presenting) and as the results of that process (e.g. a photograph, newspaper article, documentary or poem)? Who is doing the representing? And why? These are the questions that preoccupy me in this chapter. My answers to them will directly inform the arguments of the next one. They will get you thinking about who you pay attention to in your own life, and for what reasons.

Before I begin, let me make one thing clear. Our reliance on others for our understandings of 'nature' (and not just this) is hardly unique to our time. It was ever thus. However, I believe that this reliance is just as pronounced today as it was, say, four centuries ago. In the period prior to the so-called Enlightenment in Europe, a cadre of clerics, merchants and nobles exerted considerable influence on the minds of the citizenry in countries like Britain. In my view, we are today no less subject to the claims and contentions made by others than our supposedly less 'enlightened' forebears. What's arguably different is that we find ourselves influenced by a greater number and diversity of spokespersons for 'nature' (and everything else) than perhaps ever before. Today we're positioned as consumers in a *very* busy marketplace of information, ideas, incitements and experiences. As Andrew Barry notes, 'One of the characteristic features of contemporary . . . life is the extraordinary range and quantity of information that citizens ... are expected to process' (Barry, 2001: 153). What's also different today is that most of the spokespersons presenting this information belong to communities and institutions whose inner workings are hidden from view and elude our everyday understanding – this despite us living in an age of supposed 'transparency' where important information is, according to some, more readily available than heretofore.[1]

When the German philosopher Immanuel Kant famously asked the question 'What is Enlightenment?' back in 1784, he could not have foreseen a world in which so many people's knowledge would be shaped by so wide an array of organisations, groups and individuals communicating via such a multiplicity of media. In this light, the issue is not how we can become truly 'independent knowers' – because this is strictly impossible. Instead, the key question is how, precisely, should we relate to all those who seek to shape our thoughts, beliefs, norms, feelings and actions? The answer has practical implications for realising Kant's famous and stirring injunction '*aude sapere*' ('Dare to know!'). I'll attend to these implications in Chapter 3, and return to them more fully in the final chapter.

NATURE'S SPOKESPERSONS: EPISTEMIC COMMUNITIES AND EPISTEMIC DEPENDENCE

It's often said of advanced capitalist countries that their citizens are more 'educated' and 'informed' than ever before. Indeed, we're frequently told that we inhabit a 'knowledge society' and live in 'the information age' (see Box 2.1). In part, this reflects the unprecedented number of university graduates being produced, and the sheer volume, accessibility and diversity of both 'information' and 'knowledge' on offer these days. In the West, more of us possess a bachelor's degree compared with our parents' generation. More of us utilise the considerable resources of both the Internet and the media (print and broadcast, mass and niche, mainstream and 'alternative'). In addition, more of us work in jobs that involve the gathering, analysis and production of facts, evidence, ideas and various symbolic forms. Think of management consultancy, web design, magazine publishing and product advertising, for instance. But are we as educated and informed as some say we are?

My own answer is 'yes', but only if 'educated' and 'informed' mean that more of us are deriving our attitudes, opinions, preferences, prejudices and understandings from a wider range of epistemic communities than perhaps ever before. As I said previously, however much (we think) we know about the world, one cannot get around the fact that a very great deal of our understanding and (even) lived experience is 'second hand' and 'indirect'. In Steve Fuller's words, 'We have more beliefs than we can justify, and hence we must rely on . . . others to justify those beliefs' (Fuller, 2007a: 4). Consider the case of Wikileaks, which is credited with democratising knowledge and exposing corruption, malpractice and duplicity in politics, business and the media. Wikileaks, for all its revelatory benefits, in no way reduces people's dependence on unknown others for knowledge and information. Instead, it renders people less reliant on a range of 'official' sources by challenging those sources' versions of events.

BOX 2.1

'THE KNOWLEDGE SOCIETY' AND 'THE INFORMATION AGE'

These two terms have been used frequently over the past 20 years in order to describe what's distinctive about the early twenty-first century, especially in advanced capitalist countries. At a global level, the Organisation for Economic Cooperation and Development report *The knowledge-based economy* (Organisation for Economic Cooperation and Development, 1996) and the World Bank report *Knowledge for development* (World Bank, 1998) were key publications. Together they helped to popularise an emerging idea among policymakers that education (especially further and higher education) was an undervalued form of knowledge capital – undervalued in the sense that withholding it from people constitutes 'lost wealth' that would otherwise be created. The argument was that equipping more people with knowledge and skills would bring its own rewards in a hyper-competitive global economy. This has underpinned a global attempt to produce more graduates by expanding university and college enrolments. But more broadly, the term 'knowledge society' refers to two other things: first, the increased importance of various forms of knowledge as lucrative commodities (e.g. the reports of management consultants) along with the increased number of 'symbolic workers' (e.g. software designers); and second, the increased volume and availability of knowledge, notably because of the Internet.

The term 'the information age' is associated with the writings of sociologist Manuel Castells (1996). It describes an epochal increase in the volume of information available and the speed at which it moves between producers and users. Though there's no real consensus on the definitional niceties, we might think of 'information' as akin to *pieces* of knowledge (e.g. facts or instructions), and 'knowledge' as statements (e.g. narratives or theories) in which those pieces are *connected* logically or plausibly, and *contextualised* so that their meaning or significance is apparent. For information to be formative, it must usually be processed into knowledge. Equally, we might say that 'information' predominantly addresses 'what?' questions, while 'knowledge' tends to speak more 'why?', 'how?' and 'what should be done?' questions. It's been suggested that in mature democratic societies the production of knowledge has been extended and democratised in the past 30 years – notably because non-governmental organisations (NGOs), foundations, think tanks and charities now routinely produce and disseminate their own research (breaking the previous quasi-monopoly of universities).

When we say someone is an 'informed' or 'knowledgeable' person, we are pointing to their breadth and/or depth of understanding. They are the antithesis of an 'ignorant' person. Hence we talk of the 'informed consent' of medical patients, or of someone having an 'informed opinion'. But does life in a 'knowledge society' and an 'information age' promise a noticeable increase in the number of informed and knowledgeable people? The jury is most certainly out on that question. When not at work, undertaking a typically specialised (and often routine or mundane) job, the average Westerner today spends just under half of their remaining non-sleep time utilising information and communications technologies (ICTs) – notably televisions, computers, mobile phones and other handheld devices. ICTs have undoubtedly increased the potential for more people to understand far more about the world in which they live than ever before. They have made possible numerous 'social media' that both speed up and democratise the flow of information, opinion, images, sounds and ideas (think of Facebook, Twitter, YouTube, Wikipedia and Wikileaks to name five obvious cases).

However, critics make four counterpoints. First, they say that much of our use of ICTs is trivial or about ourselves rather than the wider world (e.g. texting friends on our mobiles or posting Facebook messages). Second, they argue that far too much ICT use relates to 'entertainment' or commodity consumption (e.g. viewing shopping channels on television or gambling online). Third, they suggest that even the 'serious' or edifying parts of television, radio and the World Wide Web are far too 'dumbed down', conveying simplistic and selective representations of a world that is far more complex. Finally, critics note that many people suffer from 'information overload' – they're bombarded with images, sound bites, headlines and so on, and thereby suffer an inability to select and integrate the most important information. People can simply choose from a buffet of ideas, arguments and evidence to suit their prejudices and preferences without challenging or changing them over time.

Why are we so reliant on various groups to which we do not belong for most of our understanding and experience of the world? Why are we no more free to form 'our own' beliefs, values, sentiments and preferences about things that (should) matter to us than were previous generations? Complete the task below and see what answers you come up with.

Study Task: Reflect back on your own experiences as (1) a school, and now university, student; and (2) a member of the general public interested in the world beyond your doorstep. Over your lifetime, you've learnt a lot of things, much of it now no doubt half remembered. Your understanding of the world has been organised such that you've depended, and continue to depend, on a plethora of complete strangers to enlighten you. In your view, why is there little or no alternative to this 'epistemic dependence'?

The answers to my study question are, I believe, not hard to seek. First, we live in an age of specialists. A highly elaborate division of mental (and manual) labour is a hallmark of 'developed' societies worldwide. This means that each of us can only lay claim to competence, let alone 'expertise', in relatively limited areas of knowledge and action. Pick up a *Yellow Pages* in any town or city and one is quickly reminded of this fact. The totality of information, understanding and experience that we as individuals want (or are encouraged to want) far exceeds that which we could accumulate if left purely to our own devices. As Nora Jacobson so nicely put it, today 'it's impossible for any one person to know everything – or even to know what it is that he or she doesn't know!' (Jacobson, 2007: 122).

Second, and relatedly, we live in an era where 'credentials' count for a lot. While complex divisions of labour are hardly new, the very idea of a 'knowledge society' (see Box 2.1) demands that citizens and workers not only become *more* knowledgeable than their forebears over a lifetime but *certifiably* so. In order to instruct, offer specialist knowledge to, or do particular things for other people, one is increasingly expected to possess the 'right' training and to have sufficient 'hands on' experience as a practitioner. This expectation, notwithstanding its very obvious benefits, creates barriers to thought and action. It defines 'insiders' and 'outsiders' and renders the latter 'amateurs', 'autodidacts' or even 'incompetents'.

Third, even if these barriers did not exist, it takes a lot of time and energy for any of us to become polymaths. The time and energy required increases in proportion to how detailed the division of mental and manual labour in any society becomes. It's far easier to be the recipients of others' knowledge, ideas, inventions or wisdom – though it helps a lot if we have reasons to trust those others implicitly, or to regard them as otherwise credible, admirable, likeable or talented. Of course, we do not have to be *passive* recipients, but the point is that we are not authors of that which we receive from myriad 'epistemic workers'. To summarise, in the early twenty-first-century world, we have organised the production and consumption of information, knowledge and experience so as to render 'epistemic dependence' a signature characteristic of our lives.[2]

I borrow this term from the American philosopher John Hardwig (1985), who published an essay of this name in 1985. In this context, there's something touchingly naïve about science historian John Dupré's insistence that 'our belief in . . . things should be grounded, ultimately, on experience' (Dupré, 1993: 42). Dupré is searching for a foundation upon which our knowledge or belief may rest secure, however temporarily. His empiricism recalls an 1877 essay by mathematician William Clifford, well known to philosophy students. Entitled *The ethics of belief*, the essay advanced the Humean argument that 'It is wrong, always, everywhere and for anyone, to believe anything on insufficient evidence' (Clifford, 1877/1879: 5). But because we routinely rely on others to have direct experience on our behalf, or to give it to us in some organised form (e.g. via a television programme), what Dupré really means to say is that we ground our beliefs in the experience of people who (*pace* Clifford) assemble 'sufficient evidence' for us. In effect, these people become our *proxies* or *stand ins* without us nominating them to be our representatives.[3] As I'll argue later in this chapter and later in the book (see Chapter 4), this even applies to what appears to be 'first-hand' experience, where we apprehend the world in a seemingly 'unmediated' way with our own bodily senses.

Why do I favour the term 'epistemic communities' to describe those upon which we are all so dependent, and what does it mean? It was coined by policy analyst Peter Haas (1992). Despite its cerebral overtones, it was not intended to apply exclusively to groups of professional philosophers or university academics more generally. Though Haas was interested in how 'expert' communities of government advisers – possessed of specialist (and thus scarce) knowledge – shape international political agreements, his definition of epistemic communities was somewhat ecumenical:

> Although an epistemic community may consist of professionals from a variety of disciplines and backgrounds, they have (1) a shared set of normative and principled beliefs, which provide a value-based rationale for the social action of community members; (2) shared causal beliefs, which are derived from their analysis of practices leading or contributing to a central set of problems in their domain . . . ; (3) shared notions of validity – that is, inter-subjective, internally-defined criteria for weighing and validating knowledge in the domain of their expertise.
>
> (Haas, 1992: 3)

Glossing, Haas's key point is that different epistemic communities gain their distinctiveness, and sense of self-identity, through a mixture of their value-set, ontological beliefs, questions of interest, objects/domains of concern, methods of inquiry, the criteria favoured for determining worthy ideas, knowledge or information, and their chosen genre of communication. This mixture determines both how specialised a given epistemic community is, how tight knit it considers itself to be, and how distinct it is from both other communities and the 'lay public'. As I suggested in Chapter 1, most epistemic communities also discourse with each other, and the rest of us, in distinctive languages.

Haas's epistemic communities don't do what they do only for themselves. Instead, other groups need, or feel they need, to utilise and rely on their insights or incitements. If we extend the reach of Haas's definition away from exclusively 'expert' communities of 'professionals' in the arena of policy and politics, we can usefully encompass *all* those groups whose information, knowledge, ideas and experience impacts upon us at various points in our lives. There are a great many of them. It seems to me unduly restrictive to suppose that only highly credentialised and formally trained epistemic workers should be the focus of our concern – important though these workers undoubtedly are. It leads us to suppose that what really (or only) matters is cognitive knowledge (which is simply untrue); it focuses our attention unduly on information and knowledge in the conventional sense (e.g. things written in scientific research papers or reported in news bulletins); and it may lead us to ignore some of the 'alternative' communities who seek to shape our thinking and our lives (such as those advocating the idea of 'intelligent design' as a critique of Darwin's widely accepted idea that species evolution is authorless and 'blind'). Accordingly, in this book the word 'epistemic' and its filial terms cover a wide range of mental products and media rather than the narrow philosophical focus on 'justified true belief'.

In my intentionally broad definition of the term, 'epistemic communities' can be deemed synonymous with 'knowledge communities' (though I find the latter term, which has been used in the field of business and organisational studies, somewhat bland).[4] It also has filiations with the term 'communities of practice' (Lave and Wenger, 1991), which usefully reminds us that knowledge is frequently created in order to have concrete effects and definite uses.[5] I mean 'epistemic community' to cover everything from professional counsellors to novelists to astronomers to advertising professionals to journalists to guides offering tourists an 'outdoor experience'. Depending on who we are and where we live, communities and sub-communities as broad and diverse as these together routinely seek to shape our thoughts, values, feelings and actions.[6] They're engaged in structured (though not static) forms of communication that are directed outwards towards others rather than just their epistemic peers.

Study Task: Out of the following topics, pick the one you think you know the most about: future climate change/global warming; the ecological effects of genetically modified (GM) foods; oceanic over-fishing; geoengineering the climate; water resource scarcity; peak oil; experiments on live animals; or carbon capture and storage. Now, think very carefully about precisely where your knowledge of the topic has come from. Consider as well your current beliefs about the topic. Whose epistemic claims have shaped your beliefs? To what extent are these beliefs 'borrowed' from others who you've chosen to believe? If you wanted to know more about any one of these topics, what sources would you consult and why?

Some epistemic communities actively seek us out and clamour for our attention, while we voluntarily go to others as occasion demands or opportunity arises. Some may strike us as being especially important (e.g. medical doctors), while others may seem peripheral to our lives. Some aim to persuade us and win us over, while others present themselves as being in the 'respectable' business of instruction, education or edification. Regardless of the differences between them, epistemic communities of all stripes have important features in common: they are organised and have barriers to entry (for the most part); they tend to be enduring (rather than temporary or ephemeral); they usually have a 'history' behind them (involving founding figures, key contributors and formative institutions); they 'speak' to each other and the rest of us in distinctive 'voices' and media; and they actively solicit audiences, sponsors and patrons of various kinds. Exceptions arise where a community is new (and thus nascent), where it's spontaneous or ephemeral, or where it's gradually disintegrated over time.[7] It's also important to recognise that, these days, the Internet and social media have allowed new voices to be heard that do not always belong to the epistemic communities with which we're familiar.

Having said all this, I should clarify two things. First, as my last comment implies, in order to share their representations with others all epistemic communities require material infrastructures. Depending on the context, these include laboratories, ICTs, transportation networks, university campuses, measurement devices, fibre-optic cables, gallery spaces, monitoring technologies, telescopes, electricity networks, printing presses and much more besides. The infrastructures may be funded privately, by the state, or by 'third sector' bodies. They may be highly specialised and expensive, or used relatively every day and affordable. Regardless, their role (though not exclusively so) is to sustain the production, circulation and consumption of the ideas, knowledge, images or insights characteristic of different epistemic communities – be they cognitive, moral or aesthetic (or a blend of the three). Indeed, without these infrastructures most modern epistemic communities could scarcely be said to exist at all. As David Morley notes, 'a community is not an entity that exists and *then* happens to communicate. Rather, communities are best understood as *constituted in and through their changing patterns of communication*' (Morley, 2005: 50, emphasis added).

STS scholar Paul Edwards (2010) provides a marvellous example of how infrastructures we made over time and across space. His study of how meteorologists and climatologists have come to know the global atmospheric system, appropriately entitled *A vast machine*, reveals the astonishing complexity of many knowledge infrastructures, their highly specialised character, and the tremendous effort involved in creating and then sustaining them.[8] As current ICTs demonstrate, however, once created these infrastructures allow epistemic communities to become based ever less on physical proximity. They can exist successfully as geographically dispersed communities because communication at a distance is relatively easy and affordable.

Second, and with this last point in mind, I'm well aware that the language of 'community' is potentially misleading. It suggests a spatially concentrated grouping of individuals possessed of a 'common culture' or shared interests, ones whose identities are – in Lorraine Code's sharp words – 'seemingly more coherent than they could possibly be' (Code, 2006: vii).[9] Yet, following Code, many of the communities that interest me in this book are not at all tight knit or geographically concentrated in one site. Their coherence is much looser and more complicated than this. I offer one example in Box 2.2, out of many possible ones, from the world of academia that I inhabit. This said, the *rhetoric* of 'community' (i.e. the word's routine use *as if* it refers to coherent, bounded groupings) remains important in a wide range of debates and discussions.

BOX 2.2

UNDERSTANDING 'COMMUNITY' IN EPISTEMIC COMMUNITIES: AN EXAMPLE

Like any 'community', an epistemic community doesn't exist until its members come to regard themselves as engaged in a common endeavour. There is, therefore, a *nominal* as much as *practical* element to the existence of epistemic communities. Even then, these communities may take a somewhat protean form. Academia (the community of university researchers, teachers and administrators) is a highly instructive example of just how loose and complicated membership of epistemic communities can be. At what level do epistemic communities exist in the academic world? One answer is that academics, regardless of their chosen subject areas, together form a single community in their hundreds of thousands, for two reasons. First, they all occupy the same institutional space (i.e. a university or its equivalent). Second, they share some definite values and goals (e.g. the pursuit of 'truth', 'accuracy' or 'new knowledge' in an honest, systematic and rigorous way); however, all academics are also members of definite disciplinary fields, such as history, development studies or nuclear engineering. This is reflected in the journals they choose to publish their research in, the degree programmes they contribute to and the professional conferences they attend (among other things). It is also reflected in the numerous cases of non-communication between academics: for instance physicists rarely have reason to conduct dialogue with literary critics and vice versa. But disciplinary fields, while real, are not sharply defined and nor are they internally homogenous. As Theodore Porter put it, 'In only a few disciplines is the dynamic of research activity so self-contained that interactions within the community are mainly responsible for the forms of approved knowledge' (Porter, 1995: 230). For instance, consider the sprawling subject area known as 'cultural studies', molecular genetics and the discipline of geography (my own). The first bleeds into and out of literary criticism, media studies, cultural history and cultural anthropology (among others). Molecular genetics, similarly, is now quite hard to distinguish from fields of knowledge it has both drawn upon and transformed, like informatics and biochemistry. Likewise, computer science has emerged, since 1945, from a combination of mathematics, military projects and commercial opportunism. Geographers, meanwhile, draw heavily on a very wide range of cognate disciplines for their research and teaching. What's more, while they may consider themselves Geographers (with a big G) for some purposes (e.g. delivering degree courses to undergraduates), for others they regard themselves as geomorphologists, economic geographers or political ecologists. Some of these sub-disciplinary identities connect geographers quite closely to specialist communities outside geography. For instance, Quaternary science is a recognised branch of physical geography yet the members of this epistemic community come from several disciplines.

This is one example of the many centrifugal forces that keep human and physical geographers apart, especially at the level of research, and which have led to talk of geography as a fatally 'divided discipline'. In their engaging book *Academic tribes and territories*, the sociologists of knowledge Tony Becher and Paul Trowler (2001) tried to make sense of this diversity and complexity of 'community' within the academic world, focussing in depth on several disciplines. The key point is that many academics consider themselves to be members of multiple epistemic communities simultaneously, and with varying levels of commitment to each of them. I would suggest that much the same applies to other epistemic workers. Think of journalists or nature poets or wildlife photographers. These are definite groups, yet there's a lot of internal diversity within each community. It is also perfectly possible for different epistemic communities to occupy a single institution and have a shared (though usually generic) mission – a university is a good example of this, so too is a large public broadcaster like the BBC.

RE-PRESENTING 'NATURE'

Representation, reference and representatives

With the two previously mentioned clarifications about 'community' in mind, it should be fairly obvious to readers how the ideas presented in the previous section connect to those outlined in the second half of Chapter 1. For the most part, epistemic communities operate within the wider discursive norms of any given society; however, in terms of the precise content of their knowledge, and the particular ways it's presented (e.g. in specialised vocabularies or using the mass media), these communities both develop and employ discourses in the plural, count noun sense of the word. Within these multiple discourses, 'nature' and its collateral concepts can feature in their cognitive, moral or aesthetic aspects and in complex ways. Equally, these discourses can deploy and flesh out 'ambivalent concepts' in a specific manner, thus troubling – rather than simply relying on – the antonyms listed in Figure 1.5. Indeed, some epistemic communities (or some of their members) intentionally challenge existing mindsets, values and practices: they go against the grain of convention (more on this in the latter part of Chapter 5). Regardless of the exact case, metonymic references from particular biophysical entities to a wider 'nature' are commonplace in the various discourses employed (as I suggested earlier in Box 1.2).[10]

I want now (in general terms) to consider what otherwise different epistemic communities do when they represent nature to us. The term 'representation' is both a verb and a noun: it refers to a process and its various products. It involves acts of *translation* and *replacement* in which different epistemic workers make sense of the world in various ways and then concretise these sense-making acts in forms that can be shared with others. As Michael Shapiro once said, despite frequent appearances to the contrary, 'Representations do not "imitate" reality but are the practices through which things take-on meaning and value ... ' (Shapiro, 1988: xi). These practices can be visual (book, film, map, comic strip, photograph); they can be oral and aural (lecture or guided tour); they can be quantitative (e.g. a graph of mean atmospheric temperatures or a computerised Global Circulation Model) or qualitative (e.g. the words and pictures in an issue of the *National Geographic* magazine); they can be tactile (think of an interactive museum display); they can be purely linguistic; they can involve some combination of discourse and (ostensibly) wordless communication (such as instrumental music); they can be obviously 'physical' (e.g. a sculpture); they can invite silent consumption (e.g. reading an ecotourism brochure when planning a holiday); they can demand active participation (e.g. hiking through a nature reserve with a field guide in hand); they can be communicated using a rhetoric of certainty and precision, or be hedged around with qualifications and caveats.

Some representational forms are relatively durable and can 'travel' through space and time (e.g. a book or a website). Others are relatively durable over time but fixed in space and so must physically attract consumers and users (e.g. the Eden Project in southwest England, whose miniature ecosystems represent 'real' ones). Still others are transient or relatively ephemeral (e.g. the daily news on the radio, a billboard advert or a documentary on the Discovery Channel). Some representations command small audiences, others very large ones that are spread across several countries or continents. Many representations are inconsequential, but some are hugely influential in the short or longer term.

Clearly, my definition of 'representation' is as ecumenical as my definition of epistemic communities. As David Runciman and Monica Vieira note in their book on the subject, representation 'encompasses an extraordinary range of meanings and applications, stretching from mental images ... to legal processes to theatrical performances' (Runciman and Vieira, 2008: vii). This raises a number of issues, and I'll tackle these presently. For now, though, the point I want to make is this: all representations *refer*. They are commentaries on something that they themselves are not. They can take the form of descriptions, explanations, appreciations, recreations, forecasts, assessments, disagreements, depictions, critiques and so on. They are exercises in communication about the world, and they can work in all of the registers of understanding, evaluation and affect. They're the means by which, and the media through which, meanings are attached to various portions of reality and those portions of reality themselves delimited (or, perhaps, called into question and recategorised).

Representations can refer to what appears to be 'nature', to its collateral terms and their referents, or to any of the various elements of what is considered to be 'society', 'economy', 'politics' and 'culture'. This means, in the latter cases most obviously, that representations can refer explicitly to other representations and to the things those representations themselves refer to. And, because they refer – because any given representation is not simply (or usually) about itself – representations have the potential to affect the world by changing (or affirming) how we think and feel about it. Ian Hacking (2004: 279) calls this the 'looping effect' of representation, especially of those representations whose reach or significance is relatively high compared with others.

By virtue of their difference and distance from 'outsiders', representation is the principal vehicle that epistemic communities rely upon to have a wider influence. Most epistemic communities affect us by means of persuasion not force. I'd suggest that we are, much of the time, audiences who occupy a vast theatre of representation in which different plays are endlessly performed. This assertion is consistent with my argument about 'epistemic dependence'. From time to time, we may be invited on to the metaphorical stage, while some of us (like myself) may even have the chance to script part of a scene or alter aspects of the lighting, set and music. But, for the most part, we're consumers (and users) positioned at the end of chains of representation. I say 'chains' (to use a second metaphor) because representation is about making links – between us, what is represented and those who represent the world to us. Various representations refer us to things that, in some way, 'speak' to us in the voices and idioms of the representers. As Bruno Latour says 'no being, not even humans, speak on their own, but always *through something or someone else*' (Latour, 2004: 68 'emphasis in original').

In some cases this is very obvious. Think of a Hollywood blockbuster like *The day after tomorrow* (2004) with its 'end of the world' storyline and hyper-dramatic action. It is, clearly, a morality tale about the future perils of unchecked anthropogenic environmental change. Think too of contemporary 'nature writing', such as Richard Mabey's (2005) autobiographical book *Nature cure* – a moving, lyrical and highly personal account of how the author recovered from mental illness by immersing himself in 'the living earth'. But in other cases the role played by representers is not obvious at all. For instance, we typically take the 'voice' of a scientist to be neutral, as if s/he is but a mouthpiece for the biophysical world.[11] Representations are, therefore, just that: *re*-presentations of phenomena, evidence, stories, facts, occurrences, experiences and so on, regardless of the specific context or the particular 'genre' of representation. This implies that epistemic communities always act as *both* represent*ers* (vessels/carriers) and as re-present*atives* (substitutes/replacements) at one and the same time. That they do so differently in the detail does not alter the fact of this double-headed commonality.

Representations: constructed and political

We often think that these two aspects of representation are separate, but they're not. The first, which we might associate with a journalist or an academic, involves making something present to us spatially (here) and temporally (today) via words, images and so on. Things, events, information, or knowledge from the past, or from elsewhere in the here-and-now, are brought to our doorstep as it were. Consider a daily newspaper: we are enjoined to think and feel about that which might not otherwise be in front of us, near to us or visible to us. The second aspect of representation we tend to associate with democratic political systems and most legal systems. We elect senators or MPs to represent our views and desires in our absence (hence the term 'representative democracy'); delegates and consuls are appointed to act on behalf of whole countries or their leaders; and (if in court) lawyers are our appointed defenders or prosecutors, depending. Together, they *stand for* our interests and wishes because they *stand in* for us physically and rhetorically.[12]

Here we appreciate why the two senses of representation – 'epistemic' and 'political' – are intimately connected for they *both* involve substitution, as 'iconic representations' make very clear (see Box 2.3). Representers of various kinds are doing exactly what the word implies: they are making sense of the world for themselves and us, and thus acting as representatives (whether they occupy the political–legal sphere or not). In some form then – and as I said, it's often obvious – these representatives are interposing themselves between us and that to which they refer.[13] Sometimes, there's more complexity to this than meets even those with keen analytical vision. Consider Al Gore's well-known documentary film and book *An inconvenient truth* – a science-based plea for global action to reduce greenhouse gas (GHG) emissions drastically. As one critic astutely asks, 'is he [Gore] speaking as a politician, a lay expert [on climate science], or as a spokesperson for science?' (Hulme, 2009: 81). Perhaps it's the *conflation* of all three roles that helps explain the power and popularity of *An inconvenient truth* among diverse audiences worldwide.

BOX 2.3

THE NATURE OF, AND IN, ICONIC REPRESENTATIONS

An 'icon' serves as a figurative or literal embodiment of something deemed to be of considerable significance. Icons are apprehended visually and, historically, were associated with organised religion (e.g. statues or paintings of Christ on the cross). Today, the religious association has weakened and we can think of icons as those images or artefacts that are especially symbolic in *any* area of our lives. Icons are thus a subset of representations that work metonymically (see Box 1.2). A century ago, scholars like Erwin Panofsky (1892–1968) pioneered a way of studying icons – 'iconography' – that remains relevant today. Focussing on paintings initially, Panofsky and his fellow travellers rejected the idea that the content of any icon per se was of interest (e.g. subject matter or use of colour). Instead, they focussed on 'unpacking' the meaning of icons (i.e. what their content 'said') and suggested that meaning is not intrinsic to them but arises through their interpretation by viewers (see Box 3.2). These viewers, he further argued, are necessarily situated in time and space, in culture and society. In turn, this means that icons have to be studied contextually in order to be properly understood.

Icons can reveal with especial clarity the simultaneity of the two aspects of representation we often consider separately (epistemic and legal–political). Great white sharks and great whales provide two excellent, well-known examples of this. It's not too much of an exaggeration to say that from late 1975 many Westerners came to regard *Carcharodon carcharias* as iconographic of 'nature red in tooth and claw' – for at least the next 20 years. American novelist Peter Benchley's best-selling book *Jaws* was made into the first modern 'blockbuster' by film director Steven Spielberg. The movie was a massive box office success and spawned a huge merchandising industry and three sequels. By depicting great whites as 'man eaters' to be feared, it played on a much older idea that 'nature' is cruel and indifferent to the fate of humanity: something to be avoided or tamed. Not only was this a very specific representation of sharks, but it was also very obviously one constructed by Benchley in the interests of fictional entertainment, and yet, its fictional form notwithstanding, the powerful influence of *Jaws* on popular culture stemmed from its apparent 'realism'. Benchley and Spielberg were credited with telling a story seen to be anchored in the observed behaviour of great white sharks – albeit a tiny minority of them. This evidential base lent credence to an otherwise obviously contrived representation of the 'nature' of *Carcharodon carcharias.*

In contrast, and at the same time as *Jaws* was released, Greenpeace activists staged a series of now world-famous 'image events' that made great whales iconic in a far more positive way than great whites. Photographs and film clips of large, hi-tech whaling ships harpooning helpless whales, despite the efforts of activists in small Zodiac inflatables, came to encapsulate the sensibility of the 'environmental movement', a movement that had grown steadily through the 1960s. As with *Jaws*, the viewer was invited to see nature in a particular way (in this case, as mortally threatened), and to share the view of the representative (Greenpeace). Whale hunting rapidly became emblematic of Western society's wanton cruelty and utter disregard for other sentient creatures. This reversed centuries of seeing whales as 'resources' to be harvested from the oceans. It was achieved by connecting Greenpeace's moral agenda to the 'eye-witness' images of slaughter at sea. Building on this, Greenpeace also successfully demonised Canadian seal hunters in the late 1970s by depicting seal pups as akin to small children being butchered by ruthless men.

Clearly, I'm suggesting that representations must carry some trace of their authors' habits, preferences, desires or values.[14] This means that references to 'nature' and any of its collateral concepts are always politics by other means. When we represent beluga whales, natterjack toads and Siberian tigers we are doing what politicians do for their constituents, without necessarily realising it. Or rather, others are doing it to us and for us. In saying this, I'm deliberately stretching the meaning of the word **politics** beyond those activities associated with the apparatus of elections, political parties, governments and bureaucracies – as many others in social science and the humanities have done (I'll return to this theme in Chapter 3). For me, almost *everything* that we say and do is political in the sense that it involves contestable, value-laden choices – albeit often unconscious or unthinking ones – about profound issues pertaining to what's 'normal', 'interesting', 'relevant', 'good', 'right', 'permissable' or 'moral' for us and other people. These are also choices about what *not* to represent. Some of these choices are politically and practically trivial but others, upon closer inspection, are very significant. In the domain of 'formal politics' (e.g. the British house of Commons), we debate and justify these choices explicitly. But in many other walks of life we do not. Plate 2.1 provides a rich and graphic example of how the political and epistemic senses of representation bleed into one another, even in seemingly unlikely places.

Study Task: Look closely at the image in Plate 2.1 but don't read the text beneath the title just yet. Can you identify how both aspects of 'representation' detailed in this section are operative simultaneously?

Plate 2.1 The two sides of the representational coin

This Royal Society for the Protection of Birds (RSPB) advert, placed in a British broadsheet newspaper in August 2010, illustrates well the two aspects of representation: namely, 'epistemic' (speaking of) and 'political' (speaking for). The strapline 'A million voices for nature' announces clearly the political intention of the RSPB and its members: to speak up for birds and the other wildlife species to which they relate (because birds can't articulate their own interests). In this case, the RSPB is opposing funding cuts to public programmes that aim to sustain rural biodiversity. At the same time, the advert's declaration that 'Some cuts never heal' has – in addition to its intended emotional charge – an epistemic role. It refers to the biophysical damage, to birds and other wildlife, of the spending cuts, and is presented as a broadly 'truthful' statement about species endangerment and loss. Thus is a purported 'fact' about birds used to stake a political claim about their future.

In the early years of modern European democracy, when the *ancien régime* was crumbling, the influential polymath Jean Jacques Rousseau was, famously, a critic of representation in both its senses ('speaking of' and 'speaking for'). He regarded representation as alienation because, in his view, it distances us from the world and makes us reliant on others who may not, in the end, serve us well. In his most idealistic moments, he favoured the immediate over the mediate and presence over absence because he wanted to empower us and have us take responsibility for what we think, say, feel and do. The echoes in William Clifford's and John Dupré's earlier mentioned sentiments are strong. However, this critique of representation has been brilliantly parodied in the memorable stories of the fiction writer Jorge Luis Borges (1899–1986). In *The Congress*, Borges (1979: 20–2) famously considered the 'problem' of representation identified by Rousseau:

> Don Alejandro conceived the idea of calling together a Congress of the World that would represent men of all nations ... Twirl, who had a farseeing mind, remarked that the Congress had a problem of a philosophical nature. Planning an assembly to represent all men was like fixing the exact number of Platonic types ... Twirl suggested that Don Alejandro might represent not only cattlemen but also Uruguayans, and also humanity's great forerunners, and also men with red beards, and also those who are seated in armchairs.

The story ends with an echo of another Borges tale about cartography, in which the 'perfect map' is devised: 'the College of Cartographers evolved a Map of the Empire that was of the same scale as the Empire and that coincided with it point for point' (Borges, 1981: 131).

Borges reveals the absurdity of avoiding representation and achieving 'pure presence', particularly in a world containing around 7 billion people leading highly interdependent but geographically separated lives. For the reasons I listed in the previous section, epistemic dependence is writ large in the twenty-first-century world – a world Rousseau could scarcely have envisaged when *The social contract* was published in 1762. The question then becomes: what is included in any given representation, what is excluded (i.e. rendered invisible or silent) and why? But representation is unavoidable for another reason too. We might say that representation is not only a practical necessity, but also an established social convention.

What do I mean? I mean that the *idea* of representation is part of the Western worldview, which is why I included it in Figure 1.5. This idea is predicated on a distinction between representations and representatives (on the one side), and that being represented (on the other). As Tim Ingold and other anthropologists remind us, this distinction is not a cultural universal, let alone 'natural'. But it's nonetheless real for us, a distinction we rarely (if ever) question in our everyday lives. We proceed as if the distinction is a given, and so, in some sense, it therefore *is*. Dutch philosopher Bas van Fraasen puts it like this: 'there is no representation except in the [important] sense that some things are used, made or taken to represent some things as thus or so' (van Fraasen, 2010: 511). The 'reality effect' to which Judith Butler and Tim Mitchell refer (following Roland Barthes) is achieved precisely by positing a difference between things taken as objects and their representations (even when these objects are other representations or, as with my discussion in this section, representation in general). This difference is especially important for representations that advertise themselves as being 'realistic' or 'truthful' because the world 'beyond' representation gets used as a court of appeal for their veracity (see case study 2 in Chapter 4 for more on this). So, for all their differences, this is what documentary film-making has in common with (say) entomology.[15]

This returns me to the subject of 'nature', which I've slightly backgrounded in the last few paragraphs. Consider again what the term signifies. In its four meanings, its propensity to be spatialised and temporalised in our imaginations, and its metonymic conjoining of the specific and the general, nature – to adapt Karl Marx's famous statement – 'cannot represent itself; it must be represented!' Why so? The answer is simple. As I've explained, we conceive nature as an 'other', which cannot 'speak for itself' because it cannot speak *at all* or, if it does, it 'speaks' in a foreign tongue.[16] This even applies to our own bodies, in part because the mind–body dualism in our culture has us seeing our biology as something unto itself. In David Delaney's words, for us 'the body stands for nature ... [while] the mind stands for the uniquely human traits of consciousness, subjectivity, knowledge, will, and freedom' (Delaney, 2001: 495). Indeed, we even cleave the mind in two, using the huge cognitive resources it affords us to study – in ways underdetermined by them – how those self-same resources work (which is, of course, what neuroscientists do). As Robyn Eckersley notes, according to our way of thinking, natural entities 'are unable to articulate their claims according to the canons of rational argument' (Eckersley, 1999: 37).

These entities require ventriloquists of various kinds – environmentalists, medical doctors, conservationists, anatomists, artists and all the rest. Their voices may vary greatly, but they perform the same function. As Haraway put it, somewhat hyperbolically, 'Permanently speechless, ... never forcing a recall vote, in each case the object or ground of representation is the realization of the representative's fondest dream' (Haraway, 1992: 311). Thus rendered mute, what we call 'nature' is free to be represented in all manner of different ways in a variety of arenas, media and genres. Nature's apparent ubiquity and seeming ontological givenness together make it a subject of common concern. It comprises a very large semantic (and thus material–practical) space for commentators of all kinds to speak to the rest of us about it. The number and variety of these actors has increased over the past 50 years, since the first modern wave of concern about environmental degradation in the 1960s. 'Precisely because nature is something that must be represented', writes Bruce Braun, 'the act of representation becomes that much more important, for it necessarily constructs that which it speaks for' (Braun, 2002: 260). This act, we should note, includes the ability to determine and debate what counts as nature in the first place. As Evernden wisely observes in *The natural alien*, 'the examination of "nature" must entail not simply the objects we assign to that category, but also the category itself' (Evernden, 1992: xi). The same applies to nature's collateral terms.[17]

THE NATURE OF 'REPRESENTATION': FOUR ISSUES

As I said earlier, my understanding of 'representation' is very broad indeed. Here, I take inspiration from Alexander Wilson and Jennifer Price, among several others. In *The culture of nature* (Wilson, 1992) and *Flight maps: adventures with nature in modern America* (Price, 1999) they each consider the diverse sites and situations in which what we call nature is communicated to us and encountered – for instance museum dioramas, Kate Evan's *Funny Weather* cartoon, *National Geographic* magazine, US National Public Radio's 'Living on Earth' series, or the Canadian Broadcasting Corporation's once popular documentary *The nature of things* (fronted by scientist David Suzuki). Wilson's and Price's inclusive sense of representation chimes with Stuart Hall's edited book, *Representation*. Hall has argued that all forms of representation are, in effect, linguistic: 'Spoken language uses sounds, written language uses words, musical language uses notes on a scale, the "language of the body" uses physical gesture, the fashion industry uses items of clothing ... ' (Hall, 1997: 4). His point is that, like language, even ostensibly non-linguistic media convey meaning by in some sense 'standing for' – that is to say, representing, evoking or referring to – something else.[18] What's more, in all these other media, one has various people deliberately contriving and presenting image, sight, sound and even smell towards certain semantic ends. Many others agree. For instance, in his book on music, Nicholas Cook (2000: viii) is hardly the first to insist that everything from jazz to opera can operate 'as an agent of meaning'. We might say the same about a good deal of performance art or dance, to take some other cases.

To reiterate: I'm not suggesting that representation is, in effect, everything and everywhere. Here's a mundane example of why. If I kick a football (as I do many times each week when playing 5-a-side soccer), it does not 'represent' much, if anything, except, perhaps, my brain's ability to coordinate eye, foot and manufactured spherical object. Likewise, a red traffic light is a command (stop!) not a visual representation. It's also the case that a lot of our communicative activity performs the function of maintaining relationships (as when I ask my children about their school day when I arrive home each evening). Yet *much* of what we think and do proceeds by way of representation. This is why it, as both verb and noun, has been a major preoccupation of the arts, humanities and social sciences for several decades (and usually connected to the use of important analytical concepts like 'ideology', 'hegemony' and 'power-knowledge' – more on which can be found in Chapter 6).

So far, so good. But some readers might regard my definition of representation as being too loose and insufficiently parsimonious. I say this for four main reasons, and want to offer a response in each case. They are presented in no particular order.

Representation = language?

First, some may disagree with Hall's suggestion that representation goes beyond 'language' in the conventional sense of the term: that is, beyond spoken or written words and their meanings (discourse). My response to this is straightforward. It is, I would argue, impossible to 'contain' language and thus wrong to draw overly sharp distinctions between the different registers in which we make the world meaningful to ourselves and others. Let's take vision and the act of seeing. For the vast majority of us who are not blind or sight-impaired, our eyes are perceptual instruments of the first order. However, seeing is not a purely physiological process: the human eye may be structured the same the world over, but what is seen (and how) varies. As art critic John Berger (1972) famously pointed out, there are historically and geographically specific 'ways of seeing'. Over the past 30 years, an interdisciplinary field studying **visuality** has proved him right. 'Visuality', writes geographer Fraser MacDonald (2009: 1), 'refers to the acculturation of sight'. Because one learns to use one's eyes at the same time as one learns to use a specific language, we are obliged to acknowledge 'the discursive nature of vision' (ibid: 5). The words 'image' and 'imagination' alert us to the fact that seeing is an active mental process not an unmediated receipt of externally generated visual stimuli.[19] Think about adverts on television and elsewhere: what we 'see' is both images and words, working in tandem and together generating meaning, usually after a lot of effort on the part of advertising professionals. The same applies to maps of various kinds, for instance an atlas of the world or a city street map. Most cartographic representations are littered with text or with symbols that users translate into linguistic signifiers, signifieds and referents.

Accordingly, when Audrey Kobayashi (2009: 1) defines representation as 'the practice of constructing meaning through language', she's only wrong if we take her to mean that 'language' exists in a hermetically sealed box. The famous idea that a picture is worth 'a thousand words' is true, not because we see it non-linguistically, but because images can, at a stroke, condense myriad and complex meanings that would otherwise have to be communicated at length on a page or verbally. Similarly, analyses of representation focussed principally on writing, such as the books *Writing culture* (Clifford and Marcus, 1986) or *Writing worlds* (Barnes and Duncan, 1992), should not be interpreted too literally.[20] It's surely better to think of words and their meanings as 'leaking out' into ostensibly non-linguistic registers of understanding and affect. This is, perhaps, what philosopher of mind Daniel Dennett means when he says that, 'Language infects and inflects our thought at every level' (Dennett, 1991: 330). For us, it's a signally important sense-making medium that is ultimately indissociable from all the other ways in which we make the world make sense. I thus concur with Rom Harré and his co-authors who, in their book *Greenspeak: a study of environmental discourse*, contest the idea that 'language exists as a self-contained, independent mental [phenomenon] . . . that can be studied in isolation from its use, functions, history and specific contexts of employment . . . ' (Harré *et al.*, 1999: 1).

Representation = mimesis?

Second, some may equate representation with what's called **mimesis** and 'correspondence'. This is the idea that any given representation aims to resemble or even be a 'carbon copy' of some aspect of a reality taken to exist beyond any given representation. This idea remains central, as an ideal at least, to the practice of the various pure and applied sciences. Relatedly, it remains equally important to the education system at all levels (because most educational institutions declare themselves to be enemies of 'propaganda', 'lies' and 'falsehoods', hence the recent debate in some countries over whether 'creationism' should be taught in schools). But it's important elsewhere too, notably in the news and current affairs sections of the media. Some might also say that political and legal forms of representation involve mimesis, not so much as 'correspondence' but as 'faithful' re-description and advocacy. Such 'realist' forms of representation, and the epistemic communities committed to producing them, are hardwired to the reproduction of familiar epistemic norms such as 'truth', 'objectivity' and 'fact' (versus 'opinion', 'bias' and 'fiction'). These norms are central to the identity of many epistemic communities and the principal source of whatever influence they exert and whatever social legitimacy or popularity they possess.[21] While it would be foolish to underestimate the social importance of representations that claim to hold up a mirror to the world, it would be equally foolish to equate them with representation as such. Why so?

On the one side, there are many kinds of representation that, while not claiming to be 'realistic', are most certainly intended as considered commentaries on 'reality' as it appears to those doing the representing. Consider the major exhibition staged in London's Barbican Centre in 2009: 'Radical nature: art and architecture for a changing planet, 1969–2009'. The various, striking pieces of artwork assembled, such as Henrik Hakansson's *Fallen forest*, a 4-metre-square section of tropical woodland flipped on its side and suspended above the gallery floor, are not strictly speaking attempts at mimesis. However, as the catalogue foreword written by British environmentalist Jonathan Porritt makes clear, they are direct statements about something taken to be real and actual: environmental degradation. They are not intended to be *like* the world but are assuredly commentaries *on* it. As such, they invite exhibition visitors to reflect seriously on our impact on the planet. They are, as it were, *critical* representations whose status as 'art' is not intended to detract from the 'truths' their authors seek to relate. The same 'serious' intent can be found in other ostensibly non- or quasi-realistic forms of representation, from poetry to satirical comedy to science fiction to political cartoons to graphic novels.

On the other side, even representations that bear a very distant relation to 'realism' deserve to be taken seriously. Think of advertising, which I mentioned in passing previously. A lot of advertising is overtly non-mimetic: special effects, humour, unlikely scenarios and other devices are used to communicate a message, usually about a single commodity. For instance, for many years adverts selling 4 × 4 vehicles have used various images of mountains, valleys and plains that refer to 'nature'. They do so not in an objective or scientific sense, but as a way of creating highly gendered (and often ethnicised or racialised) forms of consumer desire (see Plate 2.2). Representations like these are no less important for being so obviously 'fictional' and 'made-up'.

Study Task: Look closely at the image in Plate 2.2 but, as with the previous image, don't read the text beneath the title until later. What 'nature' is displayed in the image? What qualities does nature possess that the advertised vehicle is said to get buyers close to? What kind of potential drivers do you think this advert is appealing to with its particular choice of text and image? Going back to Chapter 1, can you identify elements of 'connotative reference' in the advert?

Indeed, if we refer back to Figure 1.5, we can see how putatively 'unrealistic' or 'creative' forms of representation, which we typically associate with the arts in their various forms, help to sustain the belief that other forms are *intrinsically* realistic. Each form implies the other: they are defined relationally, one being a conceptual shadow of the other and so not separate at all (despite appearances). I say this in a non-judgemental way: it's not necessarily a 'bad' thing. In fact, putatively 'unrealistic' forms of representation can present a useful space in which to comment in creative or critical ways on what passes for fact, truth and common sense – they can unsettle that which appears to be their very antithesis.

These are also very significant forms of media in and through which 'nature' is made sense of in the modern world (and, indeed, going back many generations). The signifieds and referents of this word and its myriad collateral terms are most certainly not the preserve of epistemic communities, whose *raison d'être* is the search for truth and who see themselves engaged in a resolutely 'non-fictional' enterprise. Our understanding and experience of nature, 'race', environment, sex and all the rest are shaped profoundly by video games, television dramas, plays, the visual arts, the plastic arts and the sonic arts – among many other things. That's why the analysis of these media has congealed into entire disciplines, like cultural studies. Such is the semantic latitude of nature and its collateral concepts that they cross every conceivable representational genre and practice. These concepts thus condition our thoughts and feelings in virtually all areas of our daily life, even if we don't always recognise it.

Plate 2.2 The staging of the 'natural environment' in outdoor vehicle advertising

In what ways are images of a rugged riverbed and forested valley slopes used to sell the vehicle? Both suggest the challenge and excitement of being 'off-road' in an environment that normal vehicles – and by implication 'normal drivers' – could not cope with. This is echoed in the vehicle's name, which refers to the famous mythical wild creature of the Himalayan Mountains. To say that the advert is staged is to make the obvious point that the car is being asked to play the role of an actor. Like an actor, its performance is intended to have effects on the 'audience' (potential buyers in this case). But, also like an actor, it becomes something altogether different off-stage. Off-stage, the Yeti is used in towns and cities for the most part, and is rarely taken off-road by owners. Yet images of day-to-day use were not favoured by the Yeti's advertisers. Instead, a strapline announcing its 'Car of the Year' status is combined with the almost wholly implausible suggestion that the Yeti might, once purchased, be driven into the wilds by its owners. One suspects that the advert is intended to appeal to men with children, flattering them with the conceit that they can combine the practicality of a family vehicle with the power and strength hitherto associated with pick-up trucks, Jeeps and Land Rovers. A classic reference point for the analysis of adverts, including those that refer to 'nature' and its collateral terms, is Judith Williamson's *Decoding advertisements* (Williamson, 1978).

A focus on representation occludes the non-representational?

A third major objection to my broad definition of representation as a social practice is that, notwithstanding my first line of defence previously, I'm underestimating the importance of the 'non-representational' element in all our lives. The social sciences and humanities have undergone something of a 'post-representational turn' in the past decade. I can't describe this turn in sufficient detail here. Glossing, it's been suggested by some critics that, through the 1980s and 1990s, too many analysts became preoccupied with discourses, signs and symbols. These analysts were charged with overlooking the practical, affective and non-cognitive dimensions of our daily lives. A related criticism was that these analysts lost sight of the material world to which the representations they were 'deconstructing' referred, including our own bodily sensorium. This world, it's been argued, tends to exceed any attempts to 'fix' its meaning or content in specific signs (words, images, scientific papers, maps and so on). Accordingly, various forms of 'non-representational theory', to use geographer Nigel Thrift's (2007) portmanteau term, have focussed on the emotional, embodied, pre/unconscious and somatic aspects of human engagement with other people and the non-human world. The central argument is that much of what matters in our lives is irreducible to the creation, circulation and effects of various representations.

This argument is undoubtedly persuasive. Yet in this book I devote little or no space to the domain of the 'non-representational'. Am I therefore at fault? Yes, or so might say the editors of a book like *Nature performed* (Szersynski *et al.*, 2003), one of several works on 'nature' that deliberately de-emphasise representation. But I'd make the following points in my defence. First, to acknowledge that the 'non-representational matters' is not to imply that representation does not. Indeed, to advocate that *more* attention should be paid to the *non*-representational is surely to imply and acknowledge that representation is socially real and materially efficacious. Second, there's the risk of creating a false dualism between the representational and putatively non-representational. As cultural geographer Catherine Nash noted, 'the turn away from language and texts in non-representational theory ... seems to require a new version of an old division between thought and action, ... mind and body' (Nash, 2000: 657). Representations of various kinds are not, or not only, the antithesis of things like emotion, affect, human embodiment or human practice. Nor, as I suggested earlier, are they synonymous with speech and sight only. On the contrary, they can profoundly shape our feelings, dispositions, and habits of action in the broadest sense. For instance, consider how the world famous nature photography of Ansel Adams has long inspired, excited and moved millions of people in the United States and beyond. Or consider how mainstream, highly normative ideas about heterosexuality lead many individuals to feel intense guilt, shame and anxiety as they come to recognise their own homosexual desires. Also, affect and emotion can be consciously manipulated by means of selective representations in advertising, zoos, magazines and other arenas. Finally, to talk and write about the 'non-representational' is, necessarily, to represent it(!). As per the previous point, this reveals the difficulty of moving entirely 'beyond' or 'outside' representation as a social practice. (For a more pointed critique of recent research focussing on the non-representational, see Hemmings, 2005.)

First-hand experience versus second-hand representation?

Fourth, it's sometimes said of life in the early twenty-first-century 'knowledge society' that it's increasingly 'mediated'. At base, this means that much of our interaction with myriad others occurs via various physical media that permit communication 'at a distance'. What's more, as I noted earlier, much of this interaction is today dispersed because, with technologies like televisions, radios, home computers, the Internet and mobile phones, those on the receiving end of mediated communications are not obliged to come together physically in order to access the information, knowledge or experiences they seek. As I also said before, this applies as well to the epistemic communities creating and disseminating this information, knowledge and experience – they themselves are both localised *and* distanciated. This permits much mediated communication to be *mass* communication. Millions of otherwise different individuals, largely unknown to one another and geographically separated, are able to receive the *same* information, knowledge or experience if they wish.

Three reasons why the mass media exist are fairly obvious. First, there's a lot of money to be made influencing the tastes of myriad potential consumers and users. Second, in democracies votes count, as does public opinion. Third, distant events these days have material consequences for our enduringly local lives. Unsurprisingly, a great deal of mediated mass communication involves representation – a daily newspaper, be it a tabloid or a broadsheet, is an obvious example of this (see Chapter 7 for an in-depth discussion of mass mediated news representations).

Notwithstanding its heightened importance compared with decades gone by, mediated mass communication has not crowded out other, seemingly different ways of interacting with people and the non-human world. Apparently 'unmediated' encounters remain intensely important. For example, think of one's daily face-to-face relations with one's parents, lovers, friends, children or workmates. Think too of everyday, first-person activities like walking to the bus stop, eating food, riding a bike to work or taking a hot shower. For all of us, these activities seem different in kind from, say, watching an episode of *Inside nature's giants* – a British science documentary series, shown on Channel 4, in which very large animals were dissected forensically. We often attach great personal significance to these sorts of 'direct' encounters. For instance, it's one thing to read about British adventurer Bear Grylls climbing Mount Everest (he was the youngest person ever to do so, in 1997) but quite another to join an expedition and do so oneself. However, are 'first-hand experiences' like the latter entirely *un*mediated?

Study Task: Think of one memorable 'outdoor experience' you may have had during your life. In my case, 'ghyll scrambling' in the English Lake District comes readily to mind. Kitted out in a helmet and full-body fleece, my son and I spent an hour and a half jumping, ducking and diving our way down a stretch of intensely cold upland stream. It was exhilarating. We scrambled with a guide and, in my case, with a long history of visits to the area to hike and camp. Whatever your memorable experience was, ask yourself this: to what extent was it a result of the things you did in that place, at that time? Did other things condition your experience and subsequent memory? If so, what were they?

The notion of (mass) mediated interaction, long discussed in the field of media and communication studies, tends implicitly to support the idea that unmediated interaction is qualitatively *distinct* from it. But the two are not as separate as they may appear. The study task is designed to get you considering this possibility. First, in general terms, it's implausible to believe that in our daily lives we maintain a Maginot Line between the two forms of interaction. Surely, the information, knowledge and experiences we accumulate in both domains leach into one another in our heads and hearts. Second, and to be more specific, many of our chosen activities can, upon a moment's reflection, be seen as an obvious blend of 'mediated' and 'unmediated' interaction. For instance, a visit to the zoo or a safari park is not mediated in the sense that watching a wildlife documentary is: we encounter animals in the flesh not through a television screen. However, echoing Tim Mitchell's analysis of the old Cairo exhibition (see Chapter 1), the 'first-hand' experience one has is still mediated to the extent that zoos and safari parks are 'constructions', in both a material and semiotic sense. They invite visitors to see, smell, hear and understand in particular ways. These direct sensations can, in turn, unthinkingly provide insights into the 'reality' of animal lives – if only by foregrounding the 'unreality' of enclosing some animals while their wild and 'happier' cousins roam free (see Plate 2.3).

They illustrate why the feminist critic Joan Scott (1991), in a brilliant essay, was right to insist that personal 'experience' should never be assumed to be the basis for 'authentic' understanding, insight or belief. If, as is conventional, such experience 'is taken to be an unquestioned given, this ... asks neither whether or not there may be another origin underlying the "original" experience, nor whether or not these underlying conditions may, in fact, be the cause of a particular experience' (Stoller, 2009: 708). Hence, Braun insists that 'Even when our relation to nature seems most immediate,

Plate 2.3 Captive animals staged as dislocated representatives of wild cousins?

This is an image of black-footed penguins seen through thick glass at Jerusalem Zoo (kindly donated by Irus Braverman). In her book *Zooland*, Braverman writes that 'The most crucial assumption underlying the entire institution of captivity is the classification of zoo animals as wild and therefore as representatives of unconfined conspecifics' (Braverman, 2012: 6). I believe she's correct. Though we, as zoo and wildlife park visitors, well understand the artificiality of the space we meander through, we also arguably see the captive species as literal and metonymic representations of whole species existing 'out there' in the 'wild'. This 'reality effect' is all the more potent for us being physically close to *living* creatures, rather than mere images of these living creatures. What's more, zoos in particular have worked hard over the past 25 years to create a more 'naturalistic' experience for visitors.

it's profoundly shaped by the narratives, knowledges and technologies that enable experience' (Braun, 2002: 15).

This said, some forms of engagement with the non-human world yield knowledge and skill that aren't 'shaped' in the sociocultural sense that Braun intends. A smallholder working the land is not, it almost goes without saying, experiencing nature in the same sense as an occasional wildlife park visitor is. But, in the West today, how many of us are relating to the non-human world in this direct, practical sense? The majority of people lead lives that offer few opportunities for sustained physical interaction with land, water or other species.

SUMMARY

This chapter has advanced three principal arguments. First, it's been suggested that we derive much (most?) of what we know, value and feel about 'natural' phenomena from various epistemic communities. In large measure, we're rendered dependent on them, so I've claimed – cognitively, morally and aesthetically. While some of these communities are tight-knit and discrete, others are looser and their boundaries more protean. We pay attention to, or ignore them, depending on a combination of things. These include their social visibility, credibility and fit with our learned priorities and interests. These communities comprise an 'epistemic ecosystem' (or 'semiosphere') in which some metaphorical species are more dominant than others (such as scientists and daily newspapers).

Second, it's been argued that these communities – be they prominent or less socially visible ones – are routinely engaged in 'representing' nature to us and, as it were, for us. All representation, it was further argued, entails a simultaneous 'speaking for' and 'speaking of' what we call nature: the two are indissociable in practice. In turn, this means that what apparently different forms of representation (e.g. scientific or artistic) have in common is that they're both constructed and political. They are 'constructed' in the sense of being purposefully fashioned by epistemic workers of various kinds, and they are 'political' in the general sense of secreting – or, in some cases, reflexively questioning – the representers' particular goals, values and preferences (ones that could, in theory, be changed). Yet the dual character of representation is not always apparent, which is to say that many people are not conscious of how it 'really' seeks to work upon them. Finally, it was argued that representation still matters in the modern world, despite an analytical focus on it glossing over some other important elements of daily existence.

Let me make one last point about representation, before moving on. In this chapter, I've discussed representation in both a 'diagnostic' and 'prescriptive' way. On the one hand, I've argued that, despite appearances, *all* acts of representation are performative: they call forth that to which they seem only to refer. They're thus acts of purposeful *invention*, not simply acts of *disclosure*. They actively engender particular forms of cognition, feeling and connection (more or less successfully, depending). On the other hand, I've argued that representation is ineluctable: there's no getting away from it, meaning that having the 'right' understanding of representation is key to achieving a better understanding of how we communicate about the world. As the British philosopher Peter Osborne once put it,

> Representation is the medium of thought, [yet] all representation is misrepresentation if by representation we mean the literal . . . re-presentation of [something] ... in some self-constituted original state. However, it is precisely the inevitable failure of any such notion of representation ... which makes a representation a representation – something, that is, which is constituted through a relation between itself and something else, which it claims to re-present.
>
> (Osborne, 1991: 208)

This 'both/and' understanding may seem contradictory. Much representation is not what it seems, I've argued, yet representation matters all the same. However, like Bruno Latour (2004) in *Politics of nature*, I think the argument is consistent. The challenge is not to do away with representation (the idea and the attendant practices understood in terms of this idea). Instead, it's to foster representational practices that do not make us passive or, worse, manipulated in a world where myriad epistemic communities and workers vie to shape our perceptions, opinions, beliefs and actions.

In the next chapter, which concludes the first part of this book, I want to build on these arguments in order to show that representations of nature are heavily implicated in the process of governing the ways in which ordinary people think, feel and act. At base, I'll argue that this is a question of personhood, freedom and democracy. It has major implications for the character and quality of our own lives, as well as all those phenomena with which we share life on Earth.

ENDNOTES

1 What's additionally different is that these spokespersons are not known to us on a personal level – they are strangers or, if familiar faces and voices, they communicate with us remotely via long-distance media. Clearly, I risk making ludicrously simplistic and over-drawn historical comparisons here. However, I do stand by the generalisation that, these days, our thoughts and actions are far more influenced by a wider variety of people who operate at a distance from us when compared with our forebears – forebears who existed in a world where long-distance communication was slow and where the division of mental and manual labour was far, far less elaborate than it is today.

2 This both mirrors, and is thoroughly woven into, our material dependence on commodities made elsewhere by myriad others within a global division of labour. Following Marx and, a century after him, Jean Baudrillard, we can say that just as commodities need to be 'defetishised', so too do 'representations' – not least because commodities are typically saturated with meaning and are themselves representations (not simply 'things'). It's frequently said that we live in an 'age of consumption' – we should remember that what is being consumed is not only material goods but a wide array of meanings 'contained' in individual commodities (e.g. cars, clothes, food). A great many of these meanings involve literal or connotative references to nature.

3 This is why Clifford's contemporary, the philosopher William James (1956), criticised the argument of 'The ethics of belief' in an essay entitled 'The will to believe'. For James, the real issue was not that we acquire direct experience pertaining to all the matters that may influence our lives, but that we get ourselves in a position to assess the *quality* of the various epistemic communities that seek to shape our beliefs. I will talk about issues of epistemic quality and the governance of epistemic communities in Chapter 3.

4 As Sue Stafford (2001: 219) defines them, 'Knowledge communities are groups of individuals drawn together over time by the sustained pursuit of knowledge creation, knowledge use, and knowledge sharing.'

5 Lave and Wenger (1991), who coined the term 'communities of practice' in their book *Situated learning*, focussed on 'applied knowledge' such as that characteristic of midwives, urban planners, architects and tailors. However, this inadvertently gives the false impression that other forms of knowledge are somehow *purely* epistemic and lack any 'practical' effects or uses at all. I should also note that, despite the resonances with literary critic Stanley Fish's (1980) notion of 'interpretive communities', my expansive definition of epistemic communities is intended to go beyond Fish's preoccupation with how any given community interprets the texts authored by its members. Equally, I note that Ludwik Fleck (1935/1979) – someone who pioneered

several ideas that, together with others, form the scaffolding of this book – coined the term 'thought collectives' back in the 1930s. This sounds a little too Orwellian in the present context for me to be comfortable using it. Perhaps closest to what I'm proposing here is historical sociologist Robert Wuthnow's (1989) notion of 'communities of discourse'. Wuthnow inquired into how specific discourses advanced by particular groups of people are able to be both *intelligible* within their wider context yet also able to *transform* that context by persuading enough other people to take the contents of the discourses seriously. In this book, I do not focus on 'transformative discourses' solely but, rather, on the full range of discourses about 'nature' and its collateral terms extant.

6 In saying this, I'm inspired by the seminal insight of SSK scholar David Bloor back in the 1970s. Bloor's 'symmetry principle' stated that ostensibly 'true' not just 'false' scientific findings invited a sociological explanation. In my treatment of epistemic communities and their diverse representations, I do not, in this book, make *a priori* determinations about which of these communities and representations are to be accorded special respect, influence or authority. Their varied role in shaping our lives arises not so much from any 'intrinsic' differences between them, but from socially created distinctions in the way knowledge is made, labelled, disseminated and used in society. It's also the case that epistemic communities vie with one another, consciously or otherwise, for *social legitimacy*. Legitimacy comprises the right to be taken seriously, or to be noticed or paid attention to, in a society. It is often hard won and takes time for epistemic communities to achieve.

7 Note that, while many epistemic communities utilise the insights of 'lay communities', in this book I'm not including the latter in my expansive definition of the former. The point about 'epistemic communities' is that, however loosely organised or complexly organised they may be, becoming a recognised or authorised member of one is rarely easy or quick.

8 For instance, Edwards recounts how long-run shipping data on weather conditions was, in the 1930s, laboriously translated by hand on to thousands of machine-readable punch cards.

9 Steven Pinch rightly describes the concept of community 'as one of the most widely used [yet] ... controversial notions in the social sciences' (Pinch, 2009). It's widely used because it describes the enduring presence of various defined social groups in the modern world. However, because the etymological root of the word 'community' takes us back to pre-industrial, rural society (in which, according to sociologist Ferdinand Tonnies, group cohesiveness was very high – 'gemeinshcaft'), it needs to be adapted to a world in which groups come in many different shapes, sizes and forms. It's also important to note that all communities are in some sense self-defined: unless and until they come to *describe themselves as such* (using the term 'community' or a similar word) they do not, in an important sense, exist.

10 An exceptionally good example of what I mean is provided by legal geographer David Delaney (2003). In his brilliant monograph *Law and nature* he explores how legal professionals (mostly in the United States) work with the conceptual dualisms listed in Figure 1.5 and, in varied ways, make legal claims and decisions in relation to an extraordinarily wide range of cases.

11 In both their own discourse and the minds of many citizens, scientists are typically regarded as the producers of representations that are – contrary to artistic representations, say – deliberately 'free' of value judgements, bias, opinion or speculation. This is precisely why the large environmental NGOs have increasingly sought to base their 'political representation' of the non-human world upon a supposedly 'non-political' cognitive representation by employing accredited scientists since the mid-1980s.

12 In most cases, this second sense of 'representation' does not involve a *resemblance* between the representative and that/those being represented. Because of this, representation in the second sense necessarily involves *relations* between representer and represented in which the former is, as it were, challenged to represent the latter in a 'suitable' way, depending on the situation and the stakes. This is what political

theorists have called a 'trustee' relationship, in which the representer is permitted to exercise judgement about the interests of those they are representing.

13 The cultural theorist Anne Freadman (writing in Bennett *et al.*, 2005: 307) expresses this very nicely indeed: 'representatives stand for their objects in some sense ... If an object needs a representation, then it is part of that logic that the representation is not its object; but if the representation is different from its object, how can it stand for it truly?' The key word here is 'truly': there are only ever partial and interested representations, no 'true' ones. This said, it's no trivial matter determining who gets to represent the world: there are 'better' and 'worse' representatives. For instance, in her marvellous book *Gender and green governance*, my colleague Bina Agarwal (2010) shows how the *presence* of women in arenas where decisions about the management of forest resources are made (in Nepal and Gujarat) matters. In complex ways, it ensures that female forest users' views get represented, redressing a democratic deficit that heretofore existed. Political theorists sometimes call this 'descriptive representation', where the representative stands for us because they are the same as (or like) their constituents.

14 In saying this, I realise that I'm not only essentialising 'representation' but also ensnaring myself in a paradox. Here is Christopher Prendergast on the matter, in his fine book *The triangle of representation*: 'The difficulty is this: in the very act of talking about representation, one effectively begs the question. Assuming you can talk about representation is the same as assuming you can represent representation ... The question then arises as to whether there is a general form of representation for Representation ... this is a theoretical problem to do with a relation between meta-language and object language. I do not propose to dwell on it, since it can rapidly lead us into logical quicksand from which we are unlikely to escape with our sanity intact' (Prendergast, 2000: 3). For the sake of my, and your, sanity I will deliberately sidestep the quicksand.

15 Mitchell's claim, which I follow here, that 'representation' is a historically *recent* invention-cum-practice conflicts with philologist Eric Auerbach's (1957) belief that it can be traced back to the ancient Romans and Greeks. I can only note this difference of perspective here, rather than explain why I think Auerbach is mistaken. Though I've argued that political–legal and epistemic representation are entangled, not separate, Lisa Disch (2008: 88) dates the former to thirteenth-century England when sovereigns first assembled 'representatives' to bind their constituents to taxes and laws. This contrasted, in some European states, with a conception of the monarch as the physical *embodiment* of his or her domain/territory. Clearly, there are complex issues of philosophy and historiography at stake here, but I stand by the argument that the two forms of representation are indissociable and enduring.

16 This idea is implicit in the RSPB advert reproduced in Plate 2.1, but here's another example out of many. In early 2012, the organisation PETA (People for the Ethical Treatment of Animals) lodged a civil rights action on behalf of five orcas that were performance animals in two Sea World theme parks in the United States (San Diego and Orlando). PETA wanted a federal court to hear arguments that the orcas' rights were breached because they were being enslaved by Sea World's owners. Relatedly, when Ecuador added nature's rights to exist to its national constitution, it was obliged to name those who could demand the recognition of those rights, all of whom were (of course) human (adult Ecuadorians).

17 Let me be clear here: neither Eckersley, Haraway, Braun or Evernden is suggesting that what we call 'nature' is *nothing but* a set of representations. If representations refer then, clearly, they must have *something* to refer to – even if that something is not natural by nature. Their point is that we apprehend the world in such a way as to make 'representation' both necessary and normal – and this includes the idea that something called 'nature' exists *prior to* and *separate from* acts of representation. By delimiting a vast terrain of reality and calling it 'nature', we create for ourselves the need and desire to decipher (or reflect upon) its meaning, function, value and qualities. This process of delimitation, to cite one significant example, underpins the academic division of

labour in both schools and universities. Take my own discipline. Human geographers, like other social scientists, study a social world already presumed to be saturated with meaning. This contrasts with physical geographers who are seen (and see themselves) as having to make sense of a biophysical world that is intrinsically meaningless until subject to careful analysis. Thus is their role defined and vouchsafed. Thus are they able to speak for nature as 'geomorphologists' or 'hydrologists' (say) – indeed, nature 'demands' that they speak because without their voice it is but 'noise' or (worse) silence.

In light of this, we can readily understand why a lot of the professional academic debate about representing nature has been not about *whether* to represent it – this is taken as a given – but about *how* it should be done and *who* should be doing it. For instance, in the laboratory and field sciences, there have long been discussions about which methods and techniques will best permit nature's truths to be revealed. If rocks and electrons 'don't talk back' then how should we make them 'talk' to us? Political theorists and ethicists have, likewise, for decades discussed how best to give everything from blue whales to human foetuses a place in our moral universe. This discussion has involved a sub-debate about whether nature has intrinsic rights, entitlements and values – and, if so, who is best placed to represent them to the rest of us (e.g. do you need to have studied and swum with dolphins for years to speak for them in the public arena?). In other words, the debates have largely been about how to configure the 'epistemic' and 'political' aspects of representation so as to achieve things like 'accuracy' and 'authenticity'. Throughout these debates, there is an assumption, which in my view needs to be problematised, that some representations and representatives operate in the realms of 'truth' and 'reality', while others are altogether different in kind because their stock-in-trade is 'fiction', 'distraction', 'faith' or 'entertainment'.

18 A familiar example that illustrates Hall's argument well is a motorcar. Take a luxury vehicle, like a Mercedes, parked on a road as you stroll down a sidewalk. You look at it, and what does it 'say'? It says, or at least tries to, that the driver is rich and successful (or aspires to be). It is probably intended by its owner to symbolise his or her social status, and has been bought as much for its sign-value as its use-value.

19 Some may assume that the senses somehow work alone, sending discrete messages into the brain. For instance, we might think that vision is about 'showing', while sound and speech is about 'telling' – two potentially discrete things. But this understanding is surely crude, even naïve. It is the interplay between information received via different senses that achieves the overall effect of permitting us to understand or be affected by the world.

20 The theoretical, methodological and empirical literature on 'visuality' and what's been called the West's 'visual culture' is now simply vast. Though old, Jessica Evans and Stuart Hall's (1999) book remains good value as a primer, while Sturken and Cartwright's (2001) *Practices of looking* is excellent on the different theories. John Hartley's (1992) *The politics of pictures* is an excellent, accessible study of how public and popular culture has become 'ocularised' in the West via the mass media and advertising. Gillian Rose's (2007) *Visual methodologies* offers an attractive blend of theory, method and numerous real world examples. I follow Rose in not separating 'discourse' from ostensibly non-discursive acts of sense-making, such as seeing. She devotes two chapters of her book to 'discourse analysis' and is insistent throughout that seeing and looking are interpretive acts that necessarily involve norms, ideas and values that are expressed (or can be expressed) linguistically.

21 From the late 1980s, many in the social sciences and humanities announced a so-called 'crisis of representation'. The crisis in question arose from a critique of the idea that many representations are mimetic, with French post-structuralist philosophers and historians leading the charge. Note that these critics did not announce the *death* of representation but, instead, its problematisation.

3 GOVERNING SOCIETY WITH REFERENCE TO THE NATURAL

Having sought to clarify how and why 'representation' matters in our lives in the second half of Chapter 2, let me now build on the early part of that chapter. I want to say more about how epistemic communities of various kinds try to exert their representational influence on us. By 'influence', as readers will see, I mean a set of affects and effects that are quite profound – either in actuality or potentially. We may think we can 'take or leave' any given message or experience because of the freedom we enjoy in our daily lives. However, I'll argue that references to nature and its collateral concepts are part of a wider process of shaping people's identities, values and behaviours. This process is inescapable, albeit complex and only partially coordinated. It encompasses seemingly 'innocent', unremarkable or everyday situations, like a weekend camping trip in the mountains or the purchase of tinned tuna caught with 'dolphin friendly' nets. Whether we realise it or not, we're routinely being invited to imbibe certain ideas, norms and practices which we then take to be our own (or not, as the case may be). This raises some key questions about epistemic communities, their representations, and the ways in which we respond to them.

These questions, I will suggest, speak to fundamental issues of liberty, control and government. References to nature and its collateral concepts, it turns out, are absolutely central in shaping what sort of people we become – beginning when we're very young. They're also central to answering what many commentators regard as the most important question of all: 'how should we live?' While this is obvious in the case of something like the science of anthropogenic climate change, I'll argue that we need to attend to the full range of arenas in which one or more of the four meanings of nature are used to elicit an intellectual, emotional or practical response from us.

HOW COMMUNICATION WORKS

Genres of communication and modes of representation

As we go about our daily business, we're all well attuned to the different **genres of communication** that are used by others to instruct, entertain, motivate, move or provoke us. 'Genre' is a literary term, but it has wide applicability. It points us to the *stylistic* differences between different representations – indeed, between all forms of communication.[1] Most people arguably take these genre differences for granted, forgetting that they are social conventions (see Box 3.1). Each genre has a history – it came into being at a certain point in the past. It also has a geography – it emerged in specific places, and then (in many cases) became adopted far more widely. From this we can conclude that different genres of communication exist because they serve certain (for now) socially recognised functions and meet felt needs. But since the functions and needs did not exist fully formed prior to the genres, we can also conclude that communicative genres help to *construct, reproduce and sometimes alter the wider context they ostensibly exist to serve.* Generalising, we might say that any given epistemic community tends to be synonymous with one or other genre (or sub-genre) of communication. We might say too that specific discourses, in the count noun sense of the term, are partly genre defined in terms of their creation and reception/consumption.

BOX 3.1

THE SOCIAL CONSTRUCTION OF COMMUNICATIVE GENRES

Genre is a slippery term but it designates something very definite. *The Oxford English Dictionary* defines genre as 'a style, especially of art or literature'. The word 'style' directs us less to the particular *content* of any given act of communication, or primarily to its *medium or mode* (e.g. visual or verbal), but more to the *kind* of communication it is intended to be. How is the content of any given communication organised (for instance, is it a continuous story narrated by a single speaker or writer, or is it a debate between opposing viewpoints)? According to what rhetorical conventions and practices are different acts of communication designed and disseminated? We are well familiar with the two 'meta-genres' of 'realist' and 'fictional' communication, so too the related (though not exactly synonymous) meta-genres of 'serious' communication versus that geared towards 'entertainment'. I call these 'meta' genres because, within them, there are numerous recognised sub-genres. For instance, popular science books and daily news broadcasts are intended to be 'realist' (i.e. the opposite of fiction), yet they achieve this in rather different ways, stylistically speaking. Likewise, superhero comic strips are fictional, but not in exactly the same way as the books and films of J. R. R. Tolkien's *Lord of the rings*. In all cases, genre differences and sub-genre variations are the products of human decision-making – in other words, the genres are not determined by the content they seek to communicate.

From time to time, people come along whose ideas or creations challenge existing understandings of how a specific genre (or sub-genre) of communication is defined, and where its boundaries lie. This highlights the capacity for genres to change over time. Take the world-famous British artist Damien Hirst. In several of his early works, he utilised the bodies of dead animals. For instance, in 1992 he famously displayed 'The physical impossibility of death in the mind of someone living', a 14-foot (4.3 metre) tiger shark immersed in formaldehyde in a vitrine. Eight years later, Marco Evaristti exhibited ten ordinary kitchen blenders inside which were goldfish – visitors were given the choice to switch the blenders on. Works like this attracted a lot of criticism from 'establishment' figures in the art world, and they also perplexed many members of the public accustomed to thinking that 'art' is synonymous with painting or sculpture. Much more recently, a similar stir was caused when the internationally prestigious Turner Prize was awarded to the Scottish artist Susan Philipsz in 2010 for her 'sound installation'. Philipsz's aural creation was a recording of her singing three versions of a Scottish folk song over the River Clyde in Glasgow. The works of Hirst, Evaristti and Philipsz made many people ask the question 'Is this art?' Outside the art world itself, what's often forgotten is that what counts as 'art', in terms of both substance and style, is ultimately decided by artists, their patrons and their audiences. In other words, there is nothing intrinsically 'artistic' – to take one famous example – about the much admired natural landscape paintings of Canada's so-called 'Group of Seven' (who worked in the 1920s). Conversely, there's nothing intrinsically 'non-artistic' about the works of Hirst, Evaristti, Philipsz or other innovators. Upon close inspection, the history of art in the West and beyond shows it to be highly changeable, notwithstanding some people's insistence that its role – or, rather, the role of 'true art' – is to communicate universal, timeless truths about life and living.

For more on genre, see John Frow's (2006) excellent book *Genre*. While I will make an analytical distinction in this chapter between the genre, the content and the mode of any given act of communication, in practice the three are inextricably intertwined. This is well illustrated in Howard Becker's (1982) wonderful study, *Art worlds*, which details the complex community of actors and the elaborate material infrastructures required to make and sustain the genre known as 'works of art'.

What's interesting, I think, is the *ease* with which we recognise that different epistemic communities are engaged in throwing different 'styles' of representation our way. We often receive the particular content of their representations with equal ease because we situate it in the context of the style. We 'know', almost unconsciously, how we 'should' respond to both genre and content in any given case (even if we choose to respond otherwise). This is because we've been socialised into expecting there to be different communicative genres from an early age. We're also well accustomed to different **modes of representation** (and communication more widely) within and between these genres. By modes I mean the *ensemble of media* in and through which representations are communicated. For instance, we've come to expect that a museum of natural history will contain lots of objects to be viewed or handled: specimens, models, drawings or video films – most with some 'serious' written or spoken text. Likewise, we'd be surprised if someone researching the unique anatomy of a deep-water fish species published their findings as a YouTube audio-visual clip rather than as a peer-reviewed academic paper in a journal like *Science*.

Study Task: This book observes the conventions of academic writing. Itemise a few of these conventions. How often, if at all, have you questioned them during your education? What sort of 'power' do you think these conventions exert on you and other students? Would you take this book's contents seriously if I chose to communicate them using a mode of representation such as a coloured picture book?

Undertaken in conjunction with reading Box 3.1, the study task is designed to highlight the *arbitrariness*, but also the *social force* of differences in communicative genres and modes. From time to time, we encounter cases of genre blending or hybridisation. For instance, as a child I recall being given a field guide to British birds which combined 'accurate' ornithological descriptions with beautiful watercolour paintings of each species printed on glossy paper.[2] In other cases, we see cross-genre collaborations, as in so-called 'Sci-Art', in which scientists work with artist partners to see what each community can gain.[3] In still other cases, we're confronted by innovations that redefine a genre from within, but these crossover and paradigm altering cases do not render moot the genre differences they blur or redefine. In each case, we're disposed to interpret representations in definite ways because we're so *au fait* with the genres and modes in question.

This is not the same as saying that we're each of us equally attentive to, or affected by, the whole spectrum of genres and modes. In part, this is because it's conventional to regard certain genres and modes as decidedly 'optional' for us (such as the innovative 'body art' of Gunther von Hagens, creator of remarkable plastinated exhibitions of human and nonhuman corpses relieved of their skins). They fall outside our daily habits and routines to the extent that they strike us as almost alien. Moreover, because many genres and modes are strongly associated with certain sections of society (e.g. 'high art' is typically seen as the preserve of a sub-group of well-educated upper income earners), many individuals pay them no heed. Equally, many people actively use their detailed knowledge of certain genres and modes as markers of distinction from others in society. In still other cases, even very visible modes and genres that we know to be socially important do not always sway us – at least knowingly. For instance, having a basic and broadly positive sense of what 'science' is doesn't necessarily mean that one pays much or especially close attention to what any given scientist says or recommends. It's highly context-dependent at the level of the individual person.

Subjects and self-hood

These caveats notwithstanding, I'd suggest (and am hardly the first to do so) that our familiarity with different genres and modes of representation tells us something important about their hold over us. I say 'hold over us' in a general sense and for good reason. Most people would acknowledge that their school experience, along with their family and friends, has had a major influence on their identity, outlook, values and aspirations. This is because, in our early lives, so much time is spent being taught by various teachers, being at home, and socialising with siblings or neighbours. But what of all the other things we learn from, say, novels, trips to the library, magazines, television dramas, web surfing, a family atlas, adverts or a visit to a gallery? Are these experiences somehow secondary in conditioning the kind of people we become? Regardless, when we eventually leave school and home, aren't we sufficiently independent and self-sufficient, both intellectually and emotionally, that we're less readily (or, we might say, less 'unreflexively') influenced by the claims or actions of other people, even ones who appear to be experts? At this point, aren't we, in short, 'mature' and able to 'think for ourselves'?[4]

These questions, if read non-rhetorically, are predicated on the idea that we can, at some point and at some level, *separate ourselves* as thinking, feeling subjects from the wider social environments in which we live. So far in this book I've occasionally implied that this idea is credible by using the term 'consumer' to describe people's engagement with these environments. In modern parlance, consumers are said to have 'freedom' and to exercise 'choice' among a range of options. But what if those environments, in significant measure, constitute and reproduce our characters, preferences and practices? What if they shape not merely what we 'know', 'feel' or 'do', but *who we are*? What if the panoply of epistemic communities we're familiar with – with their various subject matters, genres and modes of address – together provide the complex, changing and pockmarked 'grid of intelligibility' through which we come to define both ourselves and the world at large?

These questions oblige us to consider the deep connections between our individual selves and what sometimes appears to be (but is surely not) an 'external' environment comprising of multiple actors, institutions, knowledges, discourses and so on. As a corollary, and thinking back to Chapter 1, the representations of nature abroad in our world may govern us more widely and powerfully than at first meets the eye. They enter into the relational constitution of people's self-understanding and habits of action on a daily basis. In Jane Bennett and Williams Chaloupka's words, far from being something to merely contemplate, discuss, utilise, manage or journey into, '"nature" has long performed an *identity function*...' in Western society (1993: ix, emphasis added). It's a reference point through which, and against which, we come to define what sort of people (we think) we are.[5]

Clearly, I'm suggesting that epistemic communities and their various representations are part of a wider process of shaping our routines of thought, feeling and action 'all the way down'. Their effects, given sufficient time, are not superficial. At a minimum, they invite us to play temporary social roles. For instance, when I read a front-page story in *The Washington Post* on the latest scientific findings about marine species loss, I'm being **interpellated** as a 'citizen' – either of the United States or of the wider world. I'm also being addressed as an 'educated person', probably a university graduate and 'middle class' (in actuality or aspiration). I accede to these calls, consciously or not. More than this, some critics would argue that everything from the news media to artists to advertising professionals actively call us into various **subject-positions** that, together, constitute our characters.

This concept of subject-positions, which has long been a staple of critical social science, suggests that any person's "self" is neither a pure product of human biology nor necessarily coherent or indivisible.[6] As historian and geographer David Livingstone put it, 'The "self" has become increasingly fractured ... nowadays all of us occupy an immense range of different [subject-positions]. In these we act differently, adopt different personae, call on different linguistic repertoires, [and] project different "selves"' (Livingstone, 2003: 183). What we call 'the self' is thus fashioned out of multiple, but not always commensurable discourses and prompts that invite us to identify ourselves with or respond to them – and perhaps perform the practices they seek to engender. I say 'invite' because not all subject-positions have significant meaning for us, and many do not generate commensurate actions on our part. For instance, environmental social scientists are familiar with the so-called 'values–action gap' found in most Western societies. This describes a situation where people profess their 'environmental awareness and concern' – often sincerely – and yet do little to follow-through on their convictions in practice (e.g. by cycling to work as opposed to driving a 4 × 4 vehicle).

The complexity of the self, and the manifest differences between selves, reflects the sheer *variety* of subject-positions that any given individual may be invited to occupy. For instance, through a combination of structure and chance, one might be interpellated as a parent, twin, lover, worker, sister, hiker, lesbian, horticulturalist, migrant, consumer, environmentalist, feminist, holiday maker, anti-abortionist, blood donor, birdwatcher, film buff, rock climber, or (in my case) Englishman. Selves are ensembles and they are actively reproduced day in and day out. What it means, in the detail, to occupy these myriad positions depends entirely on the particular mixture of discourses, norms and experiences that any given individual has been exposed to or seeks out. But clearly there are broad patterns among otherwise different individuals depending on when and where they live – if this were not the case, any given 'society' could scarcely reproduce itself over time, notwithstanding its internal differences (see Box 3.2).[7] It's also important to note that subject-positions have physical effects on our bodies if they produce habits commensurate with those positions. They're not 'merely' mental. A graphic example is provided by food. As the 2012 BBC television series *The men who made us fat* showed, food marketing professionals are now expert at interpellating ostensibly 'healthy eaters' who are conscious about their inner health and outer appearance. The adverts for and packaging of things like organic food are designed to reproduce a particular set of positive self-conceptions among a class of consumers.[8]

BOX 3.2

SUBJECTS AND AUDIENCES

One way to think about subject-formation and subject-positions is in terms of the audiences, both real and imagined, presupposed by various different epistemic communities, institutions and their associated discourses when they address us. For over 30 years, researchers in the fields of media studies and cultural studies have examined how various audiences are, in effect, brought into being by their myriad addressees (indeed, there's now an interdisciplinary sub-field called 'audience studies'). The term **audiencing** describes this process: in short, audiences do not exist until 'called forth', more or less successfully, by those who wish to address them. A simple example is a road-users' map, which you might consult when driving a car. We know how to read the map, and indeed to trust it, because a very large community of cartographers and teachers inculcate each new generation into the 'right' way to understand maps. In this sense, **socialisation** and various routinised acts of audiencing are closely connected. In most cases, audiences are apparently 'free' in two senses: they are free to read/watch/listen/participate (i.e they can choose *not* to), and they are free to respond to or interpret what they see/hear/do in a range of different ways.

However, this suggests that in our various 'audience roles' – as high school students in a classroom, television watchers or church goers, say – we can ultimately shut out the information, knowledge, sights or sounds thrown our way. From the mid-1970s, the so-called 'Birmingham School' of media and cultural studies questioned this idea. Based in England, critics like Stuart Hall showed that audiencing is not only part of a continuous process of shaping people's identities. They also showed that even 'innocent' or seemingly 'non-serious' acts of communication, such as television soap operas, can serve to reproduce (or occasionally challenge) conventional ideas about very important subjects, such as sexuality, gender roles, terrorism, national identity or environmental change. The increasing prominence, since the turn of the twentieth century, of state education and the mass media in many countries means that, today, the process of audiencing can 'capture' millions of people at a time. In other cases, audiences are far smaller and more exclusive – especially when they're 'countercultural' in their shared beliefs, political values, interests or practices. In any given case, there's no foolproof way of gauging how 'successful' the audiencing process is. Though sensitive to the way powerful institutions can shape audience beliefs and values, Stuart Hall (in several 1980s publications) nonetheless insisted that the 'encoding' of meaning by various epistemic communities (e.g. documentary film-makers) does not always translate in the ways these communities intend when meaning is 'decoded' by their target audiences in practice. This is most obvious when decoding occurs outside the country where meaning is encoded – as revealed in Liebes and Katz's (1990) *The export of meaning*, which examined how non-American audiences interpreted the hit soap opera *Dallas*. An excellent introduction to 'audiencing' is the recent *Handbook of media audiences* by Virginia Nightingale (2011).

As the influential philosopher-cum-historian Michel Foucault rightly insisted, it would be wrong to think that the process of subject-formation is planned or orchestrated by any one actor, group or agency – even one as powerful as the national state or an organised religion like Catholicism. The sociologist Nikolas Rose put it like this in his book *Inventing ourselves*:

> Subjectification ... designate[s] all those heterogenous processes and practices by means of which human beings come to relate to themselves and others as subjects of a certain type ... [It is to be] ... understood as ... a complex of apparatuses, practices, machinations and assemblages within which human beings are fabricated ....
>
> (Rose, 1998: 25, 18)

This argument encourages us to see our identities, values, habits and beliefs as 'emergent' products of a differentiated and changeable environment in which we slowly learn to become certain sorts of people – indeed, to become different people over time.

In his late work, Foucault coined the term **governmentality** to describe this uncoordinated, multifaceted and sometimes contradictory process of **subjectification**. Because of its filiations with the more familiar word **government**, this (ungainly) first term usefully reminds us that, even when it seems *not* to be the case, various actors, groups and institutions are actively trying to govern our 'selves and souls' – even if not always consciously. I use the word 'govern' here in the original etymological sense meaning to 'steer' (or 'direct') and I use the word governmentality in a generic sense (rather than Foucault's more specific, technical and historically specific sense). In Mitchell Dean's expansive definition, it is 'any more or less calculated ... activity, undertaken by a multiplicity of authorities and agencies ... that seeks to shape conduct by working through our desires, aspirations, interests and beliefs' (Dean, 1999: 11).

There's nothing necessarily sinister about this, though nothing necessarily innocent about it either. While Foucault was given to saying that all subjects are the 'effects of social power', he did not mean that everyone from science teachers to environmental lawyers are engaged in some conscious conspiracy against the masses. Instead, Foucault enjoined us to look well *beyond* the 'obvious' institutions – for instance, schools or the family – if we're to understand fully whose ideas, values and practices determine the sort of people we become. Analytically, as Joe Hermer notes, this liberates us from 'a narrow view ... which posits a compliant "subject" that can be commanded, repressed, restricted and prohibited' (Hermer, 2002: 6). The capacity to shape and mould – including also the capacity to challenge those who currently have that capacity far more than others do – is potentially widespread rather than concentrated in the state apparatus or (say) large media organisations.[9] What's more, epistemic workers in the same genre and representational community often differ and disagree considerably, meaning that they disseminate a plurality of (often contradictory) ideas, images, facts, contentions and so on. For these reasons, as I've already said, ordinary people cannot be said to be 'subject' to any one producer of knowledge or any one communicative genre.

And yet, notwithstanding the large number and variety of epistemic communities extant in the world, I should note that it's frequently been argued by left-wing and liberal critics that too many people are these days fed a very narrow discursive and experiential diet. The attack on Fox News in the United States by the political comedian Al Franken is one example of this (Franken, 2003). Franken was hardly alone in arguing that Fox's news reporting is highly biased: it tends to align itself with the values of cultural and political conservatives, as if these are the values shared by all 'normal' and 'right-thinking' Americans. This is one example of the argument that, while power is irreducible to the words and deeds of just a few large institutions, it is nonetheless not shared equally among many. The so-called 'corporatisation of the media' – both the news and entertainment media – rightly disturbs many commentators for this reason. The worry is that the diversity of content and high quality standards one gets with most public broadcasting are sharply eroded once big private companies call all the shots. This raises wider questions about how epistemic communities govern themselves, and are governed by other responsible parties – questions I'll turn to at the end of this chapter.

EPISTEMIC COMMUNITIES FROM THE INSIDE OUT

I've argued that various epistemic communities are (wittingly or not) involved in the complex process of subject-formation – to different degrees and with different effects, depending on the case. If, as I suggested earlier, many of these communities make sense of 'nature' for us, then it follows that the concept and its referents are an aspect of governmentality (conceived here in a broad, generic way as a diffuse process of governing beyond the realms of state and formal politics per se). Working within (or sometimes challenging) the antonyms mapped in Figure 1.5, different communities' particular representations of nature compete for our attention. At the very least they can affect our thoughts, attitudes, values, feelings and actions – in the short or long term. But they can also interpellate us as subjects of a certain kind, and to that extent are involved in the making of the self. That self might be wholly conventional in some or all areas of daily life, but then again, it might not be.

Note the corollary of all this: if representations of nature are implicated in the shaping of subjects and selves, they're necessarily implicated in the way what we call nature is *itself* used, altered or protected. The governance of people and the governance of bodies and ecologies are but two sides of the same very large, intricately patterned coin. Consider again the car advert in Chapter 2 (Plate 2.2). To the extent that they succeed (and they most certainly *do*), these ads engender habits that permit mass pollution as a 'normal' practice and the mass extraction of oil from the ground. Huge environmental changes are attendant upon us being interpellated as consumers in need of private transportation.

Having so far looked at epistemic communities from the outside in, from the perspective of 'consumer' and 'audience', let me now briefly look from the inside out. This matters for two reasons. First, epistemic communities obviously socialise *their own members*. Often they engender specific **epistemic identities** that members assume when working within those communities.[10] These identities then become the basis upon which group members produce and disseminate their representations. Second, this feeds into the process whereby epistemic communities both create and then actively maintain **epistemic boundaries** between themselves, other communities and the wider public realm. This fence building and maintenance is quite central to how communities and their audiences are defined and separated. To reiterate a claim I made in passing before: over time communities actively create their own niche, and they thereby help to manufacture their various audiences. Even to insiders, let alone outsiders, this process is usually too slow for its effects to be visible. Normally, the effects are taken for granted as 'just the way things are' – as things given rather than contrived.

Let me flesh out these two points by way of brief examples, which relate to the physical sciences. Even today, if you ask people about 'science' they think of individuals (usually men) in white coats busy trying to answer questions phrased crisply as testable hypotheses. In the public's mind, scientists do not pose moral, political or aesthetic questions (because these are anterior to science). Instead, they ask value-free 'what?', 'how?' and 'why?' questions about the organic or inorganic world that are addressed by them adducing the right quality and quantity of evidence ('facts'). In the West, at least, we hold science in high esteem because it promises a kind of knowledge that's seemingly rather special in its content and character. Science is regarded as the enemy of distortion, bias, falsity, ambivalence, bad faith, opinion, metaphysics and the like. Scientists thus together comprise what's been called 'a culture of no culture'. We typically expect them to say what 'nature would say to us' if only it could speak our language. To be a 'scientist' is thus to be perceived as a truth seeker who lets the facts do the talking, not their personal biases, hunches or prejudices.

However, the public image of science has always been belied by what scientists do in practice once their activities are closely scrutinised. Since the early 1970s, a now large research and teaching field called 'social studies of science' (or science and technology studies (STS)) has opened up the 'black box' of science and peered inside. Detailed studies of pure and applied scientists in a range of disciplines, and in laboratories and the field, show that 'science' is not at all a unified thing and neither are scientists. For instance, in her germinal study of two communities of high-energy physicists in Stanford, California and Tusukuba, Japan, Sharon Traweek (1988) showed that they each had a particular 'culture' – in the same sense a Western anthropologist might characterise the norms, values and practices of a non-Western 'tribe'. Traweek's key point was that, like *all* communities, physicists actively create their own ways of describing, looking, questioning, theorising, recording, measuring, testing and evaluating. These ways are *not* 'dictated' to them by the 'laws of nature', Traweek argued. Instead, they are the socially contingent means by which those laws are (apparently) 'discovered' – if the means had been otherwise, so might the discoveries. More recent work by Karin Knorr Cetina (1999) endorses this argument, building on her earlier study (1981). Her book *Epistemic cultures* contrasts the way that high-energy physicists investigate the world with that of molecular biologists. Based on ethnographic research, Knorr Cetina shows that in the two cases, different 'scientists are reconfigured to become specific epistemic subjects' (1999: 32). Their respective sense of the 'right' way to think and act as a scientist is heavily conditioned by the particularities of their prior training (bachelor's and postgraduate degrees), their colleagues, the (often expensive) equipment at their disposal and other 'local' matters. Scientists are thus socialised into different (though sometimes overlapping) 'sub-cultures' rather than into a mythical universal called 'science' (refer back to Box 2.2).

However, the myth has its uses – and here I come to the 'boundary work' that many scientists necessarily perform (my second point). Back in 1942, the American sociologist Robert Merton famously sought to capture the *differentia specifica* of 'science'. For him there were four, namely: communalism (the common ownership of scientific discoveries); universalism (claims to truth are evaluated in terms of transcendental criteria, and not on the basis of race, class, gender, religion or nationality); disinterestedness (scientists put personal interests to one side); and organised scepticism (all hypotheses and findings must be evidenced and are subject to rigorous, structured community scrutiny) (Merton, 1942). Interestingly, Merton's four principles were not based on any systematic observations of how scientists do their science. Instead, he derived them from discussions with scientists and took the latter at their word. Arguments like Merton's, which claim to reduce science to its supposed 'essential characteristics', confirm Steve Fuller's astute observation that '*defining* science quickly metamorphoses into ... [normatively] *demarcating* ... science from its various pretenders' (Fuller, 2007b: 31). Sixty years after Merton wrote, the end of the millennium saw a large number of American scientists trying to delimit their intellectual territory once more. The impetus for their boundary work was what they perceived as the 'attacks' on science being launched by scholars like Traweek and Knorr Cetina. These supposed attacks were met with return fire in the form of publications with loaded titles such as *The flight from science and reason* (Gross *et al.*, 1997).

It seems to me that two things were going on in the so-called 'science wars' of the 1990s (see Box 3.3 for more on these wars). First, sociologists of science were deliberately having their expertise called into question by their erstwhile research subjects, namely practising scientists. In other words, the latter sought to defend their craft by arguing that the former were, at best, misinformed and, at worst, anti-science and 'politically motivated'. In Thomas Gieryn's (1999: 4) terms, the two sides were engaged in a 'credibility contest' to see who had the right to represent what science is 'really' all about. Second, in keeping with the spirit (if not the letter) of Merton, the defenders of science were also downplaying their differences in order to present a reassuringly conventional definition – and thus demarcation – of science for wider public consumption. For instance, the famous American physicist Steven Weinberg, commenting critically on the sociology of science in the *New York Review of Books*, said this:

> If scientists are talking about something real, then what they say is either true or false. If it is true, then how can it depend on the social environment of the scientist?
>
> (Weinberg, 1996: np)

Relatedly, at the same time the chemist Dudley Herschbach insisted that 'science ... exalts Nature: she is the boss; we try to discover her rules; she lets us know the extent to which she has done so' (Herschbach, 1996: 12). It is usually at moments when the existing boundaries between epistemic communities and their various audiences are being challenged that attempts are made by insiders to maintain the status quo. After all, whether we're discussing scientists or any other epistemic community, insiders usually have a lot invested in what they presently do and will, unsurprisingly, seek to defend it vigorously.[11]

BOX 3.3

THE 1990S 'SCIENCE WARS'

By the early 1990s, social studies of science (also known as sociology of science and technology or sociology of scientific knowledge (SSK)) was a mature research and teaching field. It was concentrated in several departments and units, especially in British and American universities (like Bath and Cornell). Practitioners published their research in specialist academic journals, such as *Social Studies of Science* (founded in 1971), awarded degrees in their field, and offered modules for science students studying things like mathematics, engineering or particle physics. Concerned that SSK people were 'demystifying' science to the point of undermining its societal credibility, some practising scientists sought to turn the tables. Though several SSK pioneers claimed to study science scientifically, thus interrogating it according to its own standards, by the 1990s many scientists felt that SSK was misrepresenting what they do.

The opening salvo of what came to be called the 'science wars' was fired by biologist Paul Gross and mathematician Norman Levitt in their hard-hitting book *Higher superstition: the academic Left and its quarrels with science* (Gross and Levitt, 1994). This inspired a New York Academy of Sciences conference titled 'The Flight from Science and Reason', organised by Gross, Levitt and Harvard physicist Gerald Holton in 1995. At the same time, the New York physicist Alan Sokal submitted a paper entitled 'Transgressing the boundaries: towards a transformative hermeneutics of quantum gravity' to the cultural studies journal *Social Text* (Sokal, 1996a). It was accepted and published in a special issue of the journal entitled 'Science Wars' in May 1996. Later that year, in an issue of the American literary magazine *Lingua Franca*, Sokal revealed that 'Transgressing the boundaries' was, in fact, a hoax. He had, he said, intentionally tested the intellectual rigour of a non-science academic journal to see if it would 'publish an article liberally salted with nonsense if (a) it sounded good and (b) it flattered the editors' ideological preconceptions' (Sokal, 1996b: 32). The matter became known as the 'Sokal Affair'. Subsequently, supporters of Sokal and of *Social Text* traded blows and defended their respective ideas in print and at conferences. Once the dust had settled somewhat, attempts were made to identify, in a relatively balanced way, some of the 'take home lessons' from the science wars (see, for example, *After the science wars* (Ashman and Baringer, 2001)). In all this, SSK people have been insistent that they are not 'anti-science', let alone 'anti-realists'. What's clear from the science wars is that the right to define and demarcate what 'science' is – and have one's claims in this regard seen as legitimate – remains as important as it is contested. What's also clear, as Gieryn said, is that 'science is a kind of . . . "marker" for cognitive authority, empty *until* its insides get filled and its borders drawn amidst context-bound negotiations over who and what is "science" ' (Gieryn, 1995: 405).

CIRCULATING AND MUTATING REFERENCE

In this section of the chapter, I want to conclude my discussion of epistemic communities and their representations of nature by considering how the latter 'travel'. To explain why, I need to recap some points already made. Early in Chapter 2, I mentioned briefly the material infrastructures that allow all epistemic communities to undertake their work and to distribute their various creations to diverse users, patrons, consumers, clients and audiences. In the previous sub-section of this chapter, I discussed how epistemic communities socialise their members and, if need be, defend their 'turf'. In both cases, I risk giving readers the false impression that epistemic communities, once established, are able to function *sui generis* and to control where their representations appear and how. In actuality, things are far more complicated and messy than this: there's a lot of borrowing, translation and re-representation going on. There's also a lot of mutual dependency and learning, much of it profound. In short, epistemic communities routinely import, repackage and repurpose epistemic content hailing from other communities and genres (located near and far). Christopher Prendergast calls this 'the mutually constituting circularity of representations and [associated] practices' (2000: 56), though we might want to add that the metaphorical circle can not only change size as it rolls forward, but also have its circumference perforated in places. Much of the time, therefore, we're the recipients of experience, knowledge, discourses and representations that are not so much 'second hand' (a term I used earlier) as metaphorically 'third', 'fourth' or 'fifth hand'. This is, I think, true in two senses – one general, the other far more literal and concrete.

Shared meanings and semantic repertoires

First, for there to be any communication there must be some shared concepts, meanings and values – what we usually call 'culture' in the widest sense of the term. This was my focus in Chapter 1. As I argued, the 'formations of meaning' that Raymond Williams (1976: 13) insisted we must take seriously exist 'over and above' the activities of any one epistemic community. They constitute the cultural atmosphere from which otherwise different people draw breath. Consciously or unconsciously, these concepts and meanings not only inform the work of epistemic communities; they also condition how the latter's representations are received and interpreted in the wider society. The process of subject-formation discussed in the previous pages is thus, as Foucault insisted in his early and middle writings, broader than the purposeful activities of various epistemic workers. It's at once authored and authorless because the antonyms in Figure 1.5 both precede and are invested with specific meaning by epistemic communities and their various audiences.

In her book on how the concepts of 'mother' and 'nature' take on meaning in modern American society, the feminist-environmentalist Catherine Roach (2003) provides many examples of how apparently different epistemic workers draw from a shared repertoire of recognised terms and denotations. Among other things, she analyses the content of two public information films. One was produced by the US Corps of Engineers (which, along with its other activities, constructs and maintains flood defences), the other by the Earth Communications Office (ECO, an American non-profit organisation that uses Hollywood stars to promote ecological awareness). In the first film, which focuses on the Corps' attempts to control the Mississippi River, the male narrator says this to the audience in a deep, serious voice:

> This nation has a large and powerful adversary. Our opponent could cause the US to lose nearly all her seaborne commerce, to lose her standing as first among trading nations ... We are fighting Mother Nature ... It's a battle we have to fight day by day, year by year; the health of our economy depends on victory.
>
> (Cited in Roach, 2003: 117)

In the second case, spoken words, written words and music accompany images in the ECO's 1995 release (called 'Mother') – which received an Emmy Award nomination for best in genre. The film opens with an ethereal New Age soundtrack, and there are movement-filled images of sky, water, clouds and sand dunes. A male voiceover then tells viewers:

> Long ago, before the first human was born, before the first tree began reaching for the sun, her life began. She breathes [viewers see ocean waves cresting] and grows [lush forest foliage is shown]. Her blood rushes through her veins [a gushing stream is shown] ... She is not a thing. She is the Earth, and there's a

> reason we call her mother. She has the power to give us energy [Niagara Falls is shown] and the power to make us smile [lions cubs are shown playing].
>
> (Cited in ibid.: 126–7)

There is, Roach argues, nothing natural about depicting 'nature' as either a woman or, more specifically, a mother. The reason that '"Father Nature" has no cultural meaning or resonance at all ...' (ibid.: 9) is because of a historically acquired habit of mind in which only one gender is routinely synonymised with the non-human world. In certain of their meanings and uses, 'nature' and 'woman' thus become collateral concepts. However, contra Ortner (1974), Roach shows that the 'formations of meaning' operative are complex and even contradictory – echoing Williams's point about how 'keywords' get used together in practice. Thus, the Corps' film trades on the recognised, highly negative, lay idea of a 'bad mother' who fails to meet her child's needs. By contrast, the ECO film anthropomorphises the Earth as a 'good mother' under assault from a 'bad child'. Though the films operate in the same representational genre, one is ostensibly about the need to control nature, the other about the need to treat it less instrumentally. And yet, as Roach shows, *both* films reproduce gendered ideas and assumptions about parenthood and womanhood abroad in the wider society. They arise from, and help reproduce, what Alexander Wilson (1992) called a **culture of nature**. In the United States, a certain notion of the 'good' and 'bad' mother exists that's bound up with complex histories of patriarchy, racism, colonialism and nationhood. These histories form the background to the production and reception of the films that Roach analyses.

Study Task: Can you think of any other epistemic communities that have made active use of the 'mother nature' trope? You might find an answer by typing 'mother nature' into an Internet search engine and then exploring the results. Try to find some concrete instances of references to 'mother nature' in discourse (and maybe imagery too). Then examine the content of these references closely. Do the literal and implied meanings therein echo, or depart from, those discussed immediately above?

Translating between communicative genres and sub-genres

Second, more literal acts of representational exchange and translation occur routinely between different epistemic communities. Such exchange and translation is made possible, in part, by what Bruno Latour (1987: 227) has called 'immutable mobiles'. These are transportable media (like this book or an emailable PDF) that both materially 'fix' and disseminate the representations produced by various epistemic workers. Thus, the Intergovernmental Panel on Climate Change's (IPCC) periodic 'assessment reports' are immutable mobiles just as much as a DVD of Steven Spielberg's

Jurassic Park. Even 'immutable immobiles' travel because they are represented in mobile or multisite media like television advertising (think of aquariums, for example, which need to be widely promoted in order to attract enough visitors). Immutable (im)mobiles travel from, and through, long-distance material infrastructures so that various epistemic communities and the wider public can, if they wish, encounter them. When, however, one epistemic community takes an interest in the representations and practices of another, it will necessarily make sense of them according to its own needs and the demands of the genre it operates within.

There are countless examples that can illustrate how representations of nature travel between epistemic communities, only to be reframed and reinterpreted once they've 'arrived'. Let me focus, not for the first time, on just two. In 1999, the Canadian biologist Bruce Bagemihl (1999) published a work of popular science entitled *Biological exhuberance: animal homosexuality and natural diversity*. The volume surveyed scientific studies of over 300 species and showed that homosexual and bisexual behaviours are common among animals. On this evidential basis, Bagemihl proposed a theory of sexual behaviour in which reproduction is only one of its principal functions. He hypothesised that group cohesion and the lessening of tensions between animals, seen, for example, among bonobos, are other important functions of sexual behaviour. Unsurprisingly, *Biological exuberance* garnered a lot of attention within and without the community of animal biologists. As we saw in my Preface, even today many people believe that heterosexual behaviour is 'normal' because it's 'natural' – a belief that Bagemihl's book sought to challenge on factual grounds.

Among the various places where *Biological exuberance* was reported and its claims represented were the British popular science magazine *New Scientist* and the Oslo Museum of Natural History (OMNH) in Norway. Bagemihl's book was discussed in a major article published in the former in the year of its publication (Vines, 1999). Gail Vines, the article author, writes in a 'balanced' way about the book, focussing on the evidence for and against Bagemihl's thesis. Seven years later, after the fuss about *Biological exuberance* had died down, a far smaller audience of visitors to the OMNH encountered Bagemihl's ideas anew in an exhibition entitled 'Against nature?' Unlike Vines's article, the curators wanted to provoke visitors to think explicitly about homosexuality in humans (with a view to challenging homophobia). To quote from the exhibition website:

> Sadly, most museums have no traditions for airing difficult, unspoken, and possibly controversial questions. Homosexuality is certainly such a question. We feel confident that a greater understanding of how extensive and common this behaviour is among animals, will help to de-mystify homosexuality among people. At least, we hope to reject the all too well known argument that homosexual behaviour is a crime against nature.
>
> (http://www.nhm.uio.no/besokende/skiftende-utstillinger/againstnature/index-eng.html)

The exhibition was then reported in several national newspapers, thus making this local event in Oslo more widely known. For instance, writing in the British broadsheet *The Times* in early 2007, journalist Martin Fletcher (2007) reported it as 'a new exhibition that appears to debunk the theory that homosexuality is an exclusively human preference'. Thus Bagemihl's relatively specialised, esoteric book became successively represented in different media over an eight-year period for the consumption of rather different audiences of varying sizes. Its content travelled through time and across space in ways that its author couldn't entirely have foreseen.

If this example involves representations travelling between epistemic communities located within a single genre (or family of related sub-genres), the second involves translation *between* genres. As the cinema analyst David Kirby (2008a: 51) notes, 'To be successful, a modern science-based film must adhere to a sense of scientific authenticity.' The keyword here is 'sense' because, like the US hit television show *Crime scene investigation*, movies in which science looms large usually need to represent the science in a credible way. As Kirby (2008b: 166) argues, film-makers require the assistance of what he calls 'boundary spanners' or what we might call 'knowledge brokers'. These are individuals who 'facilitate communication between two unique groups [e.g. scientists and makers of popular movies]' because they claim 'membership in both' (ibid.). As examples, Kirby cites the Munich-based science consulting agency Dox, founded by three medical scientists, and the one-woman Hollywood consultancy run by Donna Cline (who also has a biomedical background). Cline is very explicit about her role as a repackager and translator of science from one medium to another:

> I take massive amounts of technical information and possibilities, the different ways we can go. I then look at the script and distil it down to cinematographically valuable units of visual and informational material which we transform into a movie. That's my job in a nutshell.
>
> (Cited in Kirby, ibid.: 172)

Kirby summarises the chain of connections between 'real science' and cinematographic 'science fiction' in Figure 3.1. It's one of many examples of how content from one epistemic community and genre is reworked in, and for, another.[12] As I've suggested in the past few pages, the wider lesson is that,

> Different [epistemic communities, genres and] media 'feed' off each other: [for instance,] quality newspapers set the agenda for other news media; news media deploy images and metaphors 'borrowed' ... from film, literature, and popular culture; television drama and entertainment pick up and adapt issues that are prominent in news and public debate; and [product] advertising ... rework[s] the images and meanings prominent in both [news] media and public debate in consumption terms ...
>
> (Hansen, 2010: 180)

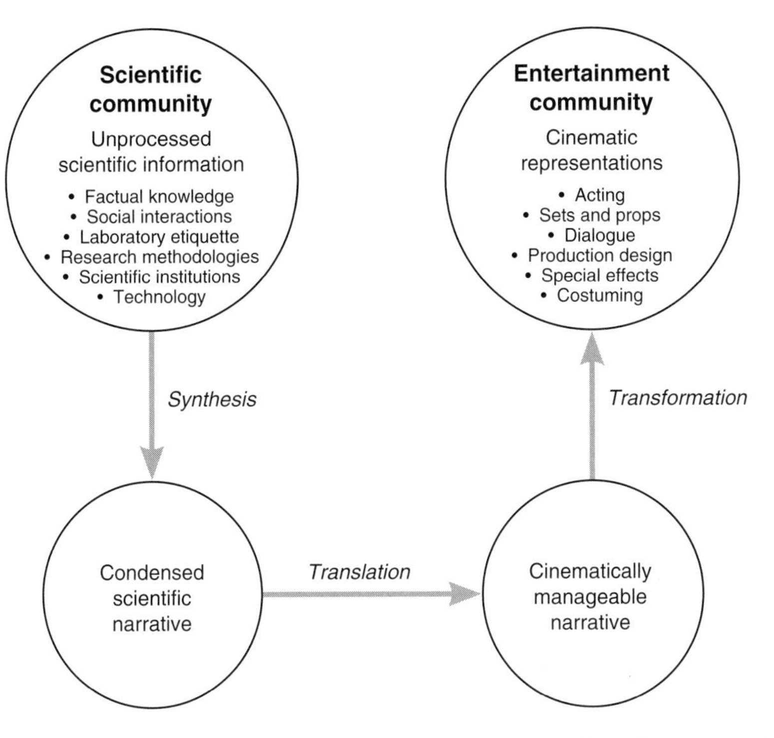

Figure 3.1 Travelling and changing representations: the role of 'boundary spanners' in translating scientific knowledge into useable material for motion pictures

'Boundary spanners' (or 'knowledge brokers') ensure information transfer without the need for different epistemic communities to communicate directly. They create a trading zone in which knowledge moves from one domain so as to be repurposed in another, with money being received in payment for the brokerage activity. (Reproduced from Kirby, 2008b: 173, with permission).

SCEPTICISM AND THE GOVERNANCE OF NATURE'S REPRESENTATIONS

In this and the remaining sections, let me conclude by touching on some very important implications of what's been argued in the previous pages. I will explore these in greater detail later in the book.

Governing epistemic dependence

If my arguments so far have been persuasive, then readers will now appreciate something that strikes me as rather important. I've suggested that, at base, references to what is said to be 'nature' and 'natural' are the means by which a panoply of epistemic workers compete for a good deal of our attention – indeed, compete to shape our sense of self and the wider world. This means that our epistemic dependence on others renders us vulnerable to manipulation, at worst. It gives certain epistemic communities, such as those comprising the mass media and mass entertainment industries, considerable power over us (my focus in Chapter 6). It also means that we may ignore, or even fail to recognise, certain epistemic communities and their products, never knowing what effects paying (more) attention to them may have had on us. The other side of the coin is this: our dependence means that we may be exposed to new ideas, evidence, values and experiences that we couldn't possibly have achieved on our own, and which – in our own view, or perhaps that of those who know us well – change us for the better (or, for that matter, worse!). Though some social scientists are wont to see relations of 'power' everywhere in modern life, people are nonetheless able to change themselves and their circumstances in ways both large and small. In short, epistemic dependence cuts two ways when it comes to representations of nature, and of much more besides.[13]

Yet, unlike the governments we elect, we don't cast a vote to authorise an epistemic worker or community to represent the natural world to us. Instead, our democratic 'choice' is exercised in different ways in different arenas. For instance, we can decide to read a book by Richard Dawkins (or not); we can decide to watch Al Gore's docufilm *An inconvenient truth* (or not); we can decide to go on a guided wilderness tour in Patagonia (or not); we can believe the latest news reports about glacial melting (or not), and so on. Though our choices are ultimately constrained, they are never predetermined. Many of them are made in the market, through us spending money on various physical and symbolic commodities. But many are not. And once we've made our choice, we have a certain freedom to make of them what we will, albeit within the wider discursive constraints of our linguistic universe.

There's no question of us eliminating or even reducing our epistemic dependence – that would be as idealistic as it is practically impossible. It also wrongly assumes that this dependence is, *ipso facto*, a bad thing. Equally, we cannot vote for a set of new epistemic workers every few years in order to keep existing epistemic communities on their toes. Aside from the practical absurdity of this idea, it wrongly presumes that elections are the only (or best) way to determine who should represent the world to us and how. Though I've argued that references to nature are politics by other means, this is not to say that all those who, in effect, represent nature 'on our behalf' should be subject to a periodic plebiscite. What, then, are the options? In general terms, there seem to me to be three.

First, if epistemic dependence usually implies an asymmetrical relationship between epistemic communities and their audiences, then a key issue is how intelligently the latter interpret the representations of the former. An audience that is unthinking, incurious and uncritical can be far more readily influenced than one that dissects what it reads, hears or sees.[14] This is partly a question of education. A public, most of whose members are trained to be sceptical, can make more considered choices about how it receives the fruits of any epistemic community's labours. By 'sceptical' I don't mean endlessly disbelieving. That would be more akin to cynicism or what we might call 'negative scepticism'.[15] I mean, instead, a disposition to be questioning, analytical and critical – something that, I readily confess, takes energy, self-confidence and a certain degree of self-discipline. Relatedly, I'd argue that a sceptical mind tolerates uncertainty, and is neither threatened nor incapacitated when confronted by a myriad of conflicting messages or opposed viewpoints. Scepticism is (or should be) a positive disposition, and a bulwark against the negative aspects of epistemic dependence. In the words of the much-quoted British political philosopher Isaiah Berlin, 'What the age calls for ... is not more faith. Rather it is the opposite: ... more enlightened scepticism, more questioning, more toleration of idiosyncracies ...' (Berlin, 1969: 92). As I've argued, one element of this should be to question the commonplace assumptions made about communicative genres.

Thus, while natural scientists apparently search for culture-free and value-neutral 'truths' about nature, we should regard their statements and findings as purposeful constructions akin (in this respect) to the work of nature poets. Likewise, just because actor-director Sean Penn's 2007 film *Into the wild* is a work of audio-visual fiction, we shouldn't assume that its potential effects on us as viewers are any less significant than the pronouncements of a television scientist like David Attenborough – mere 'entertainment' rather than something 'serious'. Taking a 'symmetrical' approach to the representations of all epistemic communities does not entail *denying* the genre differences that exist or their current social functions. Instead, it's a way of sensitising ourselves to the fact that by taking an 'asymmetrical' approach we risk placing too much trust in some communities, dismissing or ignoring others, and mistaking the actual or potential affects on us of still other ones.

Positive scepticism is, clearly, not something that can thrive when private individuals merely 'talk back' in their heads to what they read in newspapers, see on television or hear on the radio. It's more robust where it arises from, and helps reproduce, a civic and political culture that encourages debate – be it face-to-face or virtual. Such debate presumes that people's views can be changed, especially if they conduct dialogue with people holding dissimilar views and values. In this way, positive scepticism becomes a *social* disposition as much as an individual one. It leads to informed judgements, decisions and actions; it's the enemy of spontaneous 'opinion' and unexamined 'belief'. In some cases it can be used in procedures designed to allow us to get 'up close and personal' with epistemic workers whose knowledge we may need to rely on (more on this in Chapter 8).

Second, unlike elected politicians who can be voted out of office every few years, most of the epistemic communities whose representations of nature we consume are *self*-governing, and will remain so. I earlier borrowed Foucault's neologism 'governmentality' to describe their combined influence on the wider society. The term **governance**, by contrast, means two things in the present context. On the one hand, it refers us to the way in which epistemic communities consciously regulate the practices of their members and constituent institutions. The operational word here is 'consciously': governance is deliberate and, ideally, reflexive. By what values, principles, measures and techniques do epistemic communities satisfy themselves that 'good work' is being done, as well as 'innovative work' that might break the existing mould? How rigorously do these communities hold themselves to their shared standards, and how do they deal with diversity and dissent within their ranks? Indeed, how effectively do these communities permit internal debate about what the standards *should* be, and about what might *count* as 'good' and 'innovative work'? Over the past century, the visual and performance arts have been especially good at breaking the mould, and in the process challenging the wider society to examine its own values and practices – but this is not so true elsewhere, and is that a good thing?

Given the genre differences characteristic of various epistemic communities, how should any one of them deal with representations hailing from other groups? What are their responsibilities to those who consume or use their representations? These questions speak to the *quality* of self-governance: its means and its ends. Because epistemic dependence is so prevalent in the modern world, how well epistemic communities self-govern is clearly a very important issue – even if it's one that most members of the public might consider very arcane or dull.[16]

It's important here to make a broad distinction between epistemic communities who are not, in their professional activities, seeking to be partisan and those who, for various reasons, are. For instance, we understand that ENGOs (environmental non-governmental organisations) have 'agendas' in ways that most university-based geologists do not. This does not, of course, mean that some epistemic communities simply 'concoct' representations that suit their (more or less narrow) interests – though they may well do so. Nor does it mean they *necessarily* behave in nefarious or unprofessional ways (though, again, sometimes their members may do so). But it does mean they must – or should – consider the relationship between their modus operandi and how their audiences perceive them. In some situations, if they're seen as being unduly 'political', 'biased' and so on, then their ability to exert outside influence may be compromised.

Turning now to the last of the three options, in cases where self-governance is weakly entrenched, where a lot of taxpayers' money is involved, or where the epistemic stakes are very high, governance arrangements may involve **external regulation**. This is where an outside body is charged, usually by a national government, with scrutinising and controlling the conduct of the members and institutions comprising one or more epistemic communities. Together, regulatory bodies like these can have a profound affect on the particular mixture of communities whose genre-specific representations of the world become widely broadcast. In any given society today, certain communities and their representations are relatively invisible, for a variety of reasons. Others are the preserve of certain audiences and addressees, either because they occupy a very specific niche or because they peddle 'difficult' or 'subversive' subject matter. Through public education and their considerable regulatory powers, national governments (and many sub-national state bodies) have the ability to alter the diet of information, knowledge and experience that ordinary people are offered. To take a telling example, this is why many civil society groups have pressed their national governments very hard for permissive 'freedom of information' legislation. It's also why some insist that the arts and libraries should be publicly funded because leaving them to the market, it's argued, will necessarily limit the number and range of people who can benefit from exposure to them.

EPISTEMIC DEPENDENCE AND SEMIOTIC DEMOCRACY

The three options just discussed are all crucial ones for societies that claim – or aim – to be democratic in character. The word 'democracy', in simple terms, means government by and for the people: that is, 'rule by the populace over itself...' (Howard, 2011: 8). More formally, it's a mode of government in which 'the exercise of power passes through institutionalized mediums of public deliberation, giving publicly agreed norms a practical efficacy over the actions of ... [citizens]' (Barnett, 2004: 185–8). Discussion, deliberation and disagreement lead to decisions and actions that are accepted by all – though often unhappily – because they arise from a broadly inclusive and consensual political process. We often think of democracy as one of several possible political systems and, as with other political systems, equate it with such things as referenda, parliaments, election campaigns and equal voting rights among citizens. However, while it's true that any democratic polity (should) permit(s) all capable adult members the right to nominate – and remove – their political representatives, democracy is (or ought to be) much *more* than a set of formal political arrangements. As many political theorists have pointed out, this is true in three senses.

First, for citizens to hold their nominated political leaders properly to account, they need to have had the education, training and experience necessary to create confident and critical individuals (see the first option in the previous section). This education, training and experience is achieved almost entirely *outside* the formal political system in the domains of school, university, the news and educational sections of the media, churches, the family, civil society groups and so on. These domains include and sustain, but also go beyond, what's usually called the **public sphere**. This is the virtual arena in which citizens discuss matters of common concern (see Box 3.4). Second, citizens need to have the time to think and act politically. In her classic book *The human condition*, philosopher Hannah Arendt (1958) worried that modern democracies were increasingly unable to release people from the duties of home and the workplace. Relatedly, the journalist-author John Kampfner (2008) suggests that modern Westerners have traded enhanced private 'inward-facing' freedom (e.g. more shopping choices) for less public 'outward-facing' freedom (e.g. the ability to effectively challenge elected governments). Politics should not, they both argue, be the preserve of a minority of activists, political party members or pressure groups. Third, democracies emphasise the 'freedom' and 'liberty' of their members – in part to protect them from domination by others, in part to encourage diversity and novelty in ways of thinking and doing. The basic idea here is that incumbent or potential governing parties must *persuade* (not force) citizens to support them, and that part of persuasion is the obligation of those parties to debate *alternative* values, goals and policies.

BOX 3.4

THE PUBLIC SPHERE AND DEMOCRACY

According to the German theorist Jurgen Habermas (1962), in his seminal book *The structural transformation of the public sphere*, the 'public sphere' first came into existence in the fledging democracies of eighteenth-century Europe. The public sphere is not a physical place. Instead, it's a sphere of interaction that can be virtual and strung out as much as face-to-face and localised. The word 'public' has close etymological connections to the word 'people'. In the system of self-government known as a 'republic', it quite literally means self-rule rather than rule by non-elected or imposed leaders. According to Habermas, the public domain was a sphere where members of a polity could come together and consider matters of common concern on a notionally equal basis. It was irreducible to the *private domain* (the household and family), the *market domain* (the sphere of commodity production, distribution and exchange) or the domain of the *state* (elected governments, bureaucracy, the law, the police and the armed forces). Yet deliberations in the public sphere could legitimately have consequences for all three of these domains if people decided to change the law or to alter existing regulations via their elected governments. For Habermas, the public sphere overlapped with what's called **civil society**. This is that mixture of activities and organisations devoted to tackling issues that affect wide sections of society (e.g. a charity devoted to helping homeless people). It depends, he argues, on the free circulation of information, argument and opinion.

As Habermas rightly argued, a key condition of possibility for a functioning 'public sphere' is that there's such a thing as a 'public' in any given country (which itself requires that the *idea* of 'the public' is widely understood). Yet publics do not arise spontaneously. Their creation and maintenance requires considerable effort. In eighteenth-century Europe, it required revolutions, such as that in France in 1789, to unsettle the *ancien régime* and create the space for representative democracy. Citizens then had to learn, over a period of time, that it was their entitlement – some argued their *duty* – to devote energy to considering issues that were important for many other citizens. To be a member of a 'public' thus presupposes the ability of any given citizen to recognise (1) that their personal values and preferences may have to give way to those of others for the greater good, but (2) that they can learn from, but also legitimately influence, these others through open discussion, disagreement and reflection. This means that any given public is not 'a self-identical collective subject' (Barnett, 2004: 188), as if 'the people' were a homogenous mass of individuals. Instead, all publics are 'constitutively pluralised' (ibid.), united not so much about substantive political or moral issues but about the right of members to have their say about such issues. As Rebecca Ellis and colleagues put it ' "the public" as a unit does not exist ... Yet in other senses it is real because "it" ... is a functional myth, or heuristic' (Ellis *et al.*, 2010: 505). In other words, the public is a 'constitutive fiction' that does not precede its being represented.

Some believe that since the 1960s, many Western democracies have suffered the ill effects of information overload, 'public relations' and of 'populism'. The first describes the public's inability to cope with the exponential proliferation of messages and communications coming at them. The second refers to conscious efforts on the part of everyone from professional politicians to corporations to manage their image and their communications so as to avoid unwanted questions or criticism. The latter refers to views and values widely shared among members of a public but which are simplistic and unreflexive (bordering on naïve or prejudicial). Still others believe that 'opinion polls' – snapshots of people's current views – are substituting ever more for proper public debate. Meanwhile, some argue that the global expansion of commercial broadcasting and the assault on the 'free press' (courtesy of media moguls Rupert Murdoch and Silvio Berlusconi among others) are deforming the public sphere to the detriment of democracy. Elected politicians also these days work *very* hard to control their words, deeds and images – arguably, this turns formal politics into theatre, as least so far as publics are concerned. The political philosopher Nancy Fraser (1997) coined the term 'weak publics', which we might interpret to mean those where (1) the extent/volume and (2) the quality of discussion and debate in the public sphere is low. Though 20 years old now, the critical reflections on the 'phantom public sphere' contained in Bruce Robbins's book still repay close reading (Robbins, 1993).

Substantive and formal democracy

In this light, we can ask the question: how can any society that limits 'democracy' to periodic elections ultimately call itself democratic? A *substantive* – as opposed to purely formal – democracy is one in which freedom and liberty are actively cultivated across the *totality* of domains and institutions that individuals traverse weekly, monthly and yearly. As one commentator puts it, 'Democracy thrives when there are major opportunities for the mass of ordinary people to participate actively … in shaping the agenda of public life, and when they are actively using these opportunities' (Crouch, 2004: 2). Because (as I intimated earlier) freedom and liberty are not self-generating or spontaneous, they depend upon a sufficiently rich offering of information, experience and opportunity across a wide range of domains and institutions. Somewhere between the 'free' but isolated figure of Robinson Crusoe and the completely manipulated Orwellian individual lies the person for whom epistemic dependence is a powerful resource, not (or not only) a root cause of their vulnerability. It's a resource that can give individuals the capacity to think in innovative, creative and deep ways about themselves, others and the wider biophysical world we all inhabit.

As Nadia Urbinati puts it, in theory 'Democracy … makes *all* issues an object of public evaluation and all values a matter of opinion and consent' (Urbinati, 2009: 65, emphasis added). In other words, in democracies pretty much *any* question, event, activity, incident, idea, topic or proposal might reasonably become a matter of common concern and public debate.[17] We may flatter ourselves in countries like Britain, Canada, France and the United States that we have a substantive democracy, at least in comparison with autocratic states like North Korea or war-torn polities like Afghanistan; however, for many people in these countries, many (most?) issues outside the formal sphere of 'politics' are not seen as 'political'. The activity of 'being political' is thus unduly constrained in people's perceptions and practices. British political theorist Colin Crouch (2004) has coined the term 'post-democracy' to describe this: for him, many ostensibly democratic polities have seen 'democracy' reduced to periodic ballots cast by citizens. Indeed, some critics suggest that modern democracies are, in fact, studiously 'post-political' (e.g. Swyngedouw, 2007), while Urbinati prefers to apply the seemingly paradoxical term 'unpolitical democracy' (Urbinati, 2009).[18] What's sometimes called 'the crisis of democracy' thus, for some, extends well beyond inadequacies in the formal structure of political systems.

Put differently, while there are many patently undemocratic countries in the world today, some critics argue that this does not mean that the ostensibly democratic ones are necessarily living up to their ideals (see Box 3.5 for more on 'post-politics'). This contrasts with a more sanguine view, advanced by the social theorists Ulrich Beck and Anthony Giddens, that ours is the age of 'reflexive modernity' in which ordinary people actively question experts, elites and authorities of all kinds.[19] It contrasts too with the suggestion, made by several critics on the Left of the political spectrum, that our age is one of increasingly strident dissent (see, for example, Hardt and Negri's (2005) book *Multitude*). Millions of citizens worldwide, it's argued, now use ICTs to challenge strongly 'official' and mainstream views of the world. Still others take heart from citizens and non-governmental groups acting politically of their own accord, and without too much reference to the concerns of elected governments. An example is those trying to create transition towns in the United Kingdom, which are intended to have a light ecological footprint. Critics respond that these are all exceptions that prove the proverbial rule. Most citizens in democratic countries, they argue, are apathetic or cynical about politics; even political protest is seen as ineffectual, expressing 'a dull ache of frustration at power being dispensed in [elite] corridors rather than streets, at power that is ever further from [protestors'] ... grasp' (Jenkins, 2011: 33).

BOX 3.5

'POST-POLITICAL' DEMOCRACIES AND 'PUBLIC PEDAGOGY'

As we saw in the previous box, democracies in both theory and practice require a public sphere if they are to be worthy of the name; however, because any given public must function in conditions of epistemic dependence, there's always the potential for its thinking to be narrowed and shaped by powerful institutions and actors. The critical theorist Henry Giroux has, in many of his published writings, used the term **public pedagogy** to describe the way in which the citizens of modern democracies are 'schooled' by everything from soap operas, product labels and billboards to news broadcasts, television documentaries and the speeches of political leaders. The term 'pedagogy' is normally associated with the profession of teaching and with educational institutions. Giroux's key point, however, is that pedagogy is a much *wider* and *never-ending* process that transcends formal schooling. He's been especially critical of the 'weak' and 'pseudo' publics to be found in the United States. I say publics in the plural because, following Oskar Negt and Alexander Kluge's (1972/1993) critique of Jurgen Habermas's writings, Giroux argues that citizens increasingly occupy *different* 'public spheres'. The majority of people, he argues, have neither the time, inclination nor opportunity to get beyond the simplistic, superficial and often salacious diet fed to them by mainstream institutions. What's more, Giroux argues that this diet secretes a 'hidden curriculum' of norms that people are encouraged to imbibe, usually unthinkingly. This leads to 'manufactured publics' in whose name things are done as if they represented '*the* public' in the singular sense. By contrast, those minorities whose values and aspirations lie outside the mainstream may form various 'counter-publics' who either hope to challenge the mainstream or who exist, impotently, on its margins. Some of these counter-publics are part of a 'transnational public sphere' facilitated by the Internet, email, Skype and so on. But even sympathetic commentators regard these counter-publics as *visible*, but ultimately *ineffectual* in a practical sense. Unsurprisingly, left-wing academics like Giroux are part of such international counter-publics and wish them to be effective. Some, though by no means all, counter-publics aim to foster a culture of critical debate that approaches Jurgen Habermas's idealised public sphere (where sustained, non-superficial dialogue produces a robust, if impermanent, consensus).

The idea of **post-political** democracy resonates with Giroux's notion of 'public pedagogy'. It's associated with the European philosophers Slavoj Zizek, Jacques Ranciere, Alain Badiou and Chantal Mouffe. In different ways, they each suggest that modern so-called 'democracies' operate ever more by *depoliticising* critical questions and issues. Certain things become 'off-limits', that is to say unthinkable, unsayable or unimaginable. The flip side of this is that certain other things become common sense and almost beyond question. For instance, my Manchester University colleague Erik Swyngedouw argues that 'climate change' has become accepted as a 'fact' in most modern democracies – to the point where questioning it almost opens one to ridicule (Swyngedouw, 2010). In a context where capitalism appears equally beyond question, Swyngedouw argues that the idea of climate change limits political debate to 'technical solutions' to the impending 'crisis'. Not only are the claims of 'climate change sceptics' pilloried; equally, radical arguments for a future economy not based on continuous mass consumption are barely entertained in the corridors of power or everyday life. As Zizek put it sardonically, 'say and write whatever you want, on the condition that [it] ... does not effectively question the predominant political consensus' (Zizek, 2002: 544).

Clearly, in all the arguments summarised above, there's a certain idealisation of a public sphere and a form of democracy that are, respectively, 'strong' and 'deep'. However, as long ago as 1925, the early media theorist Walter Lippmann (1925), in his book *The phantom public*, worried that the ideal may *never* be realisable in practice – short of revolution and the break-up of mass democracies into far smaller polities and self-governing communities.

The value of epistemic diversity

Returning now to the subject of nature and its affiliated concepts, which representations do we – and should we – pay heed to? Which of nature's 'representatives' should be listened to, how much, and how? In what ways do we receive and consume their representations, critically or uncritically, as mere 'fiction' or something 'factual'? Which representational genres do we, as individuals, tend to favour, and which ones do we ignore or decide are not for us? How aware are we of the 'work' they perform upon us? Should we take the genre differences between these representations at face value, so too the varied social functions the genres perform? The answers to these questions lead us well beyond a concern with censorship laws and the prohibitions enforced by state-sanctioned regulators – what Luigi Pellizoni (2001: 60) calls 'power over communication'. Within the realms of what's deemed socially tolerable, how much epistemic diversity (or 'bandwidth') do we experience and which representations of nature become, in effect, ghettoised because they're perceived to be too esoteric, maverick or 'weird' by the mainstream? The question, which pertains to 'power *in* communication' (ibid.), alerts us to the quality of what Siva Vaidhyanathan (2006: 305) calls our **semiotic democracy**. To the extent that references to nature and its collateral terms play a wide role in all our lives, how can we ensure representational plurality and resist representational narrowing? How can we cultivate a 'mature' attitude towards representations, epistemic workers and genres – and so become aware of their role in shaping us as subjects?

As Vaidhyanathan argues, in the tradition of thought known as republican democracy, 'freedom' and 'liberty' do not simply consist of the right to *refuse* certain available options (since the options may be limited, similar or inferior). In their 'positive' versions, they require that a society organises itself such that *new* ideas, findings, arguments, etc. are *actively fostered and distributed*. Such a society thus decides to question its own cognitive, moral and aesthetic practices as a measure of how well it realises its self-established political ideals. Such a society may require 'positive discrimination' so that certain epistemic communities and various representations are not ghettoised, ignored or stigmatised before they've been given a chance to be more widely heard. This then obliges established or influential epistemic communities to reflect on their own practices, and it obliges the rest of us to reflect, however briefly, on our own habits of thought, feeling and action. It implies that the mere 'toleration' of unusual ideas, claims, values and so on is insufficient, at least in the first instance. For, at its worst, toleration merely entrenches mainstream beliefs, assumptions and habits, and it shelters them from any robust challenge or the glare of radical alternatives. In this sense, toleration – and its bedfellow 'freedom of speech' – can become a cover for business-as-usual, the rejection of alternative modes of being, and the refusal of a different (better?) future. Fostering epistemic diversity in claims made about 'nature' (and much else besides) counts for far less than it should if people lack the opportunity or skills to give those claims proper consideration. As Henry Giroux has argued repeatedly, creating a 'critically literate' citizenry has a pay-off only if people have a range of ideas, insights and value-claims to work with and against.

SUMMARY

In this and the previous chapter, I've suggested that the question 'what do we know, believe, feel and do?' is umbilically connected to the question 'who or what is seeking to shape our attitudes, values, emotions and actions?' Building on Chapters 1 and 2, I've argued that diverse epistemic communities invest the concept of nature with specific meanings, so too its collateral concepts. These meanings are attached by them to various material referents whose very definition is internal to discourse, genre and representational mode, which themselves are relationally defined and mutable in time and space. My major point has been that references to 'nature' and its collateral concepts are important means by which the thoughts, sentiments and activities of hundreds of millions of people are governed. Yet because nature routinely appears to be non-social and to be an ontological given, it's very easy indeed to ignore the ways in which references to it are implicated in the process of governing thought and conduct. Equally, it's easy to forget that nature is made actively knowable to us by others cognitively, morally and aesthetically. Even when we pay attention to what these others are doing, it may appear that the *last* thing any of them are engaged in is a wider process of subject-formation and the governance of belief, feeling or practice.

In the previous pages, I've used the word 'govern' in the broadest sense, just as I did the same for 'politics' in the previous chapter. To be clear: I'm not suggesting that there's always and necessarily something pernicious or malevolent about the process of governing our habits of mind or action. Though some critical social scientists regard 'power', in a pejorative sense, as a promiscuous, ever-present aspect of modern life that crowds out 'genuine' freedom, I remain deliberately agnostic on this substantive question here.[20] My (more limited) claim is that, to the extent that representations of 'nature' and its collateral terms factor in the process of governing, they can be instruments of social power, but also tools enabling resistance and opposition, fulfilment and enrichment. Equally, they can be the medium of new thoughts, feelings and experiences that unsettle received wisdom and lead to unexpected 'improvements', however defined, in our own or others' lives.

In the remaining chapters, I want to explore, by way of extended examples, how the process of governing with reference to 'nature' operates in practice. I draw no firm conclusions beyond these examples about its practical 'success' or 'failure' because these terms beg exceedingly complex questions about how one judges the effects of such a multifaceted process, let alone measure those effects. Before I proceed I should make two final points. First, it's important to read the rest of this book in light of the claims and concepts presented in this and the previous chapter. I realise that this is no mean feat because I've thrown a lot of arguments, terms and examples at readers. I will recapitulate (and elaborate) the major points in the pages to come, thereby making the reading less onerous. Second, this book is codifying, synthesising and putting to work the ideas of a loose, multidisciplinary epistemic community in order to analyse the representations of nature disseminated by other epistemic communities working outside the social sciences and humanities. In the interests of honesty and consistency, I must therefore urge readers to adopt the attitude of 'positive scepticism' towards my own arguments that I'm recommending we all adopt when dealing with epistemic workers of various kinds in our daily lives. Obviously, the framework of analysis I'm advancing in *Making sense of nature* can hardly exempt itself from its own strictures, at least if they're applied rigorously.

ENDNOTES

1 The broad understanding of 'genre' that I am advocating here is advanced by, among others, Garin Dowd *et al.* (2006) in their edited book *Genre matters* (where the latter term means both matters concerning genre and that genre, in its various forms, is important in modern societies).

2 Once, many decades ago, the difference between artistic and scientific visualisations of nature was not as clear cut as it appears to be these days: Judith Magee's (2009) *Art of nature* presents the drawings and paintings of the European naturalists who set sail during the seventeenth and eighteenth centuries.

3 For more on the interesting and growing Sci-Art movement, see Siân Ede's (2005) excellent book *Art and Science.*

4 These notions of 'maturity' and 'thinking for oneself' are, of course, central to the idea of education in Western research universities. Undergraduates move from instruction (as exemplified by the lecture as a mode of communication) to independent and critical learning in their final year(s) as students (as exemplified by the seminar as a mode of interactive education).

5 The critical theorist Timothy Luke (1999) coined the terms 'environmentality' and 'green governmentality' to describe the enhanced role of references to the non-human and of environmental spokespeople (experts, advocates, activists) in shaping people's sense of identity. Stephanie Rutherford's (2011) recent book details these references in four media (a museum, national park, theme park and a documentary). I share Luke's and Rutherford's sentiments, but would broaden the focus to include references to 'human nature' (see the third section of Chapter 4 of this book, for example). Luke and Rutherford have been inspired by Michel Foucault's writings, which I will discuss presently.

6 When first proposed in French academia in the 1960s, these ideas were seen by some as pronouncing 'the death of the subject'. This was misleading in one respect but in another accurate: the 'death' described was of a conception of the human subject as some sort of timeless essence common to all (mentally unimpaired) *homo sapiens.* If, instead, the human mind and body are seen as 'interfaces' with a wider 'ecosystem' of ideas, symbols and materials, then they're necessarily products (and producers) of a *changing history*. In social science and the humanities, this basic insight has been developed in two directions: one, which much of this book continues, explores the sociocultural ecology; the other a wider material–semiotic ecology (on the latter see, for example, Ingold (2008)).

7 Bourdieu's (1992) notion of a person's 'habitus' was intended to describe this 'complex coherence'. A habitus is an ensemble of beliefs and norms that dispose a person to act and react in regular ways to external stimuli (information, events, actions). A habitus develops slowly and is not the product of conscious decision-making or a plan on the part of the individual.

8 See http://www.bbc.co.uk/programmes/b01k0fs0.

9 One of Foucault's key insights, though hardly unique to him, was that the control of knowledge was an important source of social power as much as the control of material resources or the means of violence. Epistemologically, we may think that many people can be governed by others because they are *ignorant,* because they *lack* knowledge. Though this is certainly the case, Foucault insisted that the governance of subjects occurs, in part, through what they are invited to *know* rather than *not know*. This includes knowledge of broad social norms and goals (which may become tacit) as much as specific knowledge-content (ideas, facts, etc.). This is why he sometimes described power as 'productive' rather than simply prohibitive or punitive. This can be seen as a particular take on Francis Bacon's famous aphorism that 'knowledge is power', and it suggests that we should not focus our analytical attention only on a 'democratic deficit' (ignorance) in what people know about the world. What Foucault sometimes called 'pastoral power' emanated from the sayings and doings of diverse agencies whose task was – and remains – to measure, educate, inform and intervene rather than to allow or take individual human lives (which was the modality in which 'sovereign power' operated, the power exercised by pre-Enlightenment era monarchs in Europe). The work of these agencies blurs the boundary between the individual and the wider social environment in which they learn to become ostensibly unique selves – which was the focus of the previous endnote.

10 To return to Ian Hacking's (2004) earlier mentioned concept of 'looping effects', part of the way 'people are made up' (to use a term of Hacking's) is through the socially available categories that apply to the groups to which they belong, by choice or circumstance.

11 In Steve Fuller's view, which some might say is cynical, 'The title "science" is reserved for those who can mobilize enough personnel, resources, and clout to eliminate the competition' (Fuller, 2004: 9). This said, there are many occasions when scientists – indeed, non-scientists too – permit traffic across the 'boundaries' that separate their epistemic community from other communities and the lay public. For example, professional botanists and natural historians have a long history of interacting with the many 'amateurs' who avidly collect and record samples and sightings of wildlife. In other cases, actors outside science actively seek to use its authority by incorporating scientists into their modus operandi – this is evident in most large environmental NGOs, for example, which now employ scientists or else routinely commission scientific research.

12 In the United States, the interactions between the film industry and scientific community are facilitated by the Science and Entertainment Exchange (http://www.scienceandentertainmentexchange.org/). Frequently, epistemic communities have limited (or no) control over how other communities present their ideas, findings and activities; yet in some cases concerted attempts are made to increase the level of control. 'Science communication' is a good example. The various natural and biomedical sciences have, over the past 40 years especially, paid significant attention to how they are portrayed in the wider societies that play host to them. The reasons for this are not hard to divine. First, scientific research commands a lot of public resources, especially in Western countries. Given this, it can be especially important for scientists to explain the nature and importance of 'blue skies' or 'fundamental research', like that ongoing at the multibillion-euro Large Hadron Collider, located deep underground near Geneva, Switzerland. Second, much scientific research is ethically controversial, such as that into cloning mammals and stem cell therapy. Third, scientific research can have significant material effects upon society, as is obvious in the case of medical research into things like cancer and so-called 'genetic diseases'.

13 This may sound like I'm offering a version of the social theorists Ulrich Beck's and Anthony Giddens's well-known idea of 'reflexive modernisation', where a new, non-deferential approach to authority among lay actors and publics takes hold; however, I think that Beck and Giddens are far too sanguine about the current existence (and efficacy) of what I term 'positive scepticism' among the general population. Unlike

Hardt and Negri's (2005) hopeful thesis about the oppositional 'multitude' challenging (often nakedly corrupt) political, economic and cultural elites, I don't myself see much evidence of a power shift towards ordinary people, new social movements or oppositional non-governmental bodies. The media theorist John Thompson (2011) is right that mass mediated forms of epistemic dependence are not inferior to more face-to-face forms of epistemic dependence, but these mass mediated forms – so characteristic of our 'information age' – are only of real use if people have the means to use the informational resources they offer to conduct rich discussions and to act in politically effective ways in respect of 'public affairs'.

14 Demagogues, of course, can thrive in conditions where audiences are gullible or accustomed to favouring simple solutions over complex and conditional answers. The problem of demagoguery in democratic societies came to the fore in the 1920s and 1930s, especially in Weimar Germany. Because power cannot be exerted without popular consent, non-democrats and anti-democrats realised that the way to govern democracies was through persuasion. Adolph Hitler, taking advantage of the social and economic turmoil of the Weimar period, used his skills as a public speaker to prey on the fears and anxieties of ordinary Germans. As one of the first analysts of propaganda, Edward Bernays, realised at the time, 'The conscious and intelligent manipulation of the organized habits and opinions of the masses is an important element in democratic societies. Those who manipulate this unseen mechanism of society constitute an invisible government which is the true ruling power of our country . . . we are governed, our minds moulded, our tastes formed, our ideas suggested, largely by men we have never heard of' (Bernays, 1928: 5). Of course, Germans – and the rest of the world – *had* heard of Hitler because he aspired to become (and, of course, did become) Germany's supreme leader. But Bernays's key point was that demagogues require a wider 'invisible college' of people – especially in the mass media – who influence the public by reproducing the arguments and values of those demagogues so that they come to seem like common sense. Some critics argue that, in the early twenty-first century, demagoguery is less of a threat to modern democracies than is *distraction*. Distraction involves increasing the time ordinary people devote to entertainment, leisure and shopping so that they spend ever less time taking political affairs seriously.

15 Such negative scepticism is currently evident in many Western democracies, especially Britain. For a variety of reasons, many members of the British public are currently very cynical indeed about the claims made by their national politicians and by many business leaders. Fundamentally, *trust* has been eroded: the belief that elected politicians and many business people act in the *wider* interest, rather than their *personal* interest, has been weakened. This may, for many citizens, be linked to a dawning realisation that 'representative democracy' in large-scale, technological, complex societies renders them fairly impotent in terms of exerting any *real* influence on politics. Focussing on the United States, Jeffrey Goldfarb (1991) argued that the roots of cynicism are more sinister, lying, at least in part, in the conscious efforts of large corporations and advertising agencies and sections of the media to actively manipulate public opinion and debase public debate (see the previous endnote).

16 Interestingly, John Hartwig, the philosopher who coined the term 'epistemic dependence', places little faith in my first option and seems to favour lay individuals 'out-sourcing' reasons for their attitudes, beliefs, values and actions to various epistemic communities. While I think the quality of epistemic communities' self-governance is of critical importance, I don't share Hardwig's apparent disinterest in having a well-educated citizenry possessed of critical sensibilities. It invites the possibility, in E. E. Schnattscheider's inimitable words, that 'Democracy . . . is a form of collaboration of ignorant people and [powerful] experts' (1975: 137). One way in which political theorists have sought to address the 'democratic deficit' in erstwhile democratic societies is to find ways of increasing citizen participation *outside* elections and in the 'non-political' arenas outside the state apparatus.

17 Some might argue that this expanded conception of politics directs our attention away from the formal political sphere – detrimentally so. To be clear: I do not doubt the importance of formal politics and believe strongly that the institutional arrangements deserve the closest consideration, but so much of what is discussed in the formal political sphere is, or should be, inspired by political discussion outside it. Think, for example, of how the suffragettes in Britain (who were not permitted to be Members of Parliament) made women's inequality in the home and elsewhere a new political issue a century ago. One concern expressed by several political theorists is that 'real politics' now occurs *outside* the formal political sphere and is *disconnected* from it too. This presents us with an image of two political spheres, one involving elected governments who remain powerful, and one involving new social movements, community organisations and so on acting in local ways that may lack a wider, macro-level influence on society.

18 These criticisms rest on the idea that 'the political' is, or should be, a vibrant and sprawling field of debate, disagreement and decision-making that has, in modern (so-called) 'democracies', been *reduced* to 'politics' as a defined domain in which rulers and the ruled encounter each other on highly unequal terms that favour the former. 'Anti-political' forms of 'politics', it is argued, limit debate, choice and disagreement about person, communal or public values, goals and measures. In a series of writings, the French political philosopher Jacques Rancière has argued that 'politics' today is a form of 'policing' (note the common etymological root of the two terms). How, Ranciere and others ask, might the dialectic of the political and politics be unfrozen?

19 In several books and essays published in English through the 1990s (e.g. Beck (1992) and Giddens (1991)), it was suggested that in our age of (1) unpredicted/able 'manufactured risks' and (2) heightened uncertainty about how stable careers, place of residence, relationships and friendships will be, 'sub-politics' (Beck) and 'life politics' (Giddens) were/are assuming new importance. Sub-politics is the politicisation of events and practices within everyday life and/or in domains outside the formal political apparatus (e.g. in the activities of environmental NGOs). Life politics refers to an approach to people's daily existence that foregrounds choices, challenges and values – 'life' is something to be managed, a 'project' to be realised through conscious action to create, and then choose from, a range of options. Several critics have suggested that both Beck and Giddens were/are far too sanguine about the 'political powers' now enjoyed by ordinary people. Some have argued that neither theorist had paid sufficient attention to the enduring 'structural' constraint on people's thought, action and life chances. At worst, critics see 'sub-politics' and 'life politics' as ineffectual, as the place where 'real politics' has migrated in light of the democratic failures of official politics and elected governments.

20 In other words, I do not set out and then use a substantive theory of power in *Making sense of nature*. Instead, as will become clear in later chapters (especially Chapter 6), I discuss different conceptions of power favoured by contemporary social scientists and humanities scholars in order to show that power is undoubtedly a part of the way references to 'nature' and its collateral concepts are constructed and used.

PART II

REPRESENTATIONS AND THEIR EFFECTS

4 UNNATURAL CONSTRUCTIONS

In the previous three chapters, I've presented an analytical vocabulary that allows us to make sense of how others make sense of the world for us. Nature and its collateral terms, I've argued, are 'concepts we live by', such is their pervasiveness in our collective discourse.[1] They are, I have suggested, a major terrain upon which myriad epistemic communities operate. A metaphorical contest occurs between these communities over how the terrain is to be demarcated, partitioned and understood. In other cases, communities share and rework the epistemic products of others as part of their own practice. However, in almost all cases audiences of various kinds are invited to consume representations whose creation they know little about. These representations, operative across the full range of communicative genres, become the raw materials out of which people understand both themselves and the wider world we all inhabit. They inform our habits of action and inaction. They are part of the incessant socio-material process of 'governing' individuals, in the broadest sense of that term. They are, in other words, a key ingredient in achieving both sociocultural stability and various forms (and degrees) of change.

In this and the remaining chapters, I want to amplify and evidence the major arguments presented in the previous pages using a wide range of extended examples. Where the previous three chapters have been peppered with vignettes, the ones to come will present more in-depth case material. This material is, intentionally, drawn from field- and desk-based research of those upon whose concepts and claims I have drawn in Part 1. However, in the interests of making the chapters readable, I won't clutter the text with too many citations but, rather, let the case material do the talking. Readers can consult the research publications upon which I draw here as they wish (and I hope they do – see the References at the end of the book). I will, necessarily, recontextualise and repurpose the insights of these publications in line with the emphases of argument presented in the previous three chapters. Along the way, I'll invite you to use the arguments of Part 1 to make sense of the extended cases to be presented.

This first chapter in Part 2 has one objective. If, as I've insisted, 'nature' is not natural, then it is, in some meaningful sense, 'social'. How can what appears to be natural be shown to bear the (usually) disguised trace of people's preferences, values and assumptions? It is one thing to demonstrate that there are different beliefs, opinions or values *about* what we call 'nature'. But it's another to show that the nature being analysed, debated or evaluated is not *itself* an extra-social world lying outside social discourse and practice. I want to answer my question with reference to three cases. Between them, these cases cover the non-human world ('external nature') and human biology. They also cover 'nature' in its other principal meanings, and make plain how its collateral concepts are, similarly, anything but natural. Readers can focus on one, two or all three of the cases presented as they wish. Each of them is quite meaty, and will take some time to digest.

Those well versed in the literatures upon which *Making sense of nature* is based may regard a chapter showing that 'nature is nothing if not social' as being rather beside the point. However, I obviously beg to differ. While nature's social character may be obvious to those with a trained eye, for many people it is scarcely evident at all and nor are its important implications apparent. Equally, as I argued earlier in the book, while representations of 'nature' in art, cinema, advertising or poetry clearly express their originators' intentions, this does not mean that other genres of representation are somehow intrinsically devoid of significant social content. The three cases I choose to examine in this chapter show clearly how claims about nature can actively dissimulate socially contingent processes of creating and assigning meaning. That the cases reach back a period of years does not make them less relevant to the present: the representational lessons we can learn from them are applicable to the here-and-now. Despite some 40 years of research and writing about 'the social construction of nature', I'd suggest that the need to 'denaturalise' representations of nature remains an important one.

WHAT'S A 'FOREST'?

Globally, the surface area of forest has declined by over 50 per cent since the 1850s, and by significantly more than that in certain parts of the world (see Map 4.1). The balance between 'deforestation' and 'reforestation' is important, partly because of anthropogenic climate change: forests are a very important sink for carbon dioxide, a key greenhouse gas. But the loss or degradation of forest ecosystems matters for lots of other reasons too. Not all forests, and not all tree species, are valued equally. In many parts of the world, so-called 'old growth' forests (ones that have been undisturbed by humans) are increasingly rare. They are fringed by landscapes that have been deforested, that are reforesting (perhaps with plantation trees) or that are given over to agriculture. For these reasons, efforts are being made to protect old growth forests before they disappear altogether. These forests are valuable for aesthetic, cultural and moral reasons, as much as for economic and practical ones.

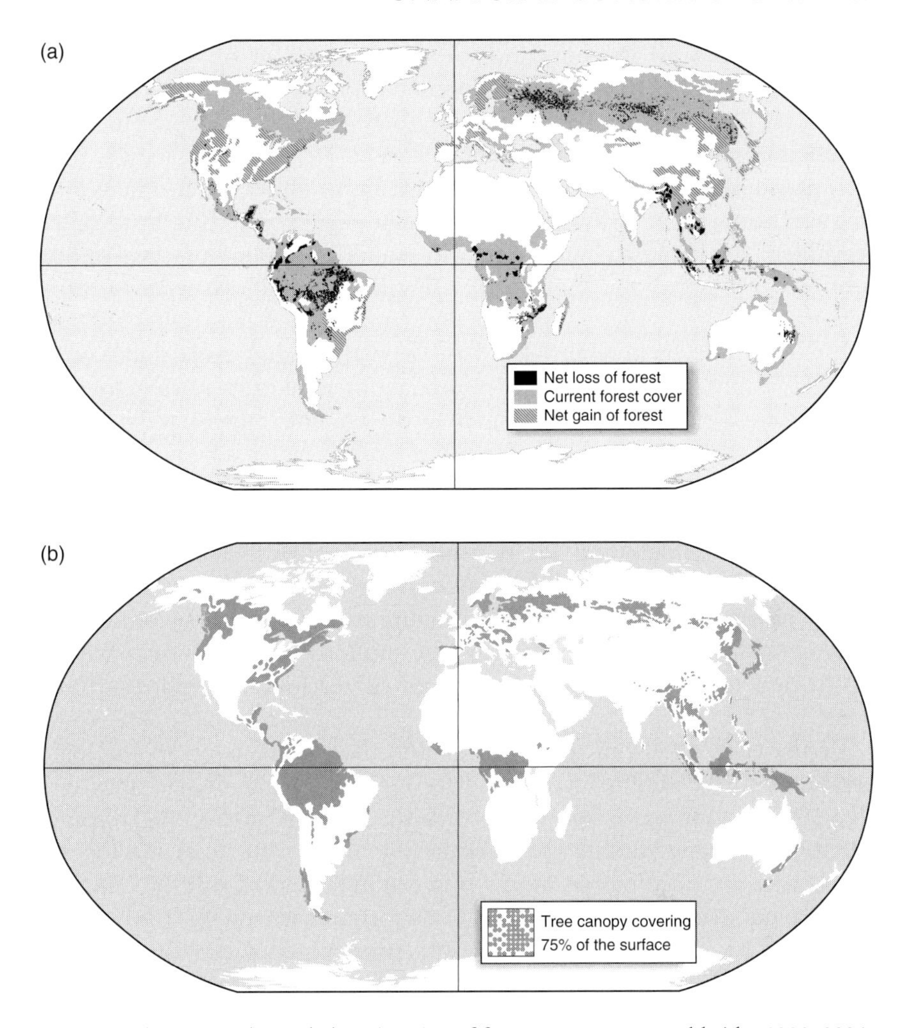

Map 4.1 The geography and changing size of forest ecosystems worldwide, 1990–2004

Adapted from United Nations Environment Programme (UNEP), 2009. These two maps show the relative size and location of the major remaining areas of forests across the globe, with the first highlighting those areas where forest loss is progressively occurring.

We often think of tropical rainforests in Brazil or West Africa when we think of tree loss; however, what are considered to be rare and special forest ecosystems exist well beyond the equatorial zone. The province of British Columbia (BC) in Canada provides a graphic example. BC is a large province: the land area is nearly 1 million square kilometres, which is four times the size of the United Kingdom. The population is just 4.5 million, and is overwhelmingly concentrated in and around Vancouver. Approximately 60 per cent of the province is forested, and an appreciable amount of this is either being logged or is reforested land (in various stages of growth). Because of its abundance of trees, the export of wood has long been a key component of the BC economy (though less so these days). The provincial government has, for decades, granted logging licences to forest products companies. These companies, such as Weyerhaeuser, harvest trees on a grand scale, but also seek to replenish timber stocks over time (under the watchful eye of provincial government regulators). This is so-called 'sustained yield forestry', and has, with some important practical modifications, proceeded on an industrial scale since the 1945 *Sloan Report*, if not before.

It's important to note that BC has, for many years, made much of its 'natural beauty', not simply the commercial potential of its forest resources. The provincial tourist board popularised the tag line 'Super natural British Columbia' some time ago, while vehicle licence plates advertise another: 'Beautiful British Columbia'. These boasts are not idle. The province is chock-full of high mountains, glaciers, lakes, valleys, rivers, streams, beaches, ocean inlets and scarcely populated islands. Seals and whales migrate along its rugged and winding coastline. This is why the province's largely urban population has used its nature parks, camping sites and trails intensively for many years. It's also why millions of tourists visit BC for sightseeing tours or the tactile pleasures of ski holidays and adventure travel.

By the early 1990s, in the south of BC, the last tracts of original temperate rainforest were being targeted for logging. As the decade drew to a close, two areas in particular had become the focus of impassioned disputes between environmentalists (intent on saving old growth trees) and logging firms (intent on turning raw timber into exportable commodities). The first was Clayoquot Sound (pronounced Clak-qot) on the western 'wild side' of Vancouver Island. In the summer of 1993, environmental activists sought to physically block vehicles carrying loggers along makeshift roads deep into the forest. By the autumn, more than 800 people had been arrested, making the blockade one of Canada's largest ever acts of civil disobedience. The Clayoquot protests became a major national news item in Canada, and to some extent internationally, and laid the basis for an organised campaign to change public and government perceptions of Clayoquot's forests in BC. This campaign, which gathered strength from 1994, fed into the politicisation (from 1997) of a second area slated for logging, 'The Great Bear Rainforest'. The evocative name, coined by environmentalists (with Greenpeace in the vanguard), covers a large area stretching from Vancouver Island across on to the coastal mainland of BC. As with Clayoquot Sound, it contains very old cedar trees and sitka spruce, and is home to black bears, grizzly bears, countless birds and insect species, and several salmon species (which spawn in rivers and streams). By the turn of the millennium, both forest regions had almost become household names in Canada, whereas a decade earlier they were simply two remote, scarcely populated parts of the country little known to non-locals (see Map 4.2).

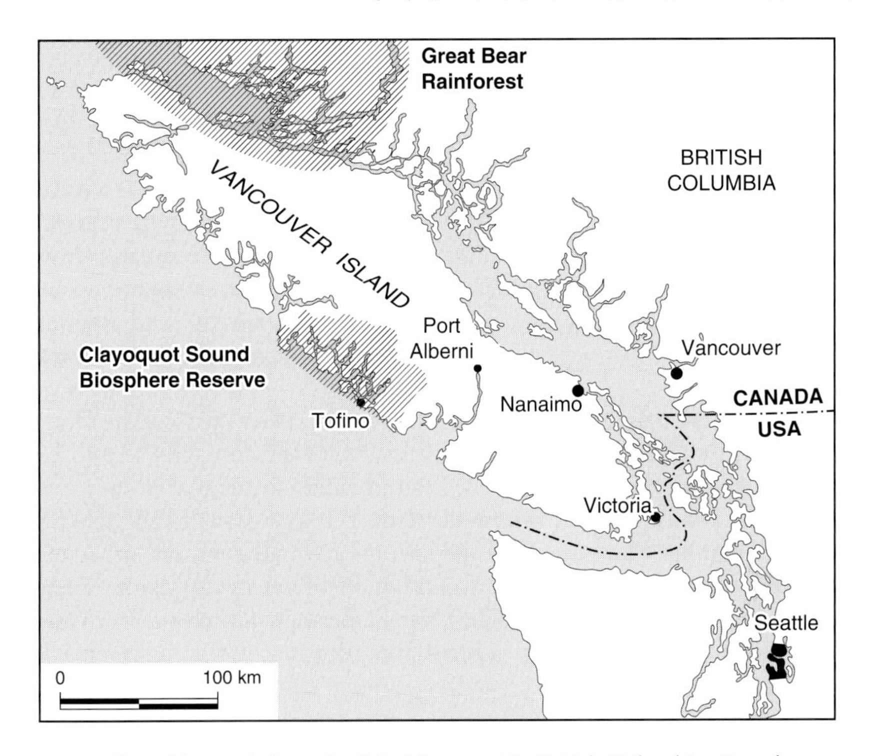

Map 4.2 Two old growth forest 'political hotspots' in British Columbia, Canada

In the 1990s, Clayoquot Sound and the Great Bear forest became the focus of environmentalist attempts to prevent the loss of temperate old growth forest.

In what ways were forests represented by the two parties involved in both conflicts? Since there are striking continuities between the Clayoquot and Great Bear cases, I will focus on the former only.

Beyond the cut

The major forest company that sought to defend its commercial interests, and reputation, in the Clayoquot conflict was MacMillan Bloedel (later taken over by Weyerhaeuser). In late 1993 and through 1994, the company ran a high-profile publicity campaign that comprised television and newspaper adverts, public relations pamphlets, news releases and even new visitor centres. Let's consider one of the early pamphlets, a mixture of glossy photographs, graphics and non-technical text (printed on high-quality paper) entitled *Beyond the cut*.

For those canny enough to notice, the title had two intended meanings. First, it was a knowing reference to the controversial practice of 'clear cut logging', where foresters used chainsaws and heavy machinery to, literally, erase sections of forest, one at a time. Second, it was a reference to what comes after logging has occurred. As the pamphlet's text made clear, fronted by a picture of a tree stump out of which grows a green shoot, MacMillan Bloedel sought to persuade readers that its approach to BC's forest resources was 'beyond' the cut in the sense that the company was a custodian of the forest on behalf of the public, not a rapacious destroyer of trees as a means to the end of money making. In his foreword, the then company president Ray Smith declared that 'We are committed to manage our forestlands in the best interests of the public.' At the heart of this rhetoric of custodianship, as Braun shows (2002: 36–41), were claims about expertise and responsibility – each of which was illustrated photographically. Let me take each in turn.

Beyond the cut was peppered with images of 'experts' at work in forests, in laboratories and in greenhouses. For instance, the company's Land Use Planning Advisory Team was profiled, and readers learnt that it comprised 'specialists in soils, wildlife, fish, water resources, and growth and yield projections'. In other places, readers saw photos of road engineers who, they were told, 'know that poor road construction practices can cause erosion and mud build-up in streams'. The overall message was clear: MacMillan Bloedel employed only highly trained and educated forest managers who planned carefully for the process of tree removal and its extended aftermath. Their planning, the pamphlet suggested, was based on careful observation and experience, both framed within the protocols of scientific rationality (see Plate 4.1).

This discourse of expert management was accompanied by one about corporate responsibility. Because BC's forest lands are public (or state owned), firms have rights of tenure for limited time periods only. In several parts of *Beyond the cut*, readers learnt that MacMillan Bloedel went well beyond its regulatory obligation to manage forest resources in a 'sustainable' way. Claims were made that areas of forest had been transferred from company tenure to become provincial nature parks, while other areas had been spared the saw. Equally, readers were told that the company worked closely with the provincial government and other stakeholders in order that 'The forests of BC no[t] . . . be decimated, or devoted exclusively to timber production'. As the pamphlet went on to say, 'As custodians of the forest, MB protects, cares for, and renews this great resource for the benefit of present and future generations . . .'.[2]

To summarise, and unsurprisingly perhaps, the forest portrayed by MacMillan Bloedel in publicity like *Beyond the cut* appeared as a stock of *potential economic resources*. The careful harvesting and renewal of these resources was legitimised on the basis of the jobs and dollars it delivered for the province. But it was recognised that forests are not only timber stocks, and that it's important to protect special areas and also permit recreational uses. However, outside these areas – whose precise size and nature was not specified in *Beyond the cut* – BC's vast forest lands were represented as the physical embodiment of potential monetary value in international wood

■ FORESTRY RESEARCH

Forestry research includes silviculture and land use. The focus within silviculture is three-fold:

■ REGENERATION

The development of techniques which best prepare growing sites and establish the next crop to ensure its survival.

■ GROWTH AND YIELD

The development of a database to realistically predict the rate of growth and the expected yield from the wide variety of growing sites. Soil conditions, weather, elevation, animal habitat are considered.

■ FERTILIZATION

Research-based decisions on the "how" and "when" of fertilization to optimize the potential for growth and yield improvement of the forest.

Landuse research concentrates on the impact of forest management practices on soil, fisheries and wildlife resources. This work contributes to the integrated management of the company's land base giving consideration to the variety of needs and demands placed on it.

15

Plate 4.1 Beyond the cut?

Above is reproduced a page from a 1993 MacMillan Bloedel public relations pamphlet distributed in British Columbia. It comprises a very particular representation of the work performed by an epistemic community of forest scientists employed by the firm. Written in matter-of-fact prose, it points to the scientists' twin focus on achieving the maximum yield of timber while protecting soils, water, fish, etc. The pamphlet avoids emotive language and makes few overt moral or aesthetic claims about the value of trees.

products markets. Managed scientifically and responsibly, they could (it was suggested) provide employment and profits well into the future.

On the wild side

What of the environmentalist riposte? What representation of Clayoquot was created and circulated by the likes of the Western Canadian Wilderness Committee (WCWC), the Rainforest Action Network, the Sierra Club of BC and Greenpeace Canada? A notable similarity and consistency was evident before and after the Clayoquot protests, feeding into the Great Bear Rainforest campaign. A popular WCWC-endorsed book of photographs,

Plate 4.2 On the wild side

This photograph by Adrian Dorst (the original is colour) is one of many that dominate *Clayoquot: on the wild side*. It represents the forest as teeming with life, as largely absent of people, and as stunningly beautiful (© Adrian Dorst).

Clayoquot: on the wild side (Dorst and Young, 1990), was, Braun argues, both symptomatic and formative. Its representations of Clayoquot resonated with those contained in a book like *The Great Bear Rainforest: Canada's forgotten coast* (McAllister *et al.*, 2007) published seven years later. The 140-plus high-quality colour images were taken by wildlife photographer Adrian Dorst, with accompanying text written by environmental journalist Cameron Young. The preface was authored by renowned Canadian wildlife artist Robert Bateman. Together, words and pictures depicted 'a sublime, enchanting landscape ... [of] powerful forces and ... intricate, even delicate, ecological relations' (Braun, 2002: 76). In contrast to *Beyond the cut*, the reader of Dorst and Young's book was enjoined to see Clayoquot as one the last redoubts of a beautiful, pristine forest ecosystem in need of preservation against the bulldozer of 'progress'.

This was achieved, in large part, through Dorst's stunning photographs, which dominated the book (see Plate 4.2). They depicted majestic, moss-covered trees, high mountains, deep valleys, rugged coves, torrid rivers, and forested islands surrounded by blue seas. Many of the views were panoramic, so that Clayoquot appeared awe-inspiring and grand. Other images were framed at ground level, training the eye on ancient, individual cedar trees. Young's text instructed the viewer on how to 'see' Clayoquot, as if instruction were needed. For instance, at one point he declared that 'This is still a virgin landscape lost in time and governed by the unequivocal laws of nature – a gift to humanity ... you feel that you are witnessing the perpetual big bang of creation ...' (cited in Braun, 2002: 78).

The unreflexive, gendered reference to a 'nature untouched' was evidenced in two ways in Dorst's beautiful images. First, most photographs contained no human presence at all, and where humans did appear, Dorst was the only one who was named. Second, in the few images containing people and their artefacts, it was aboriginals from the region who appeared (the Nuu-chah-nulth). Indeed, *Clayoquot: on the wild side* was dedicated to this local native population (numbering around a thousand), whose descendents occupied BC before it became part of the British Empire and, later, part of an independent Canada.

In a chapter on the Nuu-chah-nulth, subheadings like 'Coastal ghosts', 'A nation in distress' and 'Where worlds collide' accompanied very particular images, for instance an old lone totem pole was shown still standing in the forest; a dugout canoe appeared empty on a beach in twilight; and unnamed members of the Tla-o-qui-aht native band were pictured from a distance, silently paddling a cedar canoe across an inlet. In short, what was evoked was a pre-modern culture in seeming harmony with Clayoquot's threatened landscape. In Dorst's images and Young's text, 'first nations' people in Clayoquot were made to appear traditional and just as vulnerable to the effects of modern, commercial logging as the region's special ecology. Like Clayoquot's flora and fauna, the Nuu-chah-nulth thus, it was implied, needed 'saving', and *Clayoquot: on the wild side* sought to bear witness (through Dorst and Young) to their vanishing way of life.[3]

To summarise, the environmentalist discourse utilised in *Clayoquot: on the wild side* depicted the area less as a region of timber resources and more as *unique ecosystem* teeming with life. Placed within this ecosystem, as the symbolic opposite of MacMillan Bloedel with its heavy machinery, office-based executives and distant shareholders, were indigenous peoples. The overall affect was to both aestheticise and romanticise Clayoquot: it was presented as a source of wonder, majesty and spiritual inspiration, something too special to treat instrumentally as just another stock of lumber.

Buried epistemologies

In *Beyond the cut* and *Clayoquot: on the wild side*, we had two different representations of the same coastal forest. Both representations were 'constructed' in the obvious sense that they reflected the values and goals of their authors. They were obviously 'interested representations' and fairly typical of those produced and circulated by both sides through the 1990s. Each downplayed what the other highlighted. MacMillan Bloedel's document avoided the poetic language favoured by Young and the highly aestheticised images favoured by Dorst. Instead, it advanced a 'rational discourse' whose content and associated imagery focussed on the commercial benefits of 'sound forest management'. Dorst and Young, by contrast, presented logging as an assault on the natural world – a world whose value was irreducible to jobs and corporate profits.

Study Task: We've just seen how two different groups represented the same area of forest in contrasting ways. Focus now on what the otherwise divergent representations have in common. First, think about which of the four meanings of 'nature' itemised in Chapter 1 they share: is one most prominent? Then, consider how these meanings are spatialised and temporalised. Finally, consider if there's anything similar about how nature, so defined, is represented, notwithstanding the genre differences between *Beyond the cut* and *Clayoquot: on the wild side.*

Each document, in its own genre-specific way, made a claim to 'realism'. Both sought to reveal a 'truth' about Clayoquot Sound, albeit by using different discursive, rhetorical and visual tactics. Both sought to sway public opinion in BC and Canada, and government opinion too. Despite its emotive, moralising, pro-nature stance, *Clayoquot: on the wild side* seemed, if anything, the more authentic of the two (at least at first glance). After all, it was speaking of, and for, a forest ecosystem that was, undoubtedly, in need of representatives able to give voice to its rare and special qualities. If trees could speak, then one might reasonably presume they'd speak the language of green ethics, not the instrumentalist, anthropocentric ethics of a logging company intent on their removal. Without environmentalists like Dorst and Young, much of Clayoquot's forest could well have been clear-cut. Instead, it was made into a UNESCO World Biosphere Reserve in 2000 and continues to enjoy protection from commercial logging firms.

Interesting as all this (hopefully) is to readers, it may appear to contradict a major argument of *Making sense of nature.* It implies that, while there were different representations of nature (in this case Clayoquot Sound), nature nonetheless existed *outside* them as their referential basis. For, though *Beyond the cut* and *Clayoquot: on the wild side* differed in how they *depicted* Clayoquot, I may seem to be suggesting that 'the forest' was a pre-existing entity ontologically *available* to be re-presented in different ways. It was an 'external nature' whose essential qualities were simply being described in print and images. However, this is *not* what I am arguing. Here I follow the lead given by Braun. In his book *The intemperate rainforest* (2002), he points out that both MacMillan Bloedel and its environmentalist critics together presupposed the existence of something that their representations then failed to question. Notwithstanding their differences, what documents like *Beyond the cut* and *Clayoquot: on the wild side* shared, Braun argues, was the conviction that Clayoquot was a forest ecosystem in which people were largely absent and which, to date, had barely registered any meaningful human imprint.

This conviction was, in fact, highly contestable (though neither party hinted at it at the time). More pointedly, Braun argues that it was neo-colonial: that is, a form of colonial thinking still evident (even) after the formal end of British rule in BC many decades ago. The 'buried epistemologies' of documents like *Beyond the cut* and *Clayoquot: on the wild side* ensured that the Nuu-chah-nulth were either ignored (as in the former case) or depicted as a traditional culture sharing the modern environmentalist desire to 'respect nature' (evident in Dorst and Young, 1990). This was made possible, Braun argues, by historic processes of dispossession and relocation: when the British colonised BC from the 1850s onwards they set about controlling land and resettling indigenous bands into small 'native reserves'. This appropriation and clearing of BC's populated coasts and valleys, Braun insists, is what made it *seem* normal, a century later, to regard places like Clayoquot as essentially 'natural spaces'. What *Beyond the cut* and *Clayoquot: on the wild side* could not countenance was the suggestion that Clayoquot had been a thoroughly inhabited, utilised and culturally meaningful area prior to the arrival of British colonists. Nor could they entertain the idea that modern-day descendants of the pre-1850 Nuu-chah-nulth bear little relation to their predecessors, in terms of lifestyles, values or aspirations. In short, what the very public disagreements about the future of Clayoquot (and later, the Great Bear Rainforest) took for granted was something that BC's native bands wanted to oppose. They not only agitated for an equal say in what happened to Clayoquot; they also wanted to assert their historic rights of use and entitlement against a settler society that had taken so much away from them.

Fortunately, the Nuu-chah-nulth in Clayoquot were drawn in to the discussions and decisions about the region by the turn of the millennium. This occurred against the background of native groups across Canada making land claims cases to the courts in order to regain right of use – and indeed, ownership – of areas their forebears were displaced from. Parts of Clayoquot have in fact been logged by new firms controlled by native bands, giving the lie to the assumption that today's indigenous Canadians somehow have 'nature's interests' in their cultural genes. Though it's difficult to rectify what Braun calls the 'cognitive failures' of non-indigenous Canadians, the Clayoquot case shows why challenging seemingly unproblematic beliefs in nature's naturalness can be vitally important.

THE NATURE WITHIN: RETHINKING HUMAN IDENTITY THE GENETIC WAY

As I suggested in Chapter 1, the idea that nature is fast disappearing has given it a new saliency in public discourse and in the pronouncements of politicians, businesses and civil society organisations. The Clayoquot and Great Bear Rainforest cases remind us that much of the discussion relates to non-human nature. According to many commentators, modern capitalist societies are 'ecocidal': they're engaged in nothing less than an arms race against the living Earth. Not only are they destroying the biogeochemical diversity and complexity bequeathed by evolutionary history, they are also 'playing God' by utilising ever more intrusive technologies – like synthetic biology – whose long-run effects are impossible to predict. Degradation and unanticipated threats and risks are thus sides of the same hubristic coin. Others argue that this sort of environmentalist rhetoric is misplaced: if we want to free millions of people from grinding poverty and ensure a good standard of living for all then, it is suggested, we necessarily have to bend large parts of the non-human world to our will.

Inhabitants of early twenty-first-century Western societies frequently encounter these arguments in newspapers, on radio, on television, in magazines, in advertisements and elsewhere. They pit 'ecocentrists' committed to protecting and saving non-human nature against 'anthropocentrists' dedicated to the unfinished projects of 'development' and 'modernisation'. However, 'human nature', in the biological sense of that term, is now arguably just as much a focus of public interest as things like endangered species, melting ice sheets or organic food. This is evident in a range of arenas. I want to focus on recent 'discoveries' about human genetics and the ways these have been disseminated and received. Because these discoveries are about us, they inevitably beg, in very explicit ways, fundamental questions about our identities, potentialities and relationships to others. Representations of human genes are, I argue, an important medium through which a **re-naturalisation** of human self-understanding could be taking place. To understand why, we need first to consider the rise-and-rise of molecular biology since the 1950s.

The power of molecular biology

Molecular biology is now a large, sprawling field of basic and applied research that spans the various life sciences. It studies what some consider to be the building blocks of life: namely, those cellular level structures and process that are far too small to be observed without precision instruments. Molecular biology's numerous, highly specialised and trained practitioners occupy university departments, publicly funded research institutes and the laboratories and field stations of biotechnology companies. Their practices and discourses are highly esoteric, comprising techniques like polymerase chain reaction (PCR) and terms like 'alleles', 'introns' and 'messenger ribonucleic acid' (mRNA). Their research has had substantial wider impacts on various branches of biology (human and non-human), chemistry, information science, computational science and psychology. This research has also enjoyed significant and sustained levels of public visibility over the past 25 years.[4]

The influence and prominence of molecular biologists is not hard to explain. First, they're associated with a set of scientific discoveries that appear to be as fundamental as those of Galileo, Newton or Einstein. These include James Watson and Francis Crick's 1953 finding that deoxyribonucleic acid (DNA) contains the instructions used in the development and functioning of almost all known living organisms. Since Gregor Mendel's groundbreaking late nineteenth-century inquiries into plant heredity, biologists had wanted to understand how species achieve intergenerational consistency (structural and functional) with phenotypical variation. It was hypothesised that a 'particle' existed that, through species reproduction, was (1) passed on to each generation yet somehow itself subject to variation, or (2) able to create variation in the resulting organisms. Watson and Crick confirmed that this 'particle' was not only real but, in fact, comprised a 'double helix' in the nucleus of cells – a helix itself composed of four bases arranged along its (considerable) length.

Some decades later, molecular biologists made new and equally fundamental discoveries. From the late 1970s, they began to sequence the genome (i.e. describe the entire genetic content) of different species. Watson and Crick had only been able to identify the rudiments of DNA, namely its 'twisted ladder' structure of base pairs. It was another thing altogether to describe in detail the coupling and relative position of hundreds of thousands of bases – notably those in 'coding regions' that contain instructions vital for organism development and functioning. To date, over 100 species have been genomically sequenced, including humans. The HGP was one of two international, high-profile and exceedingly expensive attempts to produce a draft of humanity's genetic specificity, which famously came to fruition in 2000, along with a similar privately funded project by Celera Genomics. The 'complete' human genome was made public in 2003.[5]

If these and other discoveries have been exceedingly noteworthy and newsworthy, so too have a set of biotechnological tools invented by molecular biologists to intervene in the life processes they study. These tools are a second reason why molecular biology today enjoys enormous influence and prominence. The tools include the ability to isolate sections of DNA, copy it and insert it into the cell nuclei of different organisms. Without these tools, the hotly debated processes of 'genetic engineering' and 'cloning' would not be possible. Genetically modified (GM) bacteria, insects, plants and animals are now widely employed in medical, pharmaceutical and agricultural research – leading to product development, sale and use in a large and growing number of cases. The range of applications stretches from Oncomouse (a life form designed by Harvard University research biologists to be susceptible to cancer) to food crops engineered to be less vulnerable to things like drought or disease. Many GM organisms are transgenic, meaning that their DNA is a 'cut and paste' hybrid of genetic material that could not be recombined through 'natural reproduction'. In the case of cloning (i.e. replicating genetic material for the purposes of research and therapy), Dolly the sheep remains without doubt the icon. Born in 1996, she was a Finn-Dorset ewe created from nuclear DNA taken from a mammary gland cell belonging to a full-grown ewe. Her birth and growth to adulthood showed that the nucleus of any cell had the ability to 'programme' a new and complete organism in its likeness.

Study Task: I argued in Chapter 1 that 'gene' is one of nature's collateral concepts. When you think of this word, which of the four main meanings of nature come to mind? Can you think of concrete examples where a statement about genes illustrates the meaning(s) at work? The statements could be everyday ones, not necessarily ones made by scientists.

It's a measure of just how visible molecular biology now is in the public realm that its lingua franca has permeated popular discourse, almost to the point of becoming normalised. Few people are unaware, though fewer still are conscious, of the complex semantic work performed by the image of the double helix (see Plate 4.3). Terms like 'gene therapy', 'genetic database', 'genetic counselling', 'transgenic organism', 'genetic testing', 'genetic screening', 'DNA fingerprinting', 'gene patent' and 'genetic disease' are now commonplace outside the specialised epistemic communities who coined them. This is because they refer to things that have been

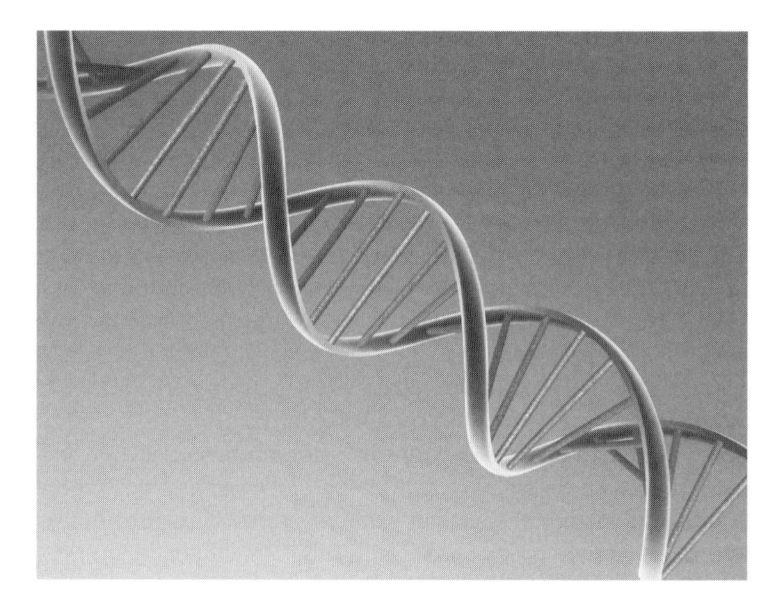

Plate 4.3 The double helix: gene as icon and metonym

What comes to mind when you see a picture of DNA? Increasingly, the 'twisted ladder' image of a gene has come to represent 'life' in general because it is perceived as *the* building block or elemental component. It signifies the biological unity of humankind, as well as the power of biological processes over human wishes and desires. It reminds us that we're all animals (whatever else we think we are), and it raises questions about group-level differences among humans (we may be genetically similar, but are we all identical?).

made to matter to ordinary people in everything from law courts to hospitals to family history websites, such as pre-implantation genetic diagnosis (i.e. embryo screening for potential health 'defects' prior to *in vitro* fertilisation). Another measure of molecular biology's prominence is that it's inspired a few successful Hollywood sci-fi films since the mid-1990s. Aside from the well-known *Jurassic Park* trilogy, these include *Gattaca* (1997), *Code 46* (2003), *Aeon Flux* (2005) and *The Island* (2005). In different ways, these films invite viewers to consider the ethical and practical implications of molecular biology's analytical and technical prowess. The same can be said of novels like Margaret Atwood's (2003) *Oryx and Crake* and the Michael Crichton books that inspired two Steven Spielberg blockbusters. They continue a long tradition in which fictional genres and modes of representation seek to shape public perceptions of scientists and the 'proper' place of science in society.

Molecularising identity

The words and deeds of molecular biologists, relayed and redescribed by myriad intermediaries, are altering our existing understandings of self and other. Their science is changing the socioeconomic context in which they operate. How? First, they are focussing their, and our, attention on entities and processes that appear to be nothing less than *definitional* of what it is to be both an individual human being and a member of a unique mammalian species. Because, to our way of thinking, we're at once natural and cultural beings, knowing *what* we are biologically speaking necessarily impinges on the question of *who* (we think) we are. Second, because sub-dermal entities and processes are invisible to the non-expert observer, then we necessarily rely on molecular biologists to, as it were, represent ourselves to ourselves. To use Evelyn Fox Keller's (2000: 10) term, 'gene talk' that can ultimately be traced back to research scientists is, these days, very hard to ignore when it comes to answering the timeless questions 'who am I?' and 'to whom am I similar, or related?' This conjugation of individual and group identity was well expressed by a Harvard geneticist, Professor Walter Gilbert, reflecting on the prospect of the fully sequenced human genome back in 1992: 'one will', he said, 'be able to pull a CD out of one's pocket and say "Here is a human being; it's me"' (Gilbert, 1992: 96).

In Chapter 2, I argued that individual and group identities are cobbled together over time as people are 'interpellated' by myriad discourses into a variety of contingent 'subject-positions'. It's the colligation of these positions that defines, in ways that vary spatially and temporally, the apparently autonomous and complete 'individuals' we typically consider ourselves to be. Rather than simply providing a 'biological foundation' upon which the house of socially constructed identities is made, we can consider human biology, or more precisely, discourses about human biology, as contributory to the process of interpellation. In respect of 'gene talk', the American anthropologist Paul Rabinow (1992) suggested that we may be entering a more 'genocentric age' in which 'biosociality' looms ever larger. **Biosociality** describes new modes of identification and association based on different people's understanding of what they are, biologically speaking, and who they might become (or themselves produce) with the assistance of biotechnological expertise. For instance, Rabinow noted that 'it's not hard to imagine [future] groups formed around chromosome 17 locus 16256, site 654376 allele variant with a guanine substitution' (Rabinow, 1992: 244). Relatedly, according to Nikolas Rose (2001), the discourses and practices of molecular biology openly invite us to become knowledgeable about our genetic connections to, and differences from, other individuals and groups. We are, he maintains, further invited to *act* on this knowledge about our 'intrinsic natures' and thus take personal responsibility for our identifications and manage whatever 'risks' (and indeed 'opportunities') our genes present to us or to our offspring (e.g. a biological disposition to have cystic fibrosis).

I don't propose to evaluate the empirical robustness of these claims here. Instead, what I take from Rabinow and Rose is the thesis that geneticised representations of human nature have a high potential to alter existing senses of 'I and you', 'me and them', identity and action. According to one commentator, these representations are part of a wider 'biologism', that is to say 'the growing ascendancy of biologically based accounts of human life' (Skinner, 2006: 461).[6] We see this in the popularity of new approaches to understanding personal, family and geographical ancestry in Western societies. These approaches, showcased in television documentaries like *African American Lives* (Public Broadcasting Service, 2006) and promoted by companies like the British-based Roots for Real (see http://www.rootsforreal.com/), use genetic tests of mitochondrial and Y-chromosome DNA to identify the probable location of a person's historic forebears.[7] We also see biologism evident in the new cottage industry of popular science books devoted to considering not *whether* but *how much* of our thought and behaviour is due to 'nature' rather than 'nurture'. These books range from the low brow, such as dietician Neal Barnard's (2001) *Turn off the fat genes,* to sophisticated interpretations of the latest scientific findings, such as Matt Ridley's (2003) bestseller *Nature via nurture.*

Study Task: Go online and spend some time examining the website of a personal genetics and family history organisation, such as *Roots for Real* (http://www.rootsforreal.com/). What overt and implicit connections are you being invited to make between your genes, your relatives and your sense of self? What actions on your part do the words and images on the site seek to engender?

In summary, it may not be too much of an exaggeration to say that 'genes' and 'the genome' have, in a variety of ways, become metonyms not only for human biology *tout court*, but for individual and group identity in many contemporary societies. What interests me about this is two things: the first is that a thing called 'genetic nature' is now widely regarded both as being real and as existing apart from (or 'below') the realms of cultural symbolism and social action; the second is that genetic researchers, advisers, technicians and counsellors are widely perceived to be in the business of sharing (or acting on) factual knowledge with the rest of us. Both beliefs, as I will now illustrate in three ways, are predicated on an implausible and remarkably old-fashioned idea that some representations have the special quality of revealing ontologically independent 'natural kinds'. Despite the 40-year project of STS, which, as I mentioned in Chapter 3, has sought to dispel a lot of the public myths about the nature of science, molecular biology's societal reception arguably demonstrates the enduring power of scientists to be taken seriously as authoritative spokespersons for a supposedly culture-free world of (human) nature.

Mapping, quantifying and differentiating genes

Genes appear to be part of our 'essence'. They are intrinsic and non-negotiable (unless we opt for biomedical procedures that alter our biological clockwork). In saying this, I don't want to underplay the considerable debates and disagreements evident in the life sciences.[8] Even so, genes and genomes are now widely understood to be discrete and elemental pieces of living nature. Their existence is taken as read and their implications, which can be interpreted in a wide range of ways, are usually regarded as being important rather than trivial. In part, this is because of the usual (and inevitable) simplifications that occur when complicated issues are reported and interpreted in the realm of everyday life. As Richard Lewontin once pointedly said, 'Measured claims about the complexity of life and our ignorance of its determinants are [just] not show biz' (Lewontin, 1991: vii). As the following trio of examples show, the (apparent) 'fact' of genomes and genes dissimulates their constitutive unnaturalness. Genetic discourse founded on scientific research is as socially conditioned as the individual and group identities that discourse is now helping to reconstitute in a biological idiom.

Cartographic metaphors

One of the principal ways in which the 'reality effect' of genetic discourse has been achieved in recent years is through the **metaphor** of mapping.[9] This metaphor has been used routinely by geneticists, as well as by a panoply of commentators reporting on and debating their research in various non-scientific arenas. To offer one example, the writer of popular science books Victor McElheny (2010) titles his recent, celebratory history of the HGP *Drawing the map of life*. As Neil Smith and Cindi Katz noted in an essay on how metaphor is used,

> it works by invoking one meaning system to explain or clarify another. The first meaning system is apparently concrete, well understood, unproblematic and evokes the familiar; in linguistic theory it is known as the 'source domain'. The second 'target domain' is elusive, opaque, seemingly unfathomable, without meaning donated from the source domain.
>
> (Smith and Katz, 1993: 61)

Gene maps, as one commentator puts it, 'represent the relative and absolute positions of genetic sequences on the 23 pairs of chromosomes that sit within the nucleus' (Hall, 2003: 152). However, they are precisely *not* the same as conventional maps in that they don't comprise graphical representations of a microscopic territory, country or city with a scale, key or grid coordinate system. Mapping, therefore, *is very much a metaphor* for representing genes, not a literal description of the representational process or eventual product. *And yet*, as Smith and Katz note, all metaphors are nonetheless intended to point to some real (or imagined) *similarity* between the source and target domains – they enable substantive comparisons to be made.[10]

In light of this, we can ask what it is about the source domain that has made the map metaphor so widely used in presentations of molecular biology leading up to and since the completion of the HGP. As several commentators have suggested, it's surely that we normally take maps to be, indeed, *positively expect them to be*, accurate depictions of landscapes that we could only perceive directly if we were flying high and looking down. In the words of one geographer,

> people rely on the map as an accurate representation of the world under scrutiny. The power that maps exert in society is bound up with the impression of exactitude and precision that they convey.
>
> (Livingstone, 2003: 154)

The metaphor of a 'genetic map' thus signifies the search for, and achievement of, truthful representation. It trades on the idea of mimesis and correspondence, and it implies facticity.[11] In contrast to conventional maps, the need for genetic maps arises not from the fact that the 'landscape' is too *large* to perceive without a representational aid, but on the contrary that it is much too *small.*

Is mapping a fitting metaphor for discussions of human genes, gene sequence and the genome? It depends on whether one chooses to take the conventional understanding of the source domain at face value. In daily life, most of us would do, wouldn't we? After all, if my A–Z city map or my car's sat-nav system get me to my destination, then I have no reason to question the assumption that maps belong to the meta-genre of 'realist' forms of representation. However, over the past 25 years, several historians and geographers have shown convincingly that, in Denis Wood's memorable words, conventional map-making '*propagandizes* exactitude as if it were the reason for its existence' (Wood, 2010: 93, emphasis added). Wood, along with Brian Harley, Matthew Edney, Mark Monmonier and Denis Cosgrove (among several others), argues that most maps fail to advertise their most important feature: namely, that they are partial and purposeful attempts to record the character and position of people, things, flows and boundaries. They are not neutral with respect to what they purport merely to record on a miniature globe or the pages of the *Times Atlas of the World*. They are, in Wood and John Fels's (2008: xv) words, 'neither what they seem nor proclaim themselves to be'.

To cite Jean Baudrillard's (1994) famously counter-intuitive observation, the map *is* the territory – not a window *on to* the territory. In this light, we can see the metaphor of genetic mapping for what it arguably is. It's an attempt, and a rhetorically successful one too, to convince us of something that's only seemingly true. This is why I placed the term discovery in scare-quotes a few pages back when describing the aims of molecular genetics. Discovery is an alluring metaphor and closely tied to the growth of mapping during the successive periods of European exploration and colonisation from the 1500s onwards. It suggests something pre-existent that's waiting to be seen, recorded objectively and positioned. In the case of genes, it gives the questionable impression that molecular biologists worldwide are engaged on a journey into the incompletely charted terrain of chromosomes, proteins, polymorphisms and haplotypes – as if the latter are all as clearly delimited as the shoreline of an island. But is there really such a thing as a 'gene' or even a whole 'genome' existing as a discrete biological entity with causal powers? And how many people outside molecular biology would ever trouble to ask the question?

Numbering the human genome

'Numbers do not make an argument.' So says one statistician (Dorling, 2011: ix). But when used *in* arguments, or in apparently matter-of-fact statements, numbers can make a huge difference. There's a strong whiff of certainty (often absoluteness) with numbers, even when expressed as probabilities, percentages or estimates. If mapping metaphors have been widely used in representations of human genes, so too have numbers – notably, those quantifying our genes relative to the gene count of non-human species. Indeed, if genes have become important to contemporary discussions of what it means to be 'human', numbers have, I would argue, been central to the claim that there's a single, universally shared genome that exists regardless of gender, 'race', ancestry or geographical origin.

Certainly, the absolute and relative numbers of human genes were highlighted persistently when the 'draft' and 'final' versions of the two genome mapping projects were unveiled early in the new millennium.[12] The 2001 special issues of the periodicals *Science* and *Nature* were devoted to presenting the two drafts of the human genome. They both led with discussions of likely gene numbers and percentage similarities to the genomes of non-human species (like thale cress). The front-page stories of almost every Western broadsheet did the same. Earlier speculation that humans might have over 100,000 genes gave way to the results of the HGP and Celera research: there appeared to be no more than 30,000, only 30 per cent more than a common worm and making us one-third genetically the same as (for example) a daffodil. Given this 'surprisingly low' difference from 'simple' species (even lower from chimpanzees), much of the subsequent media and scientific discussion centred on the *qualitative differences* between our genes and those of other species: their functional complexity rather than sheer quantity. At the same time, earlier scientific findings that all humans were no more than 0.1 per cent genetically dissimilar were repeatedly highlighted. The overall message was that humanity is one genetically distinct species, though our 'serial code' is distinguished less by its length than by its 'programming power'.

Study Task: Imagine the map of the human genome had revealed that *homo sapiens* had twice as many genes as our closest biological relatives. Why do you think scientists would have wanted to highlight this discovery? As a member of the lay public, why would this discovery have caught your attention (if at all)? What would you have inferred about humans from this discovery?

Why present information about human genetics in numerical terms? It was no doubt intended to remove, or at least reduce, ambiguity. But, intentionally or not, numbering genes links to other things too whose familiarity to us normally shields them from scrutiny. First, it invites us to regard genes as countable units of a similar size and shape (if not function) that together comprise the genome, like bricks make a house when cemented in layers. Second, it thereby implies that gene numbering is not merely an epistemological choice but, instead, an act of ontological necessity. In his classic work *The anthropology of numbers*, Thomas Crump (1990) pointed out that numbers are simply another way of describing (and intervening in) the world, no less conventional than written or spoken language. This being the case, we can ask: does mathematics comprise 'nature's own language' (as science writer Ian Stewart has argued) or is it one way we've chosen to make what we call nature speak to us? Crump is one of only a few analysts to consider the latter option in detail, and what its implications might be.[13] Most mathematicians, most practising scientists and most members of the general public tend to assume that numbers are useful because they symbolise things about the world that are real. They are not taken to be an 'imposition' on the world. But how secure is this assumption? It permits numbers to quickly assume the status of 'facts' once they're used to characterise aspects of the world, and these can then become the focus of debate – *as if what they signify depends only on the debaters' values not on the process of numbering itself.*

However, seen from the perspective of 'post-genomic' biology, representing genes like 'natural numbers' conceals a number of contestable epistemological judgements. Because numeric symbols and mathematical procedures have an abstract, formalistic quality they serve to decontextualise the things they are made to refer to and stand for. The human genome 'map' unveiled a decade or so ago was produced by a highly elaborate process of 'shredding and reassembly'. Gaps in the map were filled and smoothed by HGP and Celera biologists, and 'separate' genes identified out of the continuous (but decomposed and reconstructed) 'terrain' of the genome. Thus, genes were not, and are not, like separate beads on a continuous piece of string, at least from the perspective of post-genomic biology. Numbering them can be seen to conceal complex guesswork about where, in the human genome, 'coding regions' lie, how their boundaries are identified and what their chemical composition is.

In addition, numbering human genes (absolute quantity and relative species share) casts another complicated issue into darkness. Strictly speaking, 'the human genome' is the *entire* genetic content of *every single person* on the planet. Because it is, practically speaking, impossible to collect and 'read' the DNA of 7 billion people, sampling and selection was clearly necessary for HGP and Celera Genomics to undertake their mapping exercises. This means that the issue of a 'representative sample' of genetic material arose. How, prior to reading different people's DNA, would one know who to select for the full range of genetic differences so that a complete composite map could be created? The answer is, one wouldn't. In the event, what was called 'the human genome' by the HGP and Celera was, in fact, genetic information pertaining to a very small number of anonymous donors. Far from being a 'representative genome', it is better understood as a *reference genome* against which further and future genetic mapping projects can be compared. Yet these important and complex issues are made invisible when 'the' human genome is represented as both an absolute and relative number.[14]

Homo pluralis: dividing DNA

If the attempt to map the human genome focussed squarely on *homo sapiens'* biological similarities, the search is now on for what we might call 'human biodiversity'. These are patterns of physical difference only visible 'under the skin', but which manifest themselves phenoltypically.[15] These inner differences are known as polymorphisms, sections thereof microsatellites, and groups thereof are sometimes called haplotypes. But can human DNA really be differentiated meaningfully, on an individual and/or group basis? By what criteria does one identify 'meaningful biological differences' among people? How might those criteria be made flesh through measurable differences in a person's appearance, behaviour, performance or health?

Study Task: In light of the questions posed immediately above, think of group-level physical differences that have sometimes been said to distinguish the appearance of the following: west Africans, west Europeans and Jews. Make a list of these differences, focussing on ones that seem to be 'objectively' true. In your view, are these differences 'meaningful'? If so, can you specify the reasons why?

The questions posed above are highly sensitive ones. They open the door to some differences being judged negatively as 'threats', 'risks', 'imperfections' or 'abnormalities'.[16] This much is evident from the famous cover of *Nature* announcing the 'completion' of the human genome map (see Plate 4.4). Composed of pixels of people's faces, which together comprise the double helix of DNA, the cover was scrupulously inclusive of all humanity in its image selections. Why has the hunt for sub-dermal human difference occurred? Quite aside from scientific curiosity ('knowledge for knowledge's sake'), there are careers and fortunes to be made in what's called pharmacogenetics (i.e. drugs tailored to people's genetic needs) and the invention of other bespoke biomedical interventions. There are also lives to be improved (depending on how one defines 'improve') by identifying and then treating specific illnesses or behaviours common to subsets of people and which seem to have a genetic origin of some kind. However, as I noted in passing earlier, some critics worry that a new 'consumer eugenics' is fast coming our way – one presented in the positive language of 'liberty' and 'freedom'.

The question then arises: can one identify group-level differences in the human genome as a matter of *fact*, free from presuppositions about how the groups are defined? This question matters a great deal, and for two reasons. First, if the categories are presupposed then one may be looking for patterns of genetic difference and similarity among the wrong set of individuals. To use a classic scientific trope, the 'data cannot speak for itself' because the categorical basis for data sorting and comparison is problematic from the get-go. Second, categories of group difference have a social existence outside of science – and an often nefarious one too (think of how the referents of the term 'race' have suffered from racism in the labour market, education and other arenas besides, both now and in the past). This means that scientists interested in human biological difference must be careful to separate their own epistemic practices from those of lay actors in the wider society. Otherwise they can be accused of using categories containing cultural value judgements to search for enduring patterns of group biological diversity. This accusation remains especially relevant to molecular

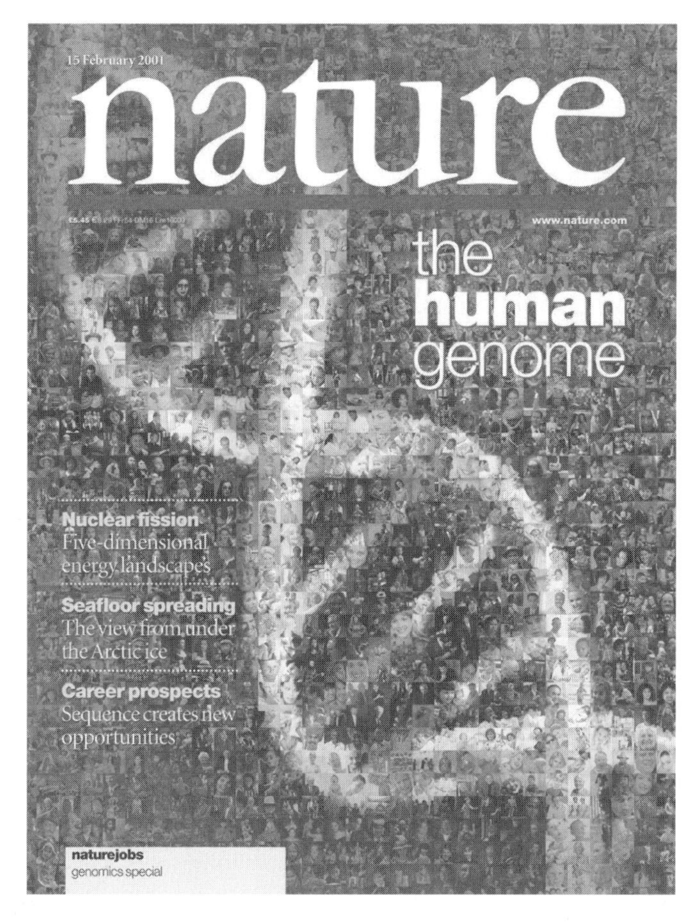

Plate 4.4 Unity trumps difference: the human genome represented by 500 unique pixel-faces from around the world (reproduced with permission from *Nature*, 15 February 2001)

This front cover image was used by *Nature* to mark the publication of the map of the human genome. With the sorry history of state-sponsored eugenics in mind, the journal's editors made a point of talking about *one* genome. They included images of phenotypically different humans who, 'under the skin', were implied to be members of a *single* biological family.

human genetics in the United States. There, particularly in the pre-1939 era, the field of human biology actively participated in what are now regarded as highly discriminatory measures against Americans of African descent (among others).

In light of these two considerations, it's no surprise that researchers looking for intra-genetic differences among *homo sapiens* have been insistent that the variations sought are real, not contrived, let alone 'tainted' by social assumptions about group difference. Let's take the subject of 'race' once more. The post-1945 consensus, at least in Western social science, has been that 'races' have no biological existence. They are instead taken to be groups identified according to perceptible characteristics that, for whatever reason, matter to people who then use those characteristics as markers of social difference. Racism, as the post-war social science orthodoxy would have it, is an act of sociocultural discrimination – even when and if the discriminating parties believe (wrongly) that racial differences are biological. It was in this context that, early in the new millennium, several human biologists claimed that racial differences were, after all, a product of natural evolution and thus physically real.

For example, the then Stanford University professor Neil Risch and his colleagues published a much discussed research paper in the online peer review journal *Genome Biology* in 2002 entitled 'Categorization of humans in biomedical research: genes, race and disease' (Risch, 2002). Risch *et al.* sought to evidence the claim that there are five human races, all with a distinct geographic origin, albeit ultimately traceable to east Africa. A not dissimilar paper was published simultaneously in the highly respected journal *Science* (Rosenberg *et al.*, 2002). This was immediately reported by the *New York Times* science writer Nicholas Wade and, a year later, the *Scientific American* magazine devoted a feature article to answering the question 'Does race exist?' (Bamshad and Olson, 2003). At the same time, the British evolutionist Anthony Edwards (2003) criticised a classic paper by Harvard's Richard Lewontin (1972), which had argued that inter-racial genetic differences were significantly smaller than other genetic differences among *homo sapiens.* A face-off with critics then ensued in the *New England Journal of Medicine* (Cooper *et al.*, 2003). Not long after, the physical anthropologist Vincent Sarich and science journalist Frank Miele published the book *Race: the reality of human differences* (Sarich and Miele, 2004). Like Risch *et al.*, they drew upon a large volume of published biological research in order to break the 'taboo' (as they saw it) about 'race' in US society. Theirs was a critique of social constructionist arguments and, as they phrased it, a long overdue attempt to look 'honestly' at 'race' and recognise its biological character. The history of racism in the United States, they argued, had made too many people afraid of acknowledging that some group differences are not just socially ascribed but scientifically demonstrable – matters of biological fact. Unsurprisingly, an often-heated debate then ensued in which Risch *et al.*, Edwards, Sarich and Miele, and like-minded researchers were reprimanded for resurrecting, or at least opening the door for the return of, a 'scientific racism' that many had thought long-dead.[17]

Though I can't recount the fine details of the debate here, what's interesting is the way that Risch *et al.*, Edwards, and Sarich and Miele prosecuted their cases. A close reading of their work reveals three understandable lines of defence against actual and potential critics. First, they all implicitly or explicitly distinguish 'race' from 'racism' (the argument being that while racism requires a belief in racial difference, 'race' is not itself a prejudicial idea).[18] Second, this distinction is bolstered by the claim that while racism today often operates on the basis of the way people's *appearance* is categorised and evaluated, genetics gets 'under the skin'. It thus (apparently) separates itself from everyday forms of racial categorisation. Indeed, making a virtue of this separation, Risch *et al.*, argued that a 'color-blind approach ... will lead to a disservice to minorities' (2002: 11). Their argument here was that *ignoring* the possibility of group-level biological differences in the name of 'political correctness' may be as harmful as falsely imputing them. Third, both these moves rest upon and reproduce a seemingly common-sense distinction between 'facts' (the domain of science and rationality) and values (the domain of culture, politics, feeling and emotion). It's not hard to see the rhetorical similarities with metaphorical mapping and numbering discussed a few pages ago.

Are these three related lines of defence strong enough to convince? Arguably not. Several sociologists of science have shown that recent research into race and genes is simply unable to maintain a Maginot Line between social and natural categories (see Key Sources and Further Reading at the end of this book). The fundamental difficulty is this: in order to identify people whose genes might be distinctive by virtue of their descent from once separate gene pools, scientists must *use categories to identify those people at the outset which cannot be purely scientific ones.* Duana Fullwiley (2007), in her ethnographic investigations into the pharmacogenetic research of a team led by Professor Esteban Burchard (in California), illustrates this difficulty well. Research subjects sampled for DNA were recruited using standard US census categories and were asked questions about their biological predecessors. When quizzed about the selection process relating to people who self-identified as Caucasian, a member of the research team revealed that 'There was really no set limit in terms of their background ... They could've been Italian, German, Jewish (...) from anywhere' (ibid.: 20). In other words, the category Caucasian, which did not come into currency in the United States until the 1910s and 1920s (the era of state-led eugenics), potentially lumps together people of widely different genetic backgrounds who happen to have a recent ancestor from Europe. It lacks granularity, and yet attains an unwarranted concreteness by organising the analytical efforts of laboratory scientists. Likewise, the category 'African American' potentially conflates groups of subtly different genetic characteristics because it pays no heed to the regional differences between African populations that go back thousands of years. In short, even when the research subjects under scrutiny get to 'self-identify' their biological provenance, they can only do so using a menu of categories that, despite their aspirations, have no clearly delineated 'natural basis'.

DOWN AND DIRTY: GETTING IN TOUCH WITH 'REAL NATURE'?

The third and final case of nature's 'construction' that I want to consider in some detail takes us back to 'external nature' of the sort discussed earlier in this chapter. Since the early 1990s, outdoor and adventure tourism has accounted for a growing share of the global market in travel and recreation. White-water rafting, wilderness camping, ocean kayaking, scuba diving and mountain biking are among the activities and experiences people will pay (often large amounts of money) for – preferably if they occur in beautiful, spectacular and remote locations at home or abroad (like Clayoquot Sound). Facilitated by more affordable and available road, water and air transportation, outdoor and adventure tourism is arguably a growth industry for two reasons.

First, the widespread belief that environmental despoliation is the rule not the exception in the twenty-first century has drawn more people to areas perceived to be 'vestiges' or 'remnants' of a once bountiful natural world now in need of protection. Second, the lifestyles of Western individuals have become ever more urbanised and sedentary, engendering a felt need for 'escape'. Over 80 per cent of people in countries like France, Australia and Germany spend most of their time in environments of concrete, brick, tarmac, stone, glass, processed wood, paint, plastic and metal where road traffic congestion is normal and ambient noise levels are often high. Because of the 'post-industrial' character of these countries' economies, most adults have office (or otherwise indoor) jobs in which they daily interact with a computer (or other technology) in heated or air-conditioned buildings with artificial lighting. In the domestic sphere, a range of electronic devices, from televisions to washing machines to food processors, conspire to keep people indoors and spare them much physical exertion. The British nature writer Roger Deakin thus opines that 'In so far as "Western" people have forgotten how to lay a wood fire, or its fossil equivalent in coal, they have lost touch with nature' (Deakin, 2007: xii).

The people Deakin describes share the sentiment. Many of them want sensory contact with its antithesis: that is, a world designed not by people but by the (seemingly) 'self-organising' forces that have, over millennia, given rise to diverse ecosystems and landforms. As Charles Louv (2005), author of *Last child in the woods*, would have us believe, many Western city dwellers are suffering from 'nature deficit disorder'.[19] Of course, the desire to make up the deficit is not just about nature 'out there' for its own sake: it's also about how walking, climbing, skiing, swimming, sailing and camping in it makes people feel about themselves, both physically and mentally.[20]

On the global stage, a relatively select number of destinations dominate the market in ecotourism and adventure travel. They include rural New Zealand, Alaska, Costa Rica, Belize, Nepal, Kerala (in India), rural Australia, Hawaii, Ecuador and Borneo. As anthropologist Adrian Peace (2001: 175) noted,

> All tourist venues are sold on the strength of their . . . distinctiveness . . . [But] eco-tourism and adventure travel especially privilege . . . the experiences that visitors might variously derive from contact with . . . [nature].

A sanctuary of peace and beauty, the Oslo Fjord region is a joy to explore. Its flat terrain makes it ideal for walking and cycling throughout those endless summer days – no other place in Norway has more sun. The area is sparsely populated so you can enjoy the tranquillity without anybody disturbing you. You'll discover beautiful towns such as historic Fredrikstad, spectacular Moss and mystical Sarpsborg – each surrounded by charming villages, rolling fields, dramatic forests and breathtaking beaches – so idyllic and so peaceful.

visitnorway.co.uk

Plate 4.5 Immersed in the non-human?

This flyer (printed in a British newspaper) is fairly typical of the way that eco- and adventure tourist destinations are promoted to Western consumers. It implies a seemingly unproblematic correspondence between representation and that which is represented. Potential British tourists are invited to see themselves paddling, walking or cycling in a sunny, 'sparsely populated', 'tranquil' Oslo fjord region – where they can, as the borrowed quote tells us, 'get close to nature'. The promise of the flyer to its readers is clear: step through the representational window to experience, in a visceral, kinaesthetic, multisensory way, what Norway's outdoors has to offer. Reproduced with permission from *The Guardian*, 15 May 2011.

Key to these experiences is the belief among tourists that they are encountering not only the non-human world 'as it really is' but also particularly special or notable elements and sites whose distinctiveness arises from their rarity: they're taken to be wonders of Earth history rather than of human artifice (see Plate 4.5). If you've never been an ecotourist or adventure traveller yourself, imagine what it might be like to dive all day in the teeming tropical waters off Palau, to swim with dolphins in Kaikoura, or to undertake an assisted ascent of K2 in the Himalayas. You'd surely be moving beyond the glossy brochures, travel websites and the guidebooks to something tangibly natural – to something too old, large, independent and ungovernable to be socially constructed or even conditioned.

Or would you be? Towards the end of Chapter 2, in my extended discussion of representation, I argued that even what appear to be 'first-hand' experiences are profoundly affected by our cultural milieux and particular actors operating according to its conventions (or, perhaps, seeking to challenge them). I now want to make good on this claim by taking the case of Fraser Island, located 190 kilometres north of Brisbane, the Australian city that's the capital of Queensland (see Map 4.3). I'll examine what, undeniably, appears to be a (tragic) case of encountering nature in the raw. However, I will show that even when one (apparently) dispenses with the representations of nature offered by others – be they molecular geneticists, environmentalists (like Dorst and Cameron) or corporations (like MacMillan Bloedel) – the nature existing 'beyond representation' proves to be rather elusive.

The nature of Fraser Island

Some 120 kilometres long and, on average, 12 kilometres wide, Fraser Island is considered to be very special in a biophysical sense. Indeed, it achieved UNESCO World Heritage status back in 1992 within the category of 'natural heritage'.[21] It's an offshore sand island and also the world's largest vegetated one. It has been formed by silt, which has been washed into the Pacific by nearby rivers and then driven northwards by powerful ocean currents. Over the centuries, the silt has accumulated and, above the water line, has been sculpted by strong winds. The resulting complex of sand dunes rises to over 200 metres above sea level. In many places on the island, dried, hardened and compacted sand in wind-created hollows has combined with plant debris. The layering of organic and inorganic matter has created a barrier that's enabled so-called 'perched lakes' to form. These exist above the water table and are typically beautiful, comprising clear blue-green water, edged by sand or subtropical vegetation.

Aside from the lakes and the fine, white sand, Fraser Island is known for its biodiversity. In some parts, imposing tropical forest can be found, while in others there's heathland (which bursts into colour each spring), marsh and coastal mangroves. Rain-fed streams give rise to corridors of water-tolerant plants, like ferns, leading down to the shoreline. There are numerous migratory bird species and several rare insect, reptile and amphibian species too. Most famously of all, there are dingoes – one of the higher mammals Australia is known for worldwide (along with the iconic kangaroo and the koala bear). At the time of writing (spring 2012), all of the main visitor websites contained prominent images of dingoes on several pages (e.g. http://www.fraserisland.net and http://www.fraserisland.info [both accessed 10 March 2012]). Dingoes arrived in Australia some 3,500 years ago, transported from southeast Asia by seafarers. In terms of natural

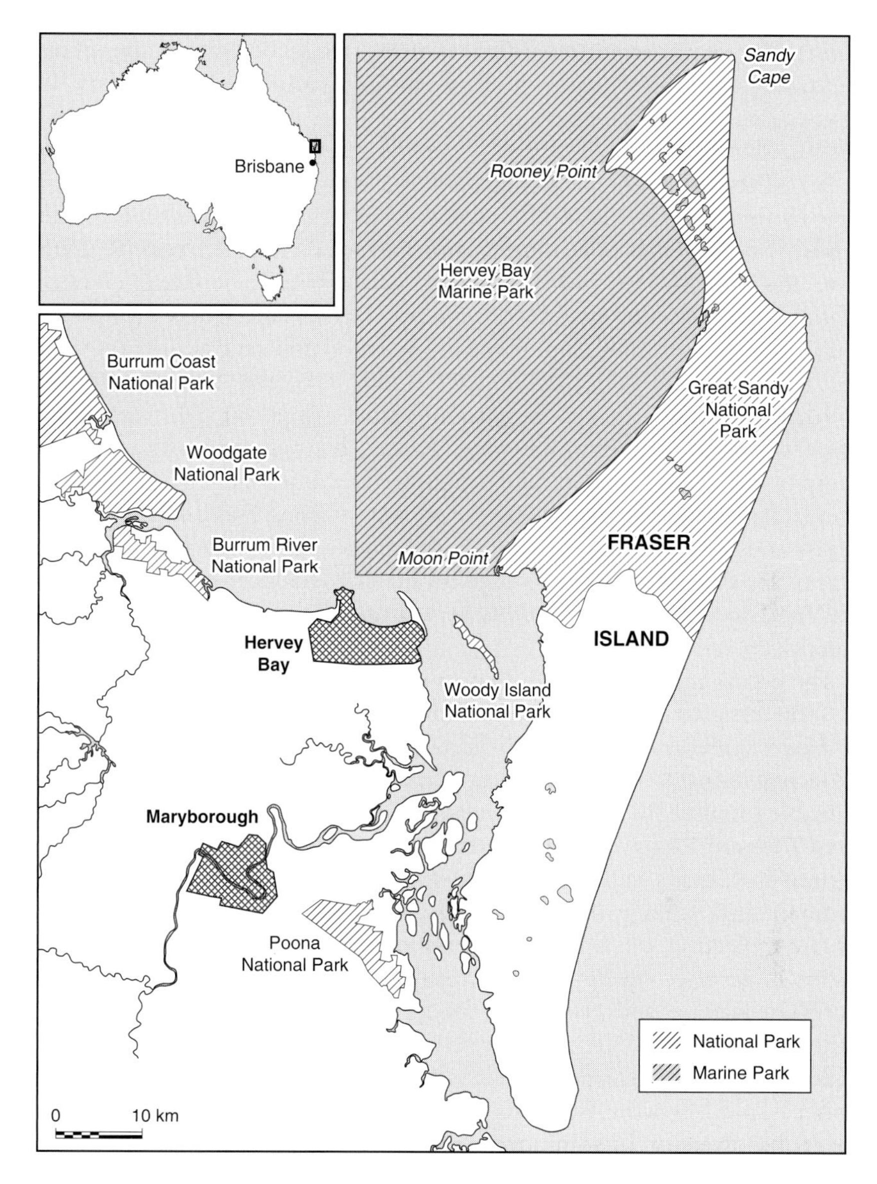

Map 4.3 Fraser Island and its region, Queensland, Australia

history, therefore, they are a relatively new species on the island, though not in terms of human history. They are triply prominent on Fraser Island. First, it's not hard for visitors to see them in the flesh. For instance, they are often spotted on beaches during daylight hours. Second, they are the *only* higher mammal to inhabit the island (apart from so-called 'brumbies', which are horses descended from working animals brought from the mainland in the late nineteenth century). This means they do not 'compete' with other large animals for visitor attention. Third, they are said to be perhaps the 'purest' remaining group of dingoes in all of Australia, attributed to the fact that many mainland dingoes have crossbred far more with domestic dogs than the island dingoes.

For all these reasons, by the late 1990s around 300,000 people each year were coming to Fraser Island for ecotourist experiences. The island was, and remains, managed by the Queensland Parks & Wildlife Service (QPWS). After the UNESCO World Heritage listing in 1992, the Service oversaw the planned expansion of facilities for visitors. Private companies were permitted to offer accommodation, food, drink, travel and recreational provision for paying tourists in designated sites. Today, eco- and adventure tourism are the major activities on the island. Visitors can go on sightseeing tours, or go whale- and dolphin-watching, fishing, hiking, and so on. They can camp or else stay in a resort, like Kingfisher Bay, built to certified ecological standards. By 2011, over 700,000 people were visiting Fraser Island for both short- and long-stay experiences. As the *Lonely planet* travel guide website says in the first sentence of its page on Fraser, 'The local Aboriginal people call the Island "K'Gari", which is very fitting as it roughly translates into paradise.'[22]

The nature that now attracts visitors in large numbers was not always seen as something to be conserved or showcased. Settlers from Europe began to harvest trees on Fraser Island from the 1860s onwards. Much later, this extractive industry was supplemented by another: sand mining (sand being a valuable resource for the construction and landscaping industries). Through the 1970s and 1980s, what happened later to the Clayoquot and Great Bear regions in Canada happened to Fraser Island: it became the focus of environmentalists seeking to protect and preserve 'natural heritage'. This was both a reflection of, and a contribution to concretising, the wider post-1950s surge of environmental concern in the West. During the era when both Greenpeace and Friends of the Earth came into existence, an NGO called Fraser Island Defenders Organization (FIDO) was founded in 1971. Through its lobbying and awareness-raising efforts, it was instrumental to Fraser Island achieving World Heritage status 21 years later. Since 1992, the island has therefore undergone a practical and presentational switch. It is (now seen as) a space of nature conservation and preservation rather than a space of resource extraction. Fortunately, logging and sand removal only affected select parts of Fraser, meaning that the QPWS has been able to focus on managing visitors so that the biophysical wonders they come to experience are not degraded by those visitors' activities.

Dingoes, danger and death

During the 1990s, there were somewhere between 180 and 220 dingoes resident on Fraser Island. The QPWS actively managed visitor behaviour in order to ensure that they were left in as wild a state as possible. This was considered good for the dingoes: it obliged them to locate their own food sources and it ensured that their social habits were not altered by contact with humans. It was also considered good for people: by encountering dingoes at a distance, Fraser Island visitors would be spared any unwanted experiences, such as the animals coming into the camp sites to steal their food. As Peace shows in his analysis of the QPWS approach to nature management in the 1990s, this separation between people and dingoes mirrored the wider separation between visitors and Fraser Island's natural environment (Peace, 2001). A battery of maps, signs, pamphlets and leaflets, in addition to physical barriers (like wood fences and marked trails), were (and still are) used to demarcate those physical spaces where people can eat, drink, sleep, walk, fish, drive, etc. from those where 'nature' prevails and requires protection from unregulated uses. Fraser Island's nature was (is), therefore, to be encountered selectively up close or else seen at a distance.

However, in the late 1990s a number of dingoes departed from their 'natural behaviour' and started to see humans as potential prey – thus crossing the 'divide' maintained between visitors and environment. First, in early 1998 a family camping on the island had their 14-month-old daughter, Kasey Rowles, snatched by a dingo from near their tent. She sustained shoulder injuries. Second, in the next three years a number of adults were attacked in broad daylight while walking or sunbathing near the ocean. Third, and most shocking of all, in 2001 a 9-year-old Brisbane boy, Clinton Gage, was killed by dingoes while playing on a beach with his younger brother. Realising they were being followed by two of the animals, the siblings started to run back to their resort but both were set upon before they could reach safety (with the younger brother suffering numerous bites and bruises, though surviving). Clinton Gage's death was the first confirmed human fatality caused by a dingo in modern Australian history.[23] Three days later, two British backpackers were also attacked by two dingoes. The QPWS response was swift and decisive: 31 dingoes were culled; visitor information guidelines were altered to stress the real threat posed by the animals; and information on how people could scare dingoes off was disseminated. Also, existing information about the fines imposed if people fed dingoes was re-emphasised and 'hard measures' relating to storing food and disposing of food waste were ramped up.

As one prominent Fraser Island visitor website explained in 2004,

> You are not allowed to feed the dingoes as this upsets the natural balance of the ecosystem where they should forage for their own food and the animals then tend to hang around campsites waiting for handouts. There have been several attacks reported over the years, and recently a tragic attack, but we have always respected the fact that though they may look friendly, they are still wild and you should keep your distance.
>
> (http://www.boxatrix.com/fraser/dingo.htm, accessed 5 July 2011)

Plate 4.6 Fraser Island's *wild* dingoes?

Attacks on people by Fraser Island dingoes in the late 1990s were repeatedly represented to Australians in ways that highlighted the animals' wildness and difference from domesticated animals.

Clinton Gage and the others bitten by dingoes before and since his death appear to have had an unwanted, and very real, experience of nature. They have encountered, it seems, something besides those parts of nature selected by the QPWS and private operators on Fraser Island. In this case, however, nature was acting 'erratically': it was a case of either a few 'rogue dingoes' or, as the website just mentioned intimates, dingoes losing their fear of humans because they were being fed by them (despite the QPWS regulations). Either way, it was dingoes' 'otherness' – their singular difference and separateness from us – that was at issue (Plate 4.6).

However, Stephen Healy's (2007) anthropological research suggests an alternative interpretation. He argues that the idea that Fraser Island contains 'wild dingoes' is something of a contrivance. It is a fiction consistent with both the environmentalist values of the FIDO, the demands of the ecotourist industry and the management practices of the QPWS. In fact, prior to the recent period, dingoes *routinely interacted* with Fraser Island's human inhabitants, albeit cautiously, and without incident. While mainland Australian dingoes were typically hunted by settlers in the past (and are still seen as pests in many rural areas), things were different on the island, largely because there were no livestock raised on farms to tempt dingoes and thus no reason for people to regard them as troublesome predators. From the 1860s, loggers began to feed dingoes food scraps (thus supplementing the animals' diet) and, as the island became more settled, twentieth-century dingoes were allowed to scavenge from public waste dumps. When logging ceased, dingoes were also allowed to hunt young, old or infirm 'brumbies' (the aforementioned imported workhorses).

The ecological management regime introduced by the QPWS changed this in the act of making Fraser Island a safe and attractive tourist destination. From the late 1990s onwards, open landfill sites were closed, fish offal had to be buried, food waste was sent for compaction to the mainland, many brumbies were removed and feeding dingoes was prohibited. As Healy points out, these measures ultimately *forced* dingoes to become what they were presented as being 'naturally', in other words 'wild' and not to be interacted with. This manufactured wildness, as several long-standing Fraser Island residents pointed out in criticisms of the QPWS management approach, placed dingoes in a very difficult situation. Starved of previous food sources, many of them had to start taking risks and attacked humans as a matter of expediency or necessity. Their behaviour, despite ecotourist rhetoric, was neither consistent with, nor a deviation from, their supposedly wild character. It was circumstantial.

IN WHAT SENSE IS 'NATURE' A 'CONSTRUCTION'?

This chapter has been a very long one, at least if you've examined all three bodies of case material presented. I'll therefore keep this penultimate section as short as I possibly can. To take stock of the lessons we can reasonably draw from the Clayoquot, new genetics and Fraser Island cases, a more philosophical discussion is now in order.

In the previous pages, I've shown in some detail that what appears to be 'natural' is indissociable from what's normally understood to be its antithesis. Disputes over old growth trees, scientific research into human genes and even 'direct' encounters with wild animals all, I have demonstrated, make reference to an asocial nature that's belied by close scrutiny of the representational practices involved. I use the term 'representational practices' because I've sought to emphasise that representations are *performative*. Even as they often appear to refer to a natural world separate from themselves, they conjure up that separateness as a rhetorical means of securing their own validity. This conjuring, allied with the specific epistemic content of any given representational act, comprises the 'force' of any representation of nature, though it's an empirical question just what specific effects follow from it.

Am I thereby arguing that representations of nature make the world in their own image? No, I'm not. To insist, as I have done, that these representations are purposeful constructions (things made), arising from equally purposeful engagements with what we call nature, is not to say that this 'nature' is constructed by us without ontological remainder. This sort of 'radical constructivism', as it has sometimes been called, has (understandably) attracted the ire of critics. It appears to imply two things that the critics regard as untenable and objectionable. First, the physicality and material agency of the non-social elements of the world seem to be discounted: these elements become a *tabula rasa* upon which societies freely inscribe their hopes, wishes and desires. Second, the door to cognitive, moral or aesthetic relativism seems to be opened: any and all representations are (apparently) as good as the next ones because the physical world's capacity to act as a 'court of appeal' – adjudicating on true and false, better and worse representations of nature – is denied.

To simplify a debate that was far more complicated than I can recount, these criticisms were voiced loudly by physical scientists and environmentalists at the turn of the millennium. For instance, in the well-known hoax he played on the cultural studies journal *Social Text*, the American physicist Sokal (1996b: 62) mocked 'those who believe that the laws of physics are mere social conventions' by inviting them to 'try transgressing those conventions from the windows of my [twenty-first floor] apartment'.[24] A year earlier, the edited book *Reinventing nature? Responses to postmodern deconstruction* was published (Soule and Lease, 1995), in which several authors took issue with the idea that environmental ethics and conservation might lack any firm anchor in the realities of the non-human world. How could nature be protected and restored, they asked, if we cannot separate facts about its current condition from falsehoods and fantasies?

Phrased like this, the criticisms created a stand off between what Kate Soper, in her germinal book *What is nature?*, called 'nature endorsing' and 'nature sceptical' positions:

> while the one party invokes nature in reference to features of ourselves and the world that are discourse independent, the other responds by querying the supposed signifier of the signified – a stance it supports by pointing to the multiple constructions placed upon 'nature' at different historical periods and in different cultural contexts . . .
>
> (Soper, 1995: 4)

Soper aimed for a *rapprochement* between the two positions, recognising that each had to imply the other for fear of being self-defeating. Thus, she argued, 'nature' could not possibly be 'constructed' if societies did not have material things, such as animals, DNA and trees, to be objects of representation and material modification in the first place. Conversely, she pointed out, for people in different times and places these things are never knowable 'in themselves', either in a purely intellectual or even kinaesthetic sense. Instead, they're always apprehended according to specific social assumptions, goals, values and desires. To suggest otherwise would make *us* the *tabula rasa* on which those things we call 'natural' inscribe *their* will. Soper's 'both/and' argument can thus be summarised as follows: what we call nature *exceeds* whatever constructions we place upon it in an epistemological or a practical sense, but we can never grasp nature's 'autonomy' or 'otherness' in a way that escapes the reach of these self-same constructions.

In this light, the three cases presented in this chapter are not attempts to deny that what we call 'forests', 'genes' and 'dingoes' possess attributes irreducible to the way they're represented by various epistemic workers. My point is simply that these workers may present particular attributes to us as if these are the main or only ones, as if they are 'objectively', 'really' or 'essentially' definitional. In the process, they make material connections – and invite *us* to make our own mental, emotional and practical connections (or, in the case of dingoes, *dis*connections) – with the physical things referred to. As I argued in Chapter 2, representation is a means by which those doing the representing aim to create particular linkages (imaginative and/or practical) between people, and between people and the world of animate and inanimate phenomena. Representations seek to engender specific modes of thought, feeling and action – more or less successfully, more or less consciously. They're not immaterial and nor are they passive in relation to the things they speak of and speak for.

I think, therefore, that (properly understood) 'construction' *is* a fitting term to describe representation because representations are both fabricated and, in turn, may shape the world – they are constructs that can themselves construct. But representation is never *sui generis*, never something fashioned out of whole cloth: there are ontological limits to the process of construction. Can these limits be specified objectively? Only if we accept that 'objectivity' is (despite the word's conventional meaning) simply one of several epistemic characteristics that we've attributed to certain forms of knowledge. All of the 'limits' that (what we call) 'nature' sets on the content and influence of our representations are not absolute, but *relative to* the social goals intrinsic to those representations (in their diverse modes and genres). This is what cultural critic Katherine Hayles (1991) meant when she coined the useful term 'constrained constructivism'.

For these reasons (and Hayles's contribution notwithstanding), some analysts have eschewed the language of 'construction' altogether when seeking to denaturalise that which is presented to us as natural. While they endorse the attention it gives to our multifarious human purposes, discourses and practices, they consider it to be too 'muscular', too inattentive to the constitutive role that 'non-social' entities play in making possible even the most technologically advanced societies. For them, this is not a question of the relative balance of influence between 'social' and 'natural' phenomena. In place of this 'two-worlds' ontology (and its associated language of 'cause and effect', 'in/dependent variables', 'interaction', 'domination' and so on) the starting point for these analysts is that the world is composed not of things (which can be neatly compartmentalised into categories like 'social' and 'natural', 'urban' and 'rural', etc.) but of myriad *relations*. These relations are more-or-less seamless, more-or-less permanent, more-or-less intricate and more-or-less stretched out in space and time (see Box 4.1).

BOX 4.1

'POST-CONSTRUCTIONIST' APPROACHES TO UNDERSTANDING NATURE

For several philosophers, theorists and empirical researchers, 'social constructionist' approaches to understanding 'nature' are as problematic as the realist–naturalist approaches opposed to them. Both approaches take for granted a social–natural distinction that, it's argued, *itself needs explaining*. What's more, both approaches are ultimately contradictory. For instance, 'constructionists' seem to accord 'society' a power and existence being denied to 'nature'. Yet, implausibly, this would mean societies are *sui generis*. What's more, even supposing that there *is* such a thing as 'society' (being related to but also ontologically distinct from the realm of 'nature'), critics argue that it's never 'society as a whole' that 'constructs' nature in any given instance. Instead, it is a set of specific actors, institutions or communities whose activities are governed by certain relationships, norms and goals. As an alternative, STS scholar Bruno Latour (1993) has famously talked of 'actor-networks', 'quasi-objects' and 'intermediaries' in his critical account of how post-Enlightenment Western societies have sought to distinguish the indistinguishable in their analytical habits of thought. For him, the categories of 'society' and 'nature', along with the other antonyms listed in Figure 1.5, are symptoms of epistemic 'purification'. They're not two different 'ontological regions' with a range of unique causal powers but instead two categories whose formation invites a contextual and historical explanation rather than mute acceptance. Relatedly, in his latest book *Being alive*, the British anthropologist Tim Ingold sees the world as a complex 'meshwork'. For him, all organisms, including us, should be understood 'not as bounded entities surrounded by an environment, but as an unbounded entanglement of lines in fluid space[-time]' (Ingold, 2011: 64). This ontology of connectivity, flows and imbrications has also been recommended by Donna Haraway (2003: 6), to cite one more key contributor, who has observed (in a typically memorable formulation) that 'Beings do not pre-exist their relatings … The world is a knot in motion' (2003: 6).

The upshot of these arguments is that analysts need to look closely at (1) the process of boundary making, marking and maintenance between the 'natural' and the 'social'; (2) how boundaries are constantly transgressed in practice; and (3) what gets lost to thought, ethics and practice when the transgressions are ignored (see Chapter 5 for more on these subjects). These ideas have found an echo in Marxist writings about nature (e.g. see Harvey, 1996: chapter 1). However, as with many advocates of **social constructionism**, these Marxists want to hold on to a conception that human action is organised in distinctive ways across space and through time, in which power inequalities play a constitutive role in structuring thought and action. 'Capitalist natures', the Marxists argue, are *hybrids* of political–economic rationalities and actions and the material properties of human bodies and non-human entities. These material arrangements and alterations of 'nature', Marxists argue, are embodiments of human intentionality and action that reflect specific social relations of class inequality. The materials enrolled, however, are not reducible to these social relations but are a means by which those relations are given form and sometimes challenged. Nigel Clark (2011) has argued that neither the Marxists nor the likes of Latour, Ingold or Haraway have yet appreciated the asymmetrical nature of many of our worldly entanglements. He comes close to a new naturalism that posits 'nature' as something separate from and far more powerful than humanity.

I'm very sympathetic to these and other attempts to alter the very basis upon which we understand our place in the world, but perhaps the differences between 'constructionist' and 'post-constructionist' approaches to understanding what we call nature (and its collateral terms) have been somewhat overdrawn. In his book *The social construction of what?*, philosopher Ian Hacking helpfully identifies different meanings of the term 'construction' in recent social science debates, criticising those that seem far fetched. He suggests we might usefully 'retain one element of its literal meaning, [namely] that of building or assembling from parts' (Hacking, 1999: 49). Just as a house is a human construction, something made intentionally according to a design, it's also constructed out of non-human things (bricks, mortar, etc.) without which the design would be mere wishful thinking. Similarly, representations of nature, be they verbal or visual, on television or in a book, are the products of our values and goals (notwithstanding that this isn't always advertised or obvious). To have any influence on us or on what they refer to, at least some of these representations must have a tangible relationship with a biophysical world that at some level exists, regardless of them – a world that knows nothing of the values and goals according to which we discuss, respond to and intervene in it.[25] For us, the world of 'nature' has agency and efficacy not 'in itself' but rather in relation to the diverse ways we think about and act towards it. As I've shown in this chapter, that relationality is not always made plain in the representations that arise from and affect it. Here I agree entirely with the likes of Latour, Ingold and Haraway (see Box 4.1): if the world's relational character *obliged us* to represent it thus, there'd be no need to question representations that cleave it into bits and pieces. These writers would be out of business, as indeed would I.

SUMMARY

In this chapter, I've shown that what we call 'nature' is shot through with the interests, aims and presuppositions of what we usually take to be its opposites. This means that claims that nature is *not* socially constituted, in significant measure, are part of the war of persuasion (if that's not too dramatic a metaphor) that *Making sense of nature* is itself explicitly participating in. Representing aspects of the world as 'natural' is a process involving myriad epistemic communities operating within and between the entire range of communicative genres known to us. The representations refer to biophysical phenomena large and small, fascinating and repulsive, cuddly and threatening. These phenomena are understood to be more-or-less 'natural' in one or more of the four principal senses identified in Chapter 1. This quartet of meanings can be communicated via one or more of nature's many collateral concepts, such as 'race', even as these concepts partake of other meanings too. The four meanings can also be evoked when any concrete instance of 'nature', 'genes', 'wildlife', etc. is highlighted because each instance is just one of many to which the meanings are conventionally attached. These meanings can be taken by their recipients at face value or with a high degree of awareness and reflexivity. Some modes and genres of representation actively encourage the latter, such as those *Making sense of nature* both instantiates and is itself an example of. I'm an epistemic worker whose discourse, fashioned from the analyses of a very particular epistemic community and concretised in these pages, is intended to make you think differently to the way other discourses about nature encourage you to think.

Clearly, I'm trying to get us away from asking questions about the 'true nature of nature' – its real character or location in space and time. Is Clayoquot Sound a rare and unique ecosystem? Is there a single human genome that's distinct from other genomes and do non-trivial group-level differences in the genome exist? Are Fraser Island dingoes wild animals and were those that killed Clinton Gage departing from their 'natural behaviour'? To my mind there's no answer to these questions that can use 'nature' as an asocial reference point, court of appeal or test bed. It's better, I would argue, to focus our attention elsewhere. Try as we might, we cannot respond to questions like these without significant reference to ourselves: why do we ask such questions in the first place, and what purposes might our answers be intended to serve?

Though I didn't announce it earlier in the book (it's been somewhat implicit thus far), I take some inspiration from the philosophical movement known as **pragmatism** – a movement that connects founding figures like John Dewey with more recent influential writers like Richard Rorty.[26] Pragmatism's central insight is easily stated, even though many have been reluctant to accept its implications: there is, it maintains, no fool-proof way, no absolute standard, that will allow us to determine whether any given representation, discourse, idea, proposition or piece of information is 'true' or 'false', 'right' or 'wrong', 'fact' or 'fiction'. Instead, these scare-quoted terms are ones we attribute to various epistemological products, *depending on their intended and actual effects in the world.*

Thus, a pragmatist would say that what we call scientific knowledge is deemed 'truthful' and 'objective' not because it actually is but because it precipitates actions, events and outcomes that many people consider to be beneficial or useful. Equally, a pragmatist might argue that nature poetry is not 'useless' by virtue of its intrinsic fictionality but because (1) nature poets are unable, for a variety of reasons, to capture the attention and affect the conduct of more than a minority of any national population; and (2) nature poetry is 'useful' in ways radically different from conventional understandings of utility (as, for example, associated with a washing machine or a pair of shoes).

So pragmatists understand all language, symbolism, denotation and the like in *practical* terms: to what extent are they able to alter human thought and action? In what ways, and to what extent, do they produce outcomes that satisfy people's 'needs' and 'wants' (as currently defined). What, in the end, do they *do*? In respect of this book's subject matter, pragmatists ask not what 'nature', 'sex', 'biology', etc. *are* but rather what follows for human in/action from understanding the world in these terms. Pragmatism is usually regarded as anti-representational and so, in one sense, it is. As Rorty once put it, 'Language cannot fail to represent accurately, for it never represents at all' (Rorty, 1999: 50). Rorty (1979) was famously critical of the idea that some knowledges possess the 'special' quality of 'corresponding' point-for-point with nature or 'revealing' timeless and elemental truths; however, he recognised that this claim to epistemic realism was, despite his criticism of it, highly efficacious in the modern world: millions of people happen to believe it, in part because things done in its name 'work' in a practical sense. He also, I would suggest, endorsed the less conventional understanding of 'representation' I argued for in Chapter 2 and reiterated in the previous section of this chapter.

From a pragmatist perspective, readers should not judge this book in terms of whether its contentions and claims are ultimately 'correct'. Instead, mine is an attempt to affect you and thereby the wider world, using all the conventions of academic writing to my advantage (a non-polemic prose style, citations, a large bibliography, etc.) and the high social status that this genre of communication still enjoys. Despite this high status, *Making sense of nature* has its work cut out because it's challenging powerful habits of thought and action. As with so much social science discourse, my own, and that of the multi-disciplinary epistemic community to which I'm affiliating myself, does not have a large or ready audience. It thus participates at the margins in the drama of nature's representation and uses that it aims to describe and explain. This illustrates why pragmatists don't worry about 'relativism'. Critics of relativism fear that it treats all perspectives and practices equally (i.e. as of equivalent worth), including iconoclastic, subversive, extreme, hurtful, demotic or objectionable ones. The pragmatist response is to say that the 'equality' is purely theoretical. In practice, only certain ways of talking about and acting in the world exert a meaningful influence, including those relating to 'nature' that I'm seeking to denaturalise in this book. To understand why, we need to comprehend how power relations and the clout of certain large institutions create inequalities in the diet of 'choices' presented to people as they consume various representations, ideas, images, values, etc. in their daily lives. Challenging these power relations is central to achieving a robust 'semiotic democracy' as described at the end of Chapter 3. I focus on the nature and effects of social power in Chapter 6, but touch upon it constantly in all the other chapters to come.

ENDNOTES

1 I borrow the scare-quoted phrase from George Lakoff and Mark Johnson's celebrated book *Metaphors we live by* (1980).
2 All the quotations from *Beyond the cut* come from Braun (2002: 36–41) and Willems-Braun (1997: 8–10).
3 This brief analysis of *Clayoquot: on the wild side* is taken from Braun (2002: 73–86).
4 Lily Kay (1996) presents a readable and authoritative history of molecular biology's early growth, focussing on the United States, where it enjoyed generous funding for many decades (and still does).
5 I introduce the scare-quotes here for two reasons: first, the two versions of the human genome were, in fact, quite divergent along many sequences; second, even if this had not been the case, geneticists have continued to 'correct' and refine the supposedly 'final' versions of the genome map. Celera Genomics withdrew from the 'race' to map the human genome after their draft was completed.
6 'Biologism' goes beyond the incitement of people to think of their minds and bodies in a 'molecular genetic' way. It also includes discourses about diet, exercise, sleep, alcohol consumption, disease prevention, parenting and the like which, together, place a major emphasis on the management of one's own corporeal existence and/or that of one's children.
7 Most human cells contain two chromosomes (except for those involved in biological reproduction). These are a union of DNA and protein, and the so-called X and Y chromosomes have a somewhat different character insofar as they regulate sex differences in human embryos.
8 Indeed, lay audiences may be surprised to learn that there is, in fact, no scientific consensus on what something as apparently 'solid', indeed foundational, as a gene is (see: Baetu, 2011; Falk's (2010) superb summary of the variety of understandings extant; and Stotz *et al.*'s (2004) Representing Genes Project, which surveys' scientists' own diverse understandings of the nature and functioning of 'genes'). What's more, molecular biology is now often said to be in a 'post-genomic' phase. The emphasis has shifted from describing genomes and identifying genes (or gene sequences) to examining how they function within the wider, complex biological entities they ostensibly create. As part of this, genes themselves are understood to change somewhat during the lifetime of an organism (the study of which is called 'epigenomics'). Post-genomic biology's roots actually go back a long way, to Francis Crick's statement of the 'central dogma' of molecular biology in 1958 (Crick, 1970). Crick, co-discoverer of DNA in 1953, suggested that it may be false to suppose that genomes and their constituent genes produce one-way, linear effects on the organisms whose structure and functioning they contain the 'programme' for. Crick was alerting molecular biologists early on to the high possibility of complex feedbacks and permeable boundaries between

microbiological and macrobiological components and processes. This recalled the holistic approaches adopted in much of nineteenth-century plant, animal and human biology. Conceptually, the central dogma was atomist and reductionist, assuming not only that organisms could be decomposed into discrete structural and functional units, but also that the microbiological components built macrobiological ones from the bottom up. Conceptual alternatives to the central dogma research practices focussed on whole biological systems and subsystems. Processes, pathways, feedbacks and the like are the focus here. Paul Griffths and Karola Stotz (2006) explain what the 'post-genomic' understanding of human genes entails. Lenny Moss (2003), in his book *What genes can't do*, maps out a similar analytical terrain. For a primer on molecular genetics, see *Introducing genetics* (Jones and van Loon, 2000).

9 There are three others that have been used widely and repeatedly over the past 15 years. The first is a *textual* metaphor, in which genes are referred to as 'the book of life' whose contents need to be translated before they can be read. For example, in his well-known work of science communication, the British geneticist Steve Jones (1993) opted for the title *The language of the genes*, implying that human biology can be analogised to a discourse. Former HGP Director Francis Collins, likewise, calls his book on personalised medicine *The language of life* (2010), while Walter Bodmer and Robin McKie preferred *The book of man* (1997). The second metaphor is *computational*: genes are referred to as containing (or being) instructional code that programmes the organism. The third metaphor is *architectural*: genes are referred to as containing (or being) a 'blueprint' for the design of a whole organism rather as an architect who designs a building in their mind (and on paper) before it can be erected in reality. Clearly, the second and third metaphors are very closely related.

10 A fine analysis of the use of metaphor in representing nature is Brendan Larson's (2011) *Metaphors for environmental sustainability: redefining our relationship with nature.*

11 This is precisely the idea I myself have traded on in using the several maps in this chapter: the earlier ones of global forest cover and of British Columbia, and the later map of Fraser Island.

12 Most of the human genome, in keeping with all life forms, is metaphorically 'junk' DNA. This means that only a relatively small portion of it 'codes' for organism development.

13 As historian Mary Poovey (1998: xi) said over a decade ago, 'numbers constitute something like the last frontier of representation' for critics like her (and me). Crump (1990) is among several anthropologists who examine numbering as a cultural act. Historians of science and technology, and their relation to statecraft, have looked at the circumstances in which numbers were attached to new 'real world phenomena' and used in new ways. In a more abstract register, philosophers of mathematics have, in different ways, sought to denaturalise numbers, numbering and various complex mathematical procedures (like calculus).

14 What molecular biologists across the world know simply as 'Anderson' was an early reference genome for mitochondrial (rather than nuclear) DNA. In a fine study, science sociologist Amade M'charek (2009) explores some of the complexities typically backgrounded in its use as a research tool – complexities attaching to its manufacture as an object of knowledge.

15 'Most of genetics is no more than a search for diversity,' observed Steve Jones (1993: 21) in his earlier mentioned work of popular science *The language of the genes.* This is illustrated by the high-profile, but ultimately failed, project to map human genome diversity led by the famous population geneticist Luca Cavalli-Sforza. The founding assumption of this project is that while *homo sapiens* were, presumably, originally more genetically uniform than today, long-run processes of genetic recombination and mutation have altered our genetic make up as our long-dead ancestors slowly migrated from east Africa. Population geneticists like Cavalli-Sforza had long maintained that biological differences existed between groups of humans, but the new genetic technologies post-1970 gave him and fellow travellers powerful new tools with which to advance their research agenda. Though the Human Genome Diversity

Project led by Cavalli-Sforza stalled, other projects focussed on group-level genetic differences proceeded. Among them was the work of the International HapMap Consortium and also of a US National Institutes of Health-sponsored project (inaugurated in the new millennium) looking into patterns of disease among different groups in the American population. Also based in the United States was the Genographic Project (2005–10) led by the National Geographic Society. Relatedly, a project at the predominantly black Howard University in the United States also sought to assemble genetic data on black Americans for health-related research and applications. It's sometimes thought that these initiatives marked a break in the analytical focus on post-1945 human biology and medicine; however, as STS scholar Jenny Reardon (2004) shows, scientists in both fields held on to the idea of group-level biological differences through the decades when it was thought that science had no truck with the 'race' idea. This challenges the belief – articulated by, among others, anti-racist theorist Paul Gilroy – that the science of human genetics provides an objective basis on which to *oppose* racism because it has long treated *homo sapiens* as one indivisible species.

16 For instance, are there identifiable and remediable causes for such things as large ears, small stature and baldness, and should such visible forms of human difference be considered more or less 'undesirable'? All humans are born biologically unique, except for identical twins. But our genetic commonality is not synonymous with genetic *uniformity*: the estimated 0.1 per cent of variation in the human genome is associated, in poorly understood ways, with a range of phenotypical differences between people. Most readers will no doubt be familiar with the idea that there are, or at least might be, 'genes for' a number of perceptible differences between individuals. Certainly, the Western news media has been happy to report on the latest findings in the fields of behavioural, medical and population genetics. While (one hopes) some of these seem rather absurd to most people (like the idea of 'criminal genes', which we might dismiss as the product of fanciful journalism), others appear to be less controversial. Perhaps the classic case is PKU (phenylketonuria), one of the first 'genetic diseases' to be identified and, as one scientist puts it, a seemingly 'uncomplicated example of an inborn error of metabolism in which a single gene carries a mutation conferring dysfunctionality . . . manifested by a fairly uniform phenotype' (Rosoff, 2010: 215). The phenotype is mental retardation, learning disabilities and seizures. PKU and other ailments that appear to have a link to genetic 'abnormalities' (like haemophilia) have encouraged geneticists, using the latest technologies, to search for a wider family of possible links between genotypical and phenotypical signatures. Perhaps, some speculate, there are genes that (alone or together) dispose certain people towards things like aggressive behaviour, fearfulness and timidity, homosexuality, psychopathy or high intelligence (to name but a few).

17 The worry is that the new genetics of 'race' may underpin a new form of 'soft scientific racism', based on the idea of the monogenesis of humans, as opposed to the 'hard scientific racism' of the eighteenth and nineteenth centuries when many scientists argued that there was more than one 'race of man' (the idea of polygenesis). In other words, this would be a racism predicated on the idea of biological *degrees* of difference not *kinds* of difference.

18 I should say here that these scientists and science writers are not insistent that 'race' be the *only* word that should be used to describe group-level genetic differences. 'Population' has been used by human geneticists for years to characterise macro-level genetic variation among large numbers of people. Where contemporary scientists do use the term 'race', they are typically very careful to ensure it is seen as a *descriptive* term, not an *evaluative* one.

19 This is the burden of Bill McKibben's bestseller *The age of missing information* (1992/2006). The book details McKibben's immersion in a full day of modern television as compared with a day atop a mountain in the Adirondacks. He subjects his mass-mediated experience of the real on television to sharp criticism and sings the praises of direct contact with the non-human world.

20 Until recently at least, quite a bit of the academic research into tourism, both 'cultural' and 'environmental', inquired into how 'authentic' it really is for the tourists involved. This is linked to a concern that the 'commodification' of tourist destinations is transforming them into something quite different to what they appear to be. For a summary and critique of the way the ideas of 'authenticity' and 'commodification' have been defined and used in tourism research, see Shepherd's (2002) essay.
21 It's one of a very small number of Australian UNESCO World Heritage sites. Others include the Great Barrier Reef and Uluru (formerly known as Ayers Rock).
22 http://www.lonelyplanet.com/australia/queensland/fraser-island, accessed 30 June 2012.
23 For this reason, it reopened a national debate about the highly controversial case of baby Azaria Chamberlain whose parents claimed she was snatched by a dingo near Uluru (then Ayers Rock) in 1980. Lindy, Azaria's mother, had been imprisoned for murder/manslaughter, even though she and her husband remained adamant that their daughter was a much loved and cherished baby. Now divorced, the Chamberlains were invited by the media to pass comment on the Gage tragedy.
24 Sokal's hoax and book *Fashionable nonsense: postmodern intellectuals' abuse of science*, co-authored with Jean Bricment (1998), were a key piece of artillery in the so-called 'science wars' that raged in the United States in the late 1990s. I referred briefly to these wars in Chapter 3.
25 Tim Mitchell (2002: 45) makes this point nicely in a critique of Karl Marx's well-known metaphor of the bee and the architect. Even architects, he reminds us, do not work *purely* in the realms of the imagination, numbers or discourse before their designs are realised in practice. Prior to drawing and planning, Mitchell notes, all architects must examine the planned construction site and consider the metals, plastics and other materials that will 'work' if their design is to be practicable. Nigel Clark, in his wonderful book *Inhuman nature* (2001), takes this further: the non-human world, he argues, is often far too powerful to be altered according to a plan or design. Vast domains of nature, he suggests, are literally beyond our power to 'construct', even in a weak sense of this term. Clark thus argues that the metaphors of 'construction' and 'hybridity', which have in their different ways dominated social science thinking about 'nature' for 25 years, only really make sense when applied to a *small aspect* of what we call 'nature'. The rest is simply beyond our reach, ours neither to make nor to 'hook up' with.
26 Rorty is now deceased, while other leading 'neo-pragmatists' (as Rorty's generation have been called) are now well into retirement. There are many good book introductions to philosophical pragmatism. My own favourite is Robert Talisse and Scott Aikin's (2008) *Pragmatism: a guide for the perplexed*.

5 ENCLOSING NATURE: BORDERS, BOUNDARIES AND TRANSGRESSIONS

In Part 1, I said more than once that what counts as 'nature' is not given in nature, just as the particular meanings we attach to things so categorised aren't natural either. The previous chapter evidenced these arguments with reference to three extended examples, which together covered a range of ways in which we've come to know nature and its collateral referents. In this chapter, I want to focus on something I left rather implicit in the previous one: namely, the process of deciding where to draw the lines between phenomena considered to be natural, and those that aren't. I'll examine how semantic divides are enforced and sometimes challenged. Even though the organising distinctions of Western thought listed in Figure 1.5 have been with us for many generations, we're constantly confronted with new situations in which we must determine when, where and how to best utilise them. By questioning the ontological solidity of these distinctions, and their usual contexts of use, we come to see that we are bound by them only to the extent that we collectively allow ourselves to be. We see too the role that various epistemic workers play in acts of mental and practical 'border enforcement', even as others intentionally try to remove, reposition or render more permeable the dividing line between 'nature' and 'not nature'.

I've already offered a simple, but I think significant, insight into the work of representational exclusion that some epistemic communities perform. In Chapter 3, I took the example of 'the science wars' and suggested that the defenders of the citadel sought to sharply demarcate their practices from those of their erstwhile critics (STS scholars). In other words, as a means of establishing their uniqueness, identity or authority, some epistemic communities – when pushed to do so – will seek to outlaw (or ridicule) 'impostors', rivals or dissenters. In this chapter, I want to focus on borders in a different (though not unrelated) way. I'm interested in how epistemic communities, as part of their ordinary practice, seek to determine where the boundary of the natural lies. How, and with what implications, are the edges of what counts as nature defined? What divisions within the realm of nature are sanctioned? Why, and to what ends, might these boundaries be moved? These are my principal questions. Unlike the previous chapter, where I examined how meaning gets actively assigned to 'natural phenomena', the pages to come look at how and why meaning gets *circumscribed*. That is, I focus on the *limits* of assignation and processes of semantic *compartmentalisation*. As analysts, the 'denaturalising' approach advocated in this book applies not only to what putatively natural entities are made to signify, but also to their demarcation from all those things listed on the right side of Figure 1.5 (their 'constitutive outsides'). To know nature, people routinely stake claims about what is *not* nature, and this has often profound implications for us and for those things we attach meaning to.[1]

BORDERS AND BOUNDARIES

What is a 'border'? Who creates and enforces borders, and why? These questions are normally thought to be the preserve of political scientists and political geographers. According to political scientist Mark Salter, 'The border is a primary institution of the contemporary state' (2011: 66). Borders define the geographical limits of sovereign nations. They enclose parcels of territory over which national governments have political and legal jurisdiction. They also mark the symbolic and material limits of state power (hence the well-known distinction between 'home affairs' and 'foreign affairs'). If a national government 'interferes' in the domestic matters of another sovereign state it may well be seen to have 'overstepped the mark'. This is most obvious in cases of a military attack or invasion, but can apply in less dramatic circumstances too. Similarly, at the sub-national level, where it's more common to talk of 'internal boundaries' rather than borders, there's usually a clear link between the partitioning of domestic territory and the delimitation of political authority and responsibility. For example, local and regional governments will raise taxes from 'their' populations and have a range of powers and duties they can (or must) discharge. It's not normally the business of other sub-national authorities to exceed their geographical remit, unless invited to do so. Political borders and boundaries can be 'hard' ones, but need not be. The intensely militarised and surveilled United States–Mexico and Israel–Palestine borders are (in)famous examples of the former, but many others are highly permeable (albeit selectively so). For example, cross-border travel within the European Union is relatively easy for its citizens, while the internal political boundaries of nation states are routinely crossed when people journey to work, make business trips or take a domestic holiday. Even so, permeability does not mean that the borders or boundaries in question don't matter. Instead, it means that while certain persons and things are allowed to cross political dividing lines with few restrictions, others (for instance, political refugees and contraband goods) encounter real barriers to movement, which cannot be ignored.

Important as political borders and boundaries are, they scarcely exhaust the subject of why, how and with what effects people segment the world in both thought and practice. Nor is this segmentation the sole preserve of politicians and bureaucrats. Geographer Reece Jones (2009) is one of several to argue that the study of borders and boundaries should be a multi-disciplinary affair that goes far beyond a focus on governments and their territorial organisation. This expanded field, he argues, can and should look for the interplay between ostensibly different arenas in which dividing lines are drawn and come to have material efficacy. Aside from state politics, these arenas include everything from commercial law, advertising, people's sense of self and other, product design, news media and much more besides. As I suggested in Chapters 2 and 3, 'politics' is present in each case (if we understand the term in an expanded sense that applies beyond the conventional domain of 'government'). The challenge, Jones argues, is to understand 'bordering' and 'bounding' as semiotic processes that have practical (never neutral) consequences.

Study Task: Make a list of some borders and boundaries that are routinely used to structure thinking and action on a day-to-day basis. Think not of the political divisions created and enforced by governments; think instead of *semantic divisions* (e.g. student versus teacher, white versus black). Consider how these divisions are used to describe, evaluate or affect the behaviour of various real world phenomena. It may help you to refer back to Figure 1.5.

Borders and boundaries are not so much things, as linguistic creations or 'conceptual cuts' that may carry sufficient weight to divide the world in their own image. While certain of these distinctions may be specific to a given arena (such as the taxonomic categories created by biologists when classifying insect species), they can also 'leak out' and influence thought and behaviour in ostensibly different domains. In this context, we can see that learning wider lessons from the study of political borders and boundaries can be more than strictly metaphorical or analogical. 'Inside–outside' distinctions are fundamental to most acts of political demarcation at the level of nation states and below. Political borders and boundaries are usually defined clearly (they are 'sharp'), and come with myriad rules, usually legal ones, that specify who or what is entitled to enter and exit. This is, in part, how 'aliens', 'illegals' and 'foreigners' come to be seen as such, and how 'deportations' and 'expulsions' become viscerally real. Considerable material resources (personnel, money, etc.) are committed to maintaining the territorial partitions that give these categories real meaning. 'Hard' borders are heavily regulated (images of fences, walls, checkpoints and patrols come to mind), while established boundaries are often difficult to break down (imagine trying to alter the political geography of any of America's contiguous states without encountering ferocious opposition from state legislatures, governors and senators). 'Soft' borders and boundaries permit movement, but only because a conscious decision has been made to treat them differently from less porous ones. Even then, certain things can appear to locals and natives as somehow 'out of place'.

Whatever the specificities of their creation and maintenance, political borders and boundaries can speak powerfully to the broader field of analysis that Jones advocates. We can inquire into whether bordering and bounding in other arenas share similar characteristics. For instance, how common is it to encounter categorical and practical distinctions that prohibit transgressions as rigorously as officials at the North Korea–South Korea border? Likewise, who or what is able to move across less solid boundaries of various kinds and what exclusions, if any, attach to these (seemingly) permeable divides? We can also inquire into whether and how categorical distinctions made in specialist discourses (administrative, legal, scientific and so on) achieve a spatial expression akin to political borders or boundaries. And we can search too for borders and boundaries that are as overtly politicised as, say, the Golan Heights, which Syria has long wished to reclaim from Israel. In each case, we can look for four things: *what* separations are made (semantically and practically), *where* are the divisions applied 'on the ground', *how* (strictly) they're enforced, and *who* (or what) is most affected by acts of bordering and bounding? What's more, we can look for substantive connections (not mere comparisons) between erstwhile different cases, and attend to the individuals, groups or institutions that engage in – or, conversely, oppose – particular acts of bordering and bounding.[2]

In the rest of this chapter, I will explore three very different cases where the borders and boundaries of the natural are delimited by different epistemic workers (in two cases legal professionals and career politicians, in the third case political activists). This involves creating conceptual quarantines for nature understood in three of its four meanings (I don't consider 'universal nature' for obvious reasons, but do include nature's collateral concepts for equally obvious reasons). It also sometimes involves the geographical expression of semantic differences, recalling the discussion of 'the spatialisation of nature' in Chapter 1. To avoid any confusion in what follows, I use the term 'borders' to refer to the line *between* what's said to be 'nature' and not-nature, while I apply the word 'boundaries' to divisions *within* the realm of what's considered to be natural.

FROM NATURE TO ARTIFICE: INTELLECTUAL PROPERTY AND BIOLOGICAL INVENTION

Patents and biotechnological manufactures

Businesses the world over have long claimed, and sought to enforce, property rights in what they take to be the unique or distinctive aspects of the commodities they sell. These property rights are prescribed in national legal systems and breaches of rights may be redressed through prosecutions and monetary charges against any offending parties. Though firms claim property rights over the *physical* commodities that they make and market (up until the point of sale), the key underpinning legal right is often purely (or largely) *discursive and ideational.* After all, commodities can be seen as material expressions of the ideas, creativity and insights of their inventors (a novel purchased in a bookstore is an obvious example). Relatedly, many commodities are marketed using certain signs and symbols (like the world-famous 'golden arches' of McDonalds). Intellectual property covers both what is 'behind' or 'inside' a commodity (but not visible to the eye), and how it appears phenomenally. In each case, the tangible commodity is separated from a realm of thoughts or designs it is taken to instantiate.[3] As such, intellectual property law, and the many thousands of legal professionals and officials involved in creating and enforcing it, is one of several technical arenas in which the many distinctions presented in Figure 1.5 are reproduced and normalised. This law covers the whole family of legal rights that can be claimed by firms and other creators of novel designs, processes, images or artefacts. The principal types of rights pertain to patents, trademarks, copyright, geographical indicators and trade secrets. Though different in the detail, each of these rights domains is intended to reward those claiming a right and to exclude those who might profit from unrestricted use of another's creations or inventions.[4]

Typically, intellectual property rights can only be claimed by real or juridical individuals (i.e. a person or an organisation). Communities or distinct categories of people (e.g. 'first nations' North Americans) must usually find some other vehicle to have whatever distinctive contributions to knowledge and practice they've made recognised. Globally, it's the United States, United Kingdom, Japan, Canada, France, Germany and several other west European countries that have done most to develop detailed, codified and enforceable intellectual property laws. Much of this endeavour has been in response to demands from entrepreneurs and firms to have their commercial interests protected. The argument has been that one cannot turn one's unique commodities into a revenue stream if rival producers are free to copy or clone them without penalty. Because national intellectual property laws and rules vary (in some cases being minimal and weakly enforced), there's been a strong push by firms and governments in export economies to create a more uniform legal and regulatory landscape overlaying international borders. The first and still most visible example of this push was the 1994 Agreement on Trade Related Aspects of Intellectual Property Rights (TRIPS). Applicable to members of the World Trade Organisation (which was created as part of the discussions that led to the TRIPS rulebook being written), the Agreement prescribes minimal intellectual property standards applicable to internationally traded goods and it contains mechanisms for redress if these standards are not adhered to.

In the past quarter century, one of the major growth areas for intellectual property claims, disputes and settlements has involved biotechnology companies. Novel techniques of the sort I briefly described in the third section of Chapter 4 (like gene-splicing) have made it possible to create new biological phenomena. These phenomena pertain to both human biology and non-human species. Agro-foods, pharmaceutical and biomedical firms, as well as universities conducting applied biotechnology research, have been seeking to protect their intellectual inventions through a proliferating number of property claims (see Box 5.1). The strongest protections are afforded by patents (in all areas of commerce, not just biotechnology). What is a patent? It's a time-limited but exclusive right to own and use information that describes a process, design, artefact or other phenomenon that's been invented and which possesses practical value. To qualify for a patent in almost all countries worldwide, this information must, to quote the World Intellectual Property Organization (WIPO),

> be of *practical use* . . . [and] show an element of *novelty*, that is, some new characteristic which is not known in the body of existing knowledge in its technical field. This body of existing knowledge is called 'prior art.' The invention must show an *inventive step* which could not be deduced by a person with average knowledge of the technical field.
>
> (WIPO, http://www.wipo.int/patentscope/en/patents_faq.html#protection, accessed 1 December 2012)

BOX 5.1

SOME RECENTLY FILED PATENTS AT THE UNITED STATES PATENT AND TRADEMARK OFFICE

The United States Patent and Trademark Office (USPTO) is the world's largest and most busy patent and trademark office. A branch of the federal government, it has dealt with over 8 million intellectual property claims in its 100-plus year existence. Below are listed just six recent patent claims in the field of biotechnology. Their titles are listed after the institutions the inventors work for and give some indication of the new sorts of things that are now subject to intellectual property claims. Some relate to technical processes, others to 'discoveries' of biophysical phenomena. It is not only inventors working for private firms who claim intellectual property rights. Research institutions do too, often with a view to earning money by 'leasing' their inventions to companies developing products on the basis of these inventions. Once granted, it is possible for a patent to be contested by a third party. Many patent cases have gone to the upper echelons of the American court system, to be adjudicated on by senior judges. This is true in other Western countries as well, and a growing number of non-Western countries.

- Merck, Sharp & Dohme Corp. RNA interference mediated inhibition of cyclin D1 gene expression using short interfering nucleic acid (siNA). US Patent 8,067,575, issued April 2011.

- US Department of Health and Human Services. Tumor suppressor gene, p471NG3. US Patent 8,067,563, issued March 2011.
- Albany Medical College. Isolated antibodies against biologically active leptin-related peptides. US Patent 8,067,545, issued December 2011.
- Cargill Inc. Method, apparatus and system for quantifying GM material in a sample. US Patent 8,150,631, issued April 2012.
- Trustees of Dartmouth College. Synthesis and biological activities of new tricyclic-bis-enones (TBEs). US Patent 8,067,394, issued May 2010.
- Kwang-Hua Development Investment Ltd. Method and composition for genetically modifying non-human cells and animals. US Patent 8,148,143, issued April 2012.

The European Patent Office (EPO) is almost as large as the USPTO and the websites of both make for fascinating reading for those who know little about the arcane world of intellectual property (see http://www.uspto.gov/index.jsp and http://www.epo.org/).

The three qualification criteria – utility, novelty and non-obviousness – must be demonstrated in words (and, where appropriate, images) in any patent claim filed with a national patent office.[5] Patent offices have elaborate and highly prescriptive procedures that any firm or person filing a patent must abide by. For instance, the *Manual of Patent Examining Procedure*, issued by USPTO (and now in its eighth edition), is a multivolume document of forbidding technical detail. The discourse of inventors (commercial laboratory and field scientists in the case of biotech firms) is thereby rendered obligatorily in the legal-procedural language of patent law. The result is a highly formalised representation that describes and explains a process, design or artefact that is claimed to be useful, novel and the result of an inventive step. I explain this in Box 5.2. However, before you read the box, complete the study task below to see if you can identify the senses in which patents 're-present'.

Study Task: As Box 5.1 shows, all patents are filed by a named inventor (or agent thereof), have a title, and comprise a detailed description of the process or thing over which a property right is being claimed. In light of this, what exactly is 're-presented' in the pages of a patent claim? Once you've attempted to answer this question, read Box 5.2.

Once a patent claim is submitted, patent officials must then check not only whether it meets the stipulated criteria, but also whether it infringes any other patents that have been granted and have not yet expired. The 'price' an applicant pays for a patent is the detailed disclosure of its content, meaning that others are free to understand the invention, if not to mimic it (at least until the patent period ends). Rejected patent claims can be pursued on appeal in all advanced and most developing capitalist countries. As we'll now see, biotechnology firms have been filing patent claims by working with, even as they seek to reposition, the distinction between 'natural biology' and 'biological inventions'. Their acts of semantic apartheid and border relocation stand to have major economic and sociocultural implications.

BOX 5.2

INTELLECTUAL PATENTS AS REPRESENTATIONS

Patents can be understood as textual and graphic representations. It was not always thus. Prior to the eighteenth century, many patent claims in Western countries and principalities were presented in a *physical* form to a representative of a sovereign. A verbal description and explanation, along with a practical demonstration, of a new device by its inventor was common. If a patent was granted, it was typically recorded in writing, but this written document did not comprise a detailed textual record of the patented invention. This changed with the intellectual property laws that came into being during the early decades of democratic, post-feudal, capitalist states like the United States (which passed a Patent Act in 1790). As Mario Biagioli argues in his history of patent law, such acts 'recast the [physical] . . . invention (which had previously been the sole instantiation of the invention) as just one of the possible . . . embodiments of . . . an idea. This was not a process of abstraction from the particular to the general, but rather one of separation between form and matter' (Biagioli, 2006: 1143). Modern patents are written and graphical representations of a process, design or artefact, not the process, design or artefact in itself. Over the decades, patent claims have become ever more detailed and precise. They 'represent' in both senses discussed in Chapter 2. Not only do they comprise an epistemic rendering of the ontological existents referred to, they also represent the achievements and commercial interests of the inventor(s). Thus, patents both 'stand for' and 'stand in for' simultaneously. Patents would be meaningless if they were not understood to be proxies for other things (i.e. inventions *and* their creators). To complicate matters further, patent officials act as representatives when assessing all patent claims. On the one hand, they are expected to uphold the public interest. Things that are either freely available or readily understandable, like oxygen or good manners, cannot be subject to patent claims because it would permit the privatisation of things that are rightly public or communal. On the other hand, patent officials are expected to uphold the private interests of current patent holders. They check against existing patents to ensure that a new patent claim represents a genuinely new invention. In each case, patent officers are stand-ins for other people as part of their normal professional practice. Their representational function is framed by the legal–moral concept of rights: property rights for patent holders and citizen rights for members of the public.

In terms of the conceptual basis on which patent claims are made, biotechnology patents by definition rest on the nature–society dualism. In effect, they aim to represent not nature but that which is demonstrably *non-natural.*

Conceptual quarantine: keeping nature out of invention

Patents, like all other forms of intellectual property, are predicated on the idea of a 'creator' or 'author'. Because creators and authors are taken to be human, patent law also rests on two assumptions: (1) what we call nature cannot itself be an inventor (or knowingly inventive); and (2) human inventors can only patent 'natural phenomena' if they're not strictly natural any more (because they display utility, novelty and non-obviousness by virtue of an inventor's inventiveness). This may seem relatively unproblematic – indeed necessary. If patent law is to do its job effectively, then it surely has to draw a line between legitimate 'invention' and the realm of the 'uninvented' and 'uninventable'. That line, it may further be thought, is no mere artifice but something real because humans routinely take products of nature and modify them.

Certainly, several early patent cases in the United States set a clear precedent in this regard. For instance, a ruling by the US Commissioner of Patents in 1889 (*Ex parte Latimer*) is often said to have codified the 'product of nature' doctrine in Western patent law. The Commissioner rejected a patent claim on wood fibre from the Southern Pine. The fibre had been isolated 'in full lengths from the siliceous, resinous, and pulpy part of the pine needles and subdivided into long, pliant filaments adapted to be spun and woven' (cited in Carolan, 2010: 113). In deciding not to grant a patent to the filing party (Latimer), the Commissioner compared the putative invention to 'wheat which has been cut by a reaper . . .' (ibid.). The logic here was that taking a 'slice' of nature does not amount to an invention. Similarly, in a 1948 ruling on a patent infringement case (involving two agro-foods companies and bacteria that allow nitrogen-fixing in plants), Justice Douglas of the US Supreme Court argued that:

> Patents cannot be issued for the discovery of the phenomena of nature. The qualities of these bacteria, like the heat of the sun, electricity, or the qualities of

> metals, are part of the storehouse of knowledge of all men [*sic*]. They are manifestations of the laws of nature, free to all [people] . . . and reserved exclusively to none.
>
> (ibid.: 114)[6]

The logic underpinning these two judgements was used once more in 1980, in what became one of the most influential patent decisions ever made. In the case of *Diamond versus Chakrabarty*, a biochemist working for the General Electric Company filed 36 patent claims pertaining to a bacterium that had been genetically modified for the purposes of breaking down crude oil. Despite Chakrabarty's bacterium being apparently useful, novel and non-obvious, the US Patent Office denied him a patent. It ruled that it wasn't possible to patent a whole organism. By a narrow majority, judges in the US Supreme Court subsequently decided that the patent should be granted. They argued that it was irrelevant whether a patent pertained to inanimate or animate entities, so long as the criteria for granting a patent could be convincingly demonstrated. For the Supreme Court, 'living nature' could be invented (including a whole organism); however, this meant that it was (and had to be shown to be) *different in kind* from other natural processes and phenomena that could not be. Since *Diamond versus Chakrabarty*, biotechnology firms such as Genentech have been able to file a great many patent claims pertaining to entire living entities or key parts thereof (such as sections of DNA).

Quite aside from the bioethical questions that arise from attempts to 're-engineer nature', biotechnology patents stir passions in at least two other areas. On the one hand, some critics argue that despite the *Ex parte Latimer* ruling and similar subsequent ones, many biotech firms are wilfully trying to profit from patenting what are, in fact, natural processes or entities. They seek to do so by *stretching* the meaning of 'invention' so that it substitutes for what are, in actuality, 'discoveries'. This reduces the size of the ontological zone taken to be natural and expands the realm of the non-natural to the commercial advantage of corporations. For instance, in 2001 the American biomedical firm Myriad Genetics successfully filed seven patents on two human genes (called BRCA 1 and 2) associated with the development of hereditary breast and ovarian cancer respectively. The patents underpinned expensive diagnostic test procedures for the cancers that had been invented by Myriad and for which it charged patients.

However, in 2009 the American Civil Liberties Union (ACLU) lodged a legal case against Myriad and the US Patent Office, seeking to make it a test case for all attempts to patent information about naturally occurring genes. In March 2010, the US District Court Judge Robert W. Sweet issued a 152-page decision in which he upheld the ACLU complaint. His ruling accepted the ACLU argument that 'isolating' genes is not an act of manufacture according to patent law. In pushing back the expanding semantic frontier of 'invention' in biotechnological patent claims, Sweet was deliberately trying hold the line for 'nature' – in this case (to use my Chapter 1 terms) the 'intrinsic' and 'super-ordinate' nature of human DNA. While Myriad claimed that the judge's ruling deprived it of the commercial rewards necessary to pay for further cancer research, the ACLU argued that information about natural processes and phenomena would (and should) now be more freely available for the use of all. Sweet's decision was subsequently overturned on appeal.

If the Myriad case represents an attempt to reduce the scope of applicability of 'invention' to biological phenomena, other critics of biotechnology patents have, conversely, sought to *widen* that scope, but in a subversive way. A great many patents conform comfortably to the legal criteria stipulated by intellectual property law, at least according to the officials assessing patent claims. Patents on various genetically modified foods are a now familiar example. Critics have questioned the way that the capacity to 'invent' is limited to corporations and their scientists in these patents. As Rosemary Coombe and Andrew Herman (2004: 561) note, the concept of 'property' is normative in that it means both that which someone owns *and* a standard of behaviour that is deemed to be 'proper' (or right). By granting patents to biotechnology companies, it is arguable that patent officials are redefining property in both senses of the term by pretending that what's said to be natural is not itself *already the result of prior human inventiveness*. Let me explain, with reference to 'invented seeds'.

For decades, government scientists and agro-foods companies have collected seeds in order to conduct experiments in controlled breeding of new crop strains for commercial use by farmers. These new strains have frequently been bred so as to necessitate farmers purchasing specialised herbicides and pesticides, and related equipment and advice, from companies supplying the new seed stock. For example, in the mid to late 1990s, Monsanto was granted US patents on seeds that produce what it called Roundup Ready corn and Roundup Ready soybeans. More recently, it has been granted patents on seeds for Roundup Ready sugar beet, cotton, canola and alfalfa. These seeds all produce crops that are resistant to 'over the top' applications of herbicide in which glyphosate is the dominant ingredient. Glyphosate kills plants by inhibiting an enzyme their cells use to synthesize amino acids (molecules used by the plants to make protein). By introducing a gene into the plant cells that is 'pre-programmed' into the seeds from which they grown, scientists have been able to produce crops that can survive the herbicide unscathed. However, the weeds that grow in and around Roundup Ready crops are not immune to glyphosate, which allows the herbicide to work selectively to the crops' (and thus farmers') advantage. In line with the logic underpinning the US Supreme Court ruling on the *Diamond versus Chakrabarty* case, Roundup Ready seeds are taken to be inventions. This implies that their creators have taken 'products of nature' (as per *Ex parte Latimer*) and then altered them inventively. But is this really so?

In Chapter 1, when I discussed the meanings of nature, I talked about 'degrees of naturalness'. I argued that we tend, almost as a matter of 'common sense', to recognise that some phenomena are far less 'pure' in their natural character than others (for instance a botanical garden as opposed to a wild orchid in a remote forest). This recognition reflects the varied extent to which human actors are seen to have modified the non-human world, intentionally or otherwise. Farmed landscapes are, of course, anthropogenic in significant measure, including those that are not the result of chemical pesticides, herbicides, commercial seeds and mechanised cultivation. For most commercially significant crops the world over, the source of new genetic characteristics (germplasm) is to be found in the tropics and subtropics (notably Central and South America, and South Asia). Firms like Monsanto routinely 'bioprospect' in these regions and use the seeds harvested from 'wild' and 'native' strains of corn, etc. as resources for their attempts to invent new varieties of the same. But are bioprospected landscapes 'natural'?

According to the patent officials and lawyers involved in granting Roundup Ready and similar patents, they are. Yet this judgement denies claims made by peasant and indigenous farmers (and their representatives) that the germplasm from which Monsanto and other agro-foods companies benefit is, in fact, the product of generations of skill and effort on their part. These farmers and their forebears have routinely crossbred strains of corn, wheat and other crops by selecting from cultivated and 'wild' plants and exchanging seeds between villages and communities. Does this not count as 'invention', even though there's no written record or single named inventor? Should a firm like Monsanto be permitted to profit from the creations of farmers in the developing world by claiming these creations are 'products of nature'? Is a genetically modified seed different in kind from the bioprospected, genetically hybridised seed stock Monsanto scientists utilised to create their Roundup brands?

In an insightful essay, Thom van Dooren (2008) has pointed out that contemporary biotechnology firms and their legal representatives answer these questions by linking the 'product of nature'/invention distinction to the notion of 'the human' (one of nature's collateral concepts) in a very self-serving way. This involves dividing the realm of the human asymmetrically and aligning this strategic distinction with the nature/not nature one. On the one side, the humans who are classified as 'inventors' are those who work *on* non-human nature, not *within* it. They take nature (seeds) as their object and analyse and experiment with it systematically so as to alter it in replicable ways. On the other hand, indigenous and peasant farmers in tropical countries have '[t]heir labour classed as mere "evolution", a part of "nature", [and] not genuinely inventive' (ibid.: 682). As I said in Chapter 1, the concept of 'the human' has always been an ambivalent one, straddling the family of antonyms arrayed in Figure 1.5. As van Dooren points out, this creates possibilities for some actors to use this ambivalence by seeking to divide up the realm of 'the human' in ways that accord with their attempts to (re)locate the boundaries of the natural. To use Michael Carolan's (2008) felicitous term, in capitalising on these possibilities biotechnology firms engage in **ontological gerrymandering**. Some humans are assimilated to nature, others said to act on external nature in ways that remake its intrinsic properties.

Degrees of 'un/natural difference': purification is complicated and contentious

If strategic delimitation of the nature/not nature border brings benefits to some at the expense of others, the same can be said about 'internal boundaries' within what's taken to be natural. Again, I'll take the case of intellectual property and agricultural biotechnology to illustrate the point. And, again, we see the importance of epistemic workers (patent officials, lawyers and judges) in attempts to draw clear (but contentious) lines around biophysical phenomena.

As we've seen above, patents on biotechnological inventions rely heavily on the idea of 'products of nature' as their antithesis. We've also seen that patents rely on a distinction between the physical expression of intellectual property and discourse (because the property right is in a description or instruction that, if acted upon precisely, produces a useful and novel process, design or artefact). Both things should be borne in mind when considering the case of *Monsanto, Canada Inc. versus Schmeiser*. The case was brought against Saskatchewan farmer Percy Schmeiser who was alleged to have grown Roundup Ready canola (RRc) in his fields illegally. Canola is a 'promiscuous crop': it pollinates openly with wind and insects producing crossbreeding between individual canola plants up to a radius of roughly 3 kilometres. Schmeiser argued that RRc must have grown in his fields accidentally as a result of natural pollination processes or because RRc seed blew on to his land from a neighbouring farmer's uncovered truck. In its 2004 ruling, the Supreme Court of Canada upheld Monsanto's complaint. Its logic was that:

> By cultivating a plant containing the patented gene and composed of the patented cells without license, the appellants deprived the respondents of the full enjoyment of the monopoly.
>
> (Supreme Court of Canada, 2011)[7]

The Court's judgement bracketed out the reasons why RRc was growing in Schmeiser's fields, and argued that Schmeiser gained commercial advantage from selling canola containing proprietary genes while avoiding the user fee charged by Monsanto. In so doing, as I'll now explain, it sought to legislate boundaries within the larger universe of canola plants – boundaries that had (and have) a geographical expression, which is politically consequential.

From the perspective of intellectual property law, anything that cannot be the subject of a patent, copyright, etc. is variously considered to be 'natural', 'public' or 'communal'. In the case of canola, no biotechnology company could ever claim to have invented pollen transfer by wind or insects, or the biochemical processes that lead to the growth of a mature canola plant. Seed patents are one example of the attempt to isolate proprietary components of living organisms from those aspects that cannot be claimed to be inventions according to Western intellectual property law. Yet the Canadian Supreme Court ruling in the Schmeiser case effectively made Monsanto's property right in the *former* a vehicle for it to claim rights over the *latter* (namely, the fullgrown RRc in Schmeiser's fields and its seeds). The ruling introduced a separation, at once semantic and physical, between different kinds of canola plants.

On the one hand were those that 'contained' Monsanto's RR genes. Given canola's promiscuity, this means that the Supreme Court decision effectively denied Canadian farmers the right to grow non-proprietary canola less than 3 kilometres from any field sown with GM canola. This is a non-trivial outcome, especially for farmers trying to grow crops that are certified as 'organic'. On the other hand, there were those canola plants that contained either no proprietary genes or only those of a single seed manufacturer with whom a farmer had a contract. In order to make this distinction stick, the Court had to finesse the distinction between 'products of nature' and 'inventions'. While all crops can be argued to contain processes and characteristics that are irreducible to human 'inventions' of any kind, the Court implied that some crops are more reducible than others. Though all canola has aspects that are not patentable, the Court ruling differentiated the crop according to which 'unnatural' components it contains. In this (awkward) way, the 'product of nature' doctrine was upheld, even as full-grown canola was separated into categories according to whose intellectual property it embodied.

The Court's decision sets a precedent that gives biotechnology companies reason to believe that patent rights are, in fact, rights over the entire living organism to which, technically, the patent does not apply.[8] It also places considerable responsibility on farmers of all stripes to demonstrate to potentially litigious firms that their crops are either 'organic', or else do not contain any unpaid for proprietary genes. Meanwhile, unless they claim intellectual property rights of their own and/or have the resources to pay for legal aid, farmers find it difficult to undo the legal boundary work that is placing Monsanto and similar companies in such a commercially advantageous position.

BEASTLY BEHAVIOUR: CROSSING THE HUMAN–ANIMAL DIVIDE

I want now to move away from intellectual property claims to parts of the non-human world and examine the case of bestiality (or what's more politely known as zoophilia, with which it's not entirely synonymous).[9] This will allow us to explore how making sharp distinctions between nature and its antonyms is central to the formation of personal and group identities, and also attempts to undo or breach those distinctions. It will allow us to talk again about 'human nature' (as we did in Chapter 4), and illustrate the way that attempts to achieve semantic hygiene are integral to the process of **moral regulation** – a process in which references to nature and its collateral terms have long been (and remain) very important.

The conventional, highly negative attitude towards humans having a physical relationship with animals is graphical testament to how central the division between 'us' (people) and 'them' (non-humans) is in Western society. As a boundary transgressing act, bestial behaviour makes obvious the role that this and allied distinctions play in governing our self-understanding and our actions. In using the term 'governing', I mean to emphasise the normative and practical force of these distinctions – their capacity, when reproduced by authoritative epistemic workers, to normalise thought and behaviour. Though commonly portrayed as 'unnatural' (not to mention immoral and disgusting), bestiality is better understood as unconventional – as something that both reveals and challenges the norms of social discourse and practice. It's one of many fronts in a semantic struggle over what it means to be properly 'human' – a 'fit person', a 'decent individual' (see Box 5.3).

BOX 5.3

MORAL REGULATION AND SEMANTIC BORDER CONTROLS

If you think about it for a minute, there are a great many examples of how references to nature and its collateral terms are used by some to govern 'proper' thought, identity and behaviour. Such references also often secrete particular conceptions of what is 'beautiful', 'desirable', 'perverse', 'ugly' or 'disgusting' that are taken to be normal (and are thus normative). Consider the following:

- The well-known critic Francis Fukuyama (2002) expressed concern that biotechnology, by permitting society to manipulate a person's genetic offspring and their own body, will create a 'post-human future'. The risk, Fukuyama argued, is that by interfering with 'natural evolution' and life's 'genetic lottery', a new 'underclass' may appear of the sort depicted in the sci-fi films *Gattaca* (1997) and *The Island* (2005). This would be a group of people identified biomedically as having 'faulty' genes, even if this was not apparent to the naked eye (i.e. phenotypically).

- In many parts of the world, sex between consenting males or between consenting females is considered a 'sin', a taboo and even a crime, punishable by law (be it customary or otherwise). Refer back to the Preface for a recent example.
- In the 1990s and early 2000s, many Western environmentalists labelled genetically modified crops as 'Frankenfoods' in order to condemn biotechnology and agro-food companies' development of putatively 'unnatural' crop varieties.
- In countries that are today thoroughly multicultural and ethno-nationally diverse (like the United States), it took many decades for intimate relationships between 'whites', 'blacks' and other ostensibly 'different' people to be culturally sanctioned. For many years, such emotional and sexual unions were seen to lead to 'miscegenation', an ugly term that's rightly considered offensive today.
- Though it's perfectly legal in many countries, people who alter their bodies to conform to their 'transsex', 'inter-sex' or hermaphroditic identities are routinely regarded as 'odd', 'weird' or 'subversive' by heterosexuals (and a few homosexuals too). Some medical professionals have sought to 'medicalise' sex- and gender-ambiguous bodies and personalities (see, for example, the essays in Barrett, 2007). (I say much more about 'trans' individuals later in this chapter.)
- While animals have been granted rights – and legally enforced ones too – in many countries worldwide, few people accept the argument that certain 'higher' animals (such as apes) might be accorded additional rights akin to the great many enjoyed by people (an argument made, for example, in Stephen Wise's (2000) *Rattling the cage* and in the more recent book *Zoopolis* by J. Donaldson and W. Kymlicka (2011)). It appears to some to bring animals *too close to us*, to treat them *as too like us.*

The Enumclaw case: outlawing animal intercourse

In mid-2005, a 45-year-old engineer who worked for Boeing (the world-famous aircraft manufacturer) died shortly after being admitted to hospital. His death became headline news because of its highly unusual nature. Kenneth Pinyan had suffered a fatal infection after his colon had been perforated during sexual intercourse with a stallion. It turned out that Pinyan had paid another man (James Michael Tait) for the pleasure of being penetrated anally by the horse on a farm in rural Washington State (near the town of Enumclaw). Apparently, Pinyan was not alone: many others had paid Tait and his associates in order to engage in sexual acts with farm animals (many of which were filmed). Realising that Pinyan had suffered an internal injury, Tait had dropped him off at Enumclaw's community hospital before absenting himself. Although the local coroner pronounced the death accidental, its unusual cause incited intense discussion about the propriety of sex with animals – both in Washington and across the United States. As Michael Brown and Claire Rasmussen note in their analysis of how Washingtonians responded to the Enumclaw incident, it 'provoked near universal disgust and disapproval . . . ' (2010: 159).[10] Few lamented Pinyan's accidental death, while many decried his sexual choices. As David Delaney observes, 'The zoophile crosses – and in so doing, erases – the boundary [between humans and animals]. He climbs over the barrier of disgust that marks off the civilized from the wild' (2003: 258). In less than a year, legislators in the state passed a law that made bestiality illegal, punishable by imprisonment for up to five years. What's remarkable about this rush to legislate is two things: not only was the Pinyan–Tait case the exception to the rule (i.e. bestiality's invisibility in the public domain), but no zoophiles had been agitating openly for bestiality to be legal in the first place. In this context, not only was an anti-bestiality law superfluous but it also stood little chance of discouraging zoophiles (given that they are highly secretive and rarely caught). Yet it was proposed, debated and passed all the same.

The law (RCW 16.52.205, entitled 'Animal cruelty in the first degree') came into force in 2006. It was drafted, discussed and agreed with little or no opposition among local politicians (a Republican, Pam Roach, sponsored the bill). It contains a fairly stringent definition of human–animal sex. I quote:

> 'Sexual contact' means any contact, however slight, between the mouth, sex organ, or anus of a person and the sex organ or anus of an animal, or any intrusion, however slight, of any part of the body of the person into the sex organ or anus of an animal, or any intrusion of the sex organ or anus of the person into the mouth of the animal, for the purpose of sexual gratification or arousal of the person.[11]

Defined thus, deliberate 'sexual contact' is, according to RCW 16.52.205, a form of animal cruelty. So too is the transmission of any images of sexual contact or of 'sexual conduct' (defined in the bill as 'any touching or fondling by a person, either directly or through clothing, of the sex organs or anus of an animal or any transfer or transmission of semen by the person upon any part of the animal, for the purpose of sexual gratification or arousal of the person'). In this regard, Washington State has aligned itself with many other US states and numerous countries around the world.

The speed with which Washington legislators moved to outlaw bestiality after Pinyan's death is symptomatic of a much older and wider attempt to enforce a human–animal divide. This divide is intended to regulate activity on either side of, and across, the putative human–animal border. While all sorts of human–animal interactions are seen as highly positive or else argued to be necessary (e.g. testing drugs or cosmetics on mammals), some incite great moral opprobrium and physical revulsion. Bestiality, even when there's no apparent harm to an animal, is one such interaction. Along with physical (and, where applicable, emotional) cruelty to animals, it's something of a taboo. Not only do most of us feel very uneasy about it: we'd prefer not to have to talk about it, unless forced to do so by circumstances. The fact that Washington lacked an anti-bestiality law prior to 2006 indicates just how effective this combination of emotional discomfort and willing silence has been in governing thought and behaviour.[12] Such learnt behaviour meant that zoophiles like Pinyan and Tait were (and are) not only small in number but also largely invisible. They kept (and still keep) their unusual sexual behaviour a secret because of the force of convention, not the threat of prosecution. Their intercourse with animals had to occur in remote and enclosed spaces, and at times when 'normal' people could not witness their bestial behaviour. In this light, the passing of RCW 16.52.205 can be seen as reflecting a moral panic: Pam Roach and her bill's supporters felt that it was important to set an example, lest political inaction be interpreted as a tacit endorsement of sex with animals. The act formalised and sought to justify moral convention, applying new penalties to convention breakers like Pinyan and Tait.

Study Task: In your view why do so many people recoil when they hear about bestiality? Is it 'natural' to be disgusted at the thought of human–animal sex? Why would most people find it hard to accept that 'zoophilia' is a legitimate practice, perhaps even the basis for identity and community among zoophiles?

The reasons why bestiality is morally outlawed are many and varied, as you may have discovered by answering the study question above. Though many do not stand up to close scrutiny on logical grounds, all rely on ideas of ontological borders that ought not to be crossed. This became clear in the way that legislators reacted to the Pinyan–Tait case, as I'll now detail (drawing upon the research of Brown and Rasmussen, 2010).

Respectable identities and proper behaviour: dividing sex in two

In the political debates leading to RCW 16.52.205, a number of arguments against animals having sex with humans were voiced repeatedly. These arguments united political conservatives like Pam Roach with animal rights groups in Washington State and the wider United States. One argument focussed on the concept of consent. On different occasions, Roach was quoted as making the following claims when advocating RCW 16.52.205:

> Animals are innocent. They cannot consent. It's wrong ...

> It's really a bill that will protect animals, who are innocent, by the fact that they can't consent. We have a good deal of our population wanting to protect innocent animals from cruelty.
>
> Our state has no ... deterrence. Animals are stimulated to do this. This is not something a stallion wants to be involved in.
>
> (ibid.: 102)

Another argument, building on the lack of consent argument, was that sex acts with animals are forms of cruelty, equivalent to inflicting physical pain on any sentient creature. For instance, Robert Reder, regional director of the Pacific Northwest Humane Society, argued in 2005 that 'The people who engage in this behaviour are victimizing animals, using their power to subject [them] . . . to this . . . harmful behaviour' (ibid.: 163). Though the horse involved in the Pinyan–Tait case was not physically hurt, the suggestion was that the other animals that might be the focus of bestial acts (e.g. chickens or cats) could be. A third, and again related, argument was that bestiality was akin to child abuse. To quote Roach once more: 'Like a child, an animal cannot consent. No animal is a willing participant' (ibid.: 164). Indeed, Roach went on to link bestiality to paedophilia: 'The studies people have sent me', she said in 2005, 'show how abusers develop by starting with something helpless, an animal; next is a child. These are patterns that develop' (ibid.). In this way, Roach categorised bestiality as the thin end of a dangerous wedge leading to the sexual exploitation of babies or toddlers.

As Brown and Rasmussen have shown, none of these arguments are very compelling. The consent argument fails in several ways, one being that we routinely slaughter animals for meat without their approval (so why not use them for sexual gratification too?). The victimisation argument treats all sexual contact with animals as emotionally or physically abusive, as if sex was a monolithic thing. However, not only do we know that sex acts can be beautiful and life affirming; according to our conventional conception of animals, we would also never know if they felt abused by bestialists. Finally, there's no convincing evidence that bestialists tend to become paedophiles. The logical and evidential weakness of the anti-bestiality case is rather beside the point – we need to focus simply on its intended effects: to outlaw bestiality and punish offenders. Roach and her supporters needed to find reasons to justify their moral stance on, and emotional reactions to, Pinyan's sex act, however weak these reasons were. Their aim was not simply to discourage bestiality among its very small number of practitioners, but more importantly RCW 16.52.205 was intended to reinforce norms about 'appropriate sexual behaviour' among the vast majority of Washingtonians. As philosopher Peter Steeves noted in a different context,

> The process of becoming an adult is a process of conforming, of learning our concepts, our categories and our places within them ... The body must suppress ... 'animality'. Whatever is *wild* must go.
>
> (Steeves, 1999: 1)

Relatedly, another philosopher (Martha Nussbaum) has observed that 'We need a group of humans to bound ourselves against, who will come to exemplify the . . . line between the truly human and the basely animal' (Nussbaum, 2001: 29). For all the talk of *animal* rights and welfare in the debates leading up to RCW 16.52.205, the bill was really about *people* and their (mis)behaviour.[13] In the name of animal protection, Roach's law could not entertain the possibility that bestiality might be considered a legitimate practice.

This move rested on an implicit bifurcation of sex acts into two classes upon which 'facts' moral reasoning was constructed. On the one hand, there was sex that we can regard as loving, cultured, sensual and ultimately 'human'. On the other hand, there was sex we can regard as instrumental, carnal, lustful and ultimately 'animalistic'. The first kind of sex is located in a wider set of human virtues that are taken to be irreducible to one's inner urges. The second kind of sex is elemental and asocial – the virtues defining 'good sex' were, for Roach and her supporters, absent or suppressed by bestialists.[14] This 'bad sex' appears wild and ungoverned: pure (aberrant) instinct. In Pinyan's case, this may have seemed especially true to critics of his behaviour; after all, he submitted to being penetrated by a creature seven or eight times his own size, rather than penetrating the creature himself. He thus apparently made himself into the *object* of a horse's desire, albeit a horse that was manually stimulated to perform a sex act.

Study Task: In Chapter 1 and the previous section of this chapter, I referred to 'ambivalent categories' that cross the putative ontological divides mapped out in Figure 1.5. From the paragraph immediately above, can you see how the ambivalent category 'human' was linked to nature's collateral concept 'sex' by Pam Roach and her supporters? Was 'sex' itself made into an ambivalent category in this case?

Despite criticisms of his predilections, by all accounts Kenneth Pinyan was an intelligent, educated, professionally successful man who loved animals. He was, it seems, a zoophile (or a 'zoo'). As part of his relationship with animals, he pursued non-violent sex in which he, not the animal, was sodomised. He evidently derived a certain pleasure from this, despite the pain he endured (and the risks he took, which ultimately cost him his life). If he had survived his mid-2005 night in Enumclaw Community Hospital, would Pinyan have argued that his bestiality was a kind of 'good sex'? Would he have argued the zoophile case and contended that love of animals – emotional *and* physical – was part of who he was as a person? Would he, in short, have contested the logic of Pam Roach's argument that sex with animals crosses the line between respectability and deviance?

In an oblique way, all these questions were raised in a 2007 documentary film about the Pinyan–Tait case entitled *Zoo* (directed by Robinson Devor, see Plate 5.1). This film sought to represent the Enumclaw incident

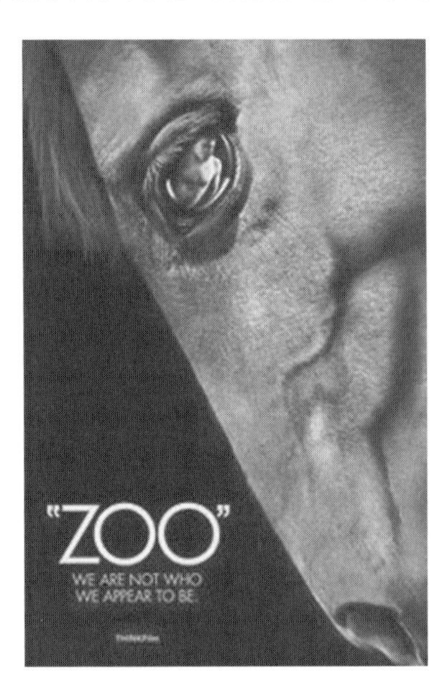

Plate 5.1 Is bestiality unnatural and immoral?

In an attempt to challenge one-sided, pejorative views of bestiality, the 2007 docu-film *Zoo* attempted to represent bestial desire and behaviour from the zoophile perspective. The movie poster's tag line was arguably intended to have a wider significance. The 'we' in 'We are not who we appear to be', it seems to imply, extends well *beyond* bestialists. In a range of ways, societies too often force 'alternative' identities and practices underground, and judge them harshly when, as in the Enumclaw case, they become visible in the mainstream.

from the zoophile perspective, without explicitly advocating for zoophilia or bestiality. It received critical acclaim and featured in two internationally prestigious film festivals.[15] *Zoo*, and the still underground world of zoophilia, can reasonably be seen as analogous to where the gay rights movement and the gay community were in most Western countries 40 years ago. Man–man and woman–woman sex acts were once routinely outlawed and considered 'disgusting', 'perverted' and 'unnatural' as recently as the 1960s in the United States, United Kingdom and elsewhere. Yet, over time, the boundary between 'good' and 'bad' interpersonal sex was relocated in both Western culture and law. What's more, it became accepted that homosexuality is (or can be) part of a person's *identity*, not simply a preference or a type of behaviour. In recalling this process of slow acceptance, I'm not suggesting that zoophilia will become normalised in the years ahead. Instead, I'm saying that because all attempts to define 'proper' identities and behaviour are underpinned by struggles over meaning, the losers in those struggles usually end up as outcasts or operate below the radar of social visibility. When the winners get to reposition the border between what's acceptable and what's not, they remind us that all borders (and boundaries) are mutable – even when seemingly unproblematic references to 'nature', 'sex' and other collateral terms are part of the justification for their relocation.

LEAKS AND TRANSGRESSIONS: RESISTING THE COMPULSIONS OF 'EITHER/OR'

In the previous pages, I've explored two cases where well-resourced, influential or authoritative social actors try to 'fix' the location of the lines between what's natural and what's not, as well (in the first case) as the 'internal boundaries' of what's taken to be natural. I've looked both at non-human nature but also at what's said to be 'normal' human behaviour and identity. I've looked also at who loses out, perhaps temporarily but possibly in the long term, in these border- and boundary-fixing projects. In this final section, I want to consider a case where those outside the cultural, moral and political mainstream have attempted to publicly challenge the discursive norms of society (unlike zoophiles, who remain largely invisible). In so doing they have – unusually and, for many, controversially and confusingly – intentionally rendered *permeable* numerous taken-for-granted distinctions, thus questioning the ontological security of phenomena normally perceived as beyond question. These transgressors have thus sought to shine a positive light on ambiguity, ambivalence and contradiction rather than see them as problematic aberrations from an otherwise orderly world. For them, 'purity is the end of possibility' (Dodge, 2000: xx).

Beyond binaries in sex and gender: representing 'trans' identities and politics

The negative reactions to bestiality in Washington State are just one of many examples of how unwelcome transgressive ideas and actions usually are. A battery of charged words attests to the moral opposition that transgressors of all kinds face in society: think of 'perverts', 'outlaws', 'whores', 'faggots', 'monsters', 'criminals' and 'freaks', for example. However, even among individuals and groups whose words and deeds go against the social grain, the normalising force of conventional distinctions can serve to stigmatise or exclude others. Put differently, those who agitate to make their own transgressions an accepted part of society can easily fail to understand or approve of other transgressors. This can make them convention-enforcers despite their erstwhile radicalism. The example I want to consider here relates to how certain transsexual and transgender individuals have been understood by some members of the feminist, gay and lesbian communities (never mind the wider society).

The terms 'transsexual' and 'transgender' are both rather baggy ones that refer to individuals with a wide range of somatic and subjective characteristics. The first is the older of the two categories. It includes hermaphrodites ('inter-sex' persons born with reproductive organs that are neither strictly male nor female) and persons who don't identify with the particular sexed bodies (male or female) they were born with. Many of the latter are transvestites who prefer to wear attire normally associated with the opposite sex. The second term, which originated in the early 1990s, refers to individuals who elect not to play one of the two established gender roles and who don't, accordingly, simply regard themselves as either a 'male' or a 'female' in a subjective sense. Transgendered persons include androgynes, drag queens, drag kings, butches, lady boys, effeminate men and many transvestites. They may identify as heterosexual, homosexual, bisexual or asexual. Unlike many transsexuals, transgendered individuals don't necessarily seek to alter their sexual appearance and functioning physically (through surgery) or biochemically (through pharmaceuticals). This said, it's not uncommon to find transsexuals who are also transgender in appearance, behaviour and self-understanding.[16]

In a political sense, transsexuals and transgender persons began to find their voice in California from the late 1980s. They were inspired by the post-1968 successes of both the feminist and the gay and lesbian movements in the United States. But some pioneering commentators also defined transsexual and transgender politics in partial *opposition* to both movements, which they saw as containing exclusionary elements. Leading feminist and 'queer' thinkers had succeeded in distinguishing between and arguing against the causal coupling of anatomical sex, sexual preference and gender. They denaturalised the latter two, with Judith Butler (1993) later arguing that even anatomical sex could not be understood as a natural substrate upon which sexual preferences and gender roles and identities were built contingently (see Box 5.4). This helped feminists and queer activists to resist any normative regulation of their constituents' behaviour based on supposedly natural 'imperatives' or on correcting 'deviations' from psychosomatic norms; however, it also led some among them to take a dim view of 'trans' people.

BOX 5.4

DENATURALISING THE SEX–GENDER CONNECTION

In Box 5.3, 'moral regulation' was the focus. Such regulation can greatly influence people's sense of self and the actions they undertake publicly and privately. A powerful example of this relates to 'heteronormativity'. As noted by Michael Warner (1991) over 20 years ago, it involves a presumed *causal alignment* of biological sex, sexuality, gender identity and gender roles. In turn, this alignment is said to be 'normal' and forms a template for young people as they mature into adults. Indeed, Butler has talked about *compulsive* heteronormativity to emphasise its social force. The presumed causal alignment mentioned above can be explained as follows: first, most people are born with male or female sex organs (their 'biological sex' is thus either/or); second, this disposes them to find members of the opposite sex desirable (their 'sexuality' is thus hetero); third, this then conditions their sense of self (they identify as 'man' or 'woman' in various ways); and finally, this sense of self conforms to and reproduces socially recognised subject-positions (i.e. specific 'gender roles' that people are invited to play).

One of the great intellectual breakthroughs of the feminist and queer movements of the 1970s was to show the *non-necessity* of causal alignment. It was argued that biological sex did not determine any sexual preference, gender identity or gender roles. Instead, belief in the necessity of this causal alignment was shown to be a powerful social convention. It was a convention that had caused many people great unhappiness in their lives. It obliged them to play gender roles and engage in sex acts that were not those they wished for. The argument was that while biological sex may be natural, sexual preference, gender identity or gender roles are all socially contingent, variable and malleable.

Notable here was Janice Raymond. Her book *The transsexual empire: the making of the she-male* argued that transsexuals, especially those who have medical help to 'reassign' their sex, were both victims and agents of patriarchy (Raymond, 1979). According to Raymond, the gender ideals of many male-to-female transsexuals reflected regressive stereotypes that did little to empower women. By placing such emphasis on *physical* change (superficially clothing but, more profoundly, sex reassignment surgery and pharmacology), many transsexuals, in Raymond's view, perpetuated the mythic, unidirectional link between anatomical sex, gender roles and self-identity. Indeed, such emphasis gave encouragement to a medical establishment that had long classified transsexualism as a 'condition' that could be 'cured' through psycho-surgical interventions.[17]

In what became a foundational text in the then nascent 'trans' political movement, Sandy Stone's essay 'The empire strikes back: a post-transsexual manifesto' (Stone, 1991) challenged Raymond's uncompromising interpretation. Stone criticised many of the transsexual autobiographies upon which Raymond's representation of the 'trans community' was based. These self-descriptions, Stone argued, reproduced the belief that transsexualism is (and should only be) a *temporary* state of affairs before an individual 'passes' to one or other sex-gender location.[18] Relatedly (and ironically), a rather **essentialist** belief that it should be possible to identify a 'real' man or woman underpinned the expulsion of a post-operative male-to-female transsexual, called Nancy Burkholder, from the Michigan Women's Music Festival in 1991. This lesbian–feminist event ejected Burkholder on the grounds that she was 'actually' a man, and in the process conflated and simplified anatomical sex, sexual preference and gender identity (contrary to the academic work of theorists like Butler). Subsequently, Trans Camp was set up in 1994 adjacent to the Festival to challenge such exclusionary practices and to valorise the 'incomplete', 'mixed-up' appearance, preferences, self-understandings and desires of many 'trans' individuals. Intellectually, it was inspired by Stone's manifesto and an equally trenchant pamphlet by Leslie Feinberg (1992), entitled *Transgender liberation: a movement whose time has come*. These epistemic workers and their written works defended the right of trans individuals to refuse assimilation to gender and sexual norms (even ostensibly 'liberated' feminist and queer ones), and to be comfortable living unusual lives. They fed in to a wider political movement, with strong academic underpinnings, that grew in size and strength throughout the 1990s in the United States (and, to a much lesser extent, in some other Western countries).[19]

Genre blending and bending

This movement has not only faced the obvious challenge of getting the voices and concerns of trans people heard in public life; it's also had to confront prejudices far more visceral than those evidenced by Nancy Burkholder's expulsion from a women's music festival. Physical attacks against and murders of trans people continue to this day. This is true even in countries like the United States where many academic and political representatives have done much to destigmatise them. As Susan Stryker notes of the 1999 Hollywood film *Big Boys Don't Cry*, which depicted the life and brutal death in 1993 of a transgender youth Brandon Teena,

> Those who commit crimes against transgender people routinely seek to excuse their own behaviour by claiming that they've been unjustly deceived by a mismatch between the other's gender and genitals.
>
> (Stryker, 2006: 10)

This plea of having been 'misled' fails to consider the way in which dominant norms about gender roles and sexual anatomy suffocate those who wish to breathe in a less tightly regulated cultural atmosphere. In Kate Bornstein's (1994: x) words, it could even be seen as a form of 'gender terrorism'. It's an open question whether sympathetic portrayals like *Big Boys*, and the professional success of trans entertainers such as Eddie Izzard, Lily Savage and RuPaul, have done much to alter mainstream views of trans people. It's very possible that today, in erstwhile 'liberal' societies, trans people are perceived as exotic (at best) – they are curiosities who pose no threat because they're so few in number. Let us remember that it was only a generation ago when Jonathan Demme's hugely successful film *The Silence of the Lambs* could pathologise a trans person without offending most viewers, namely the fictional character and serial murderer Jame Gumb (Buffalo Bill). Gumb is portrayed as a violent man who kills women for their skin, which he pieces together to create a new 'body suit' so as to transform his appearance into that of a female. The paradox is that he behaves in a stereotypically 'male' way – by using physical force without any moral restraint or compassion – against the very sex and gender he himself aspires to become. By fixating on women's skin, he's seeking to appropriate part of their bodily nature in order to counterfeit his own. In a macabre way, Demme's film thus reproduces conventional ideas that sexual and gender identities are ultimately rooted in corporeal differences between men and women.[20]

Challenging dominant representations of trans people, either in mainstream society or in politically visible (and ostensibly radical) sub-communities, has been a slow and difficult task. The recent success of 'trans' studies in the social sciences and humanities arguably reflects the values of tolerance and free speech that, unusually compared with many other arenas of life, characterise the academy (however imperfectly). But what sorts of communicative media and representational genres have been available to trans individuals to build a better understanding of – and more respect for – their preferences and aspirations among non-academic audiences? And how influential have they been?

One answer to the first question is autobiography. The prime example remains *Stone butch blues: a novel* (Feinberg, 1993, reissued 2003), written by the influential transgender activist and left-wing campaigner Leslie Feinberg (author of the above-mentioned pamphlet *Transgender liberation*). Originally circulating only within the trans community in the United States, within a few years the book's readership extended to people of all stripes who, for whatever reason, were interested in unfreezing many established sociocultural norms. Its sub-title announces its fictional status, contrary to conventional autobiography, which is, of course, a sub-genre of a large family of texts that report on 'real' events and people, be they past or present.[21] This sub-genre has, historically, been very important for transsexuals. It's allowed many of them to record in great detail their experiences, emotions and aspirations when inhabiting bodies that felt somehow 'wrong'. These personal stories have presented themselves as 'truthful' and 'authentic', reproducing the convention that autobiographies don't involve having to represent a life story from another person's perspective (as in biography). However, the American cultural theorist Bernice Hausman (1995) was one of several critics to point out that 'conventional' transsexual autobiographies were, in fact, complicit with biomedical discourses prevailing between the 1950s and 1990s. Knowing what evidence they needed to present to medical professionals in order to undergo sex reassignment, many transsexual autobiographers, Hausman and others argued, told selective stories that conformed to a 'biomedical script'. 'What is lost', Sandy Stone argued,

> is the ability to represent authentically the complexities and ambiguities of lived experience . . . Instead, authentic experience is replaced by a particular kind of story, one that supports the old constructed positions.
>
> (Stone, 1991: 295)

Given this, it's perhaps no wonder that many gays and lesbians were suspicious of transsexuals. By 'authentic' Stone was referring to autobiographies that might represent transsexual experience and aspirations in a different way. *Stone butch blues* met her call within three years. Feinberg's insistence that it's a *fictional* autobiography was arguably significant: it was an intentional subversion of the claims to authenticity of putatively 'factual' transsexual autobiography. 'Never underestimate the power of fiction to tell the truth,' Feinberg intoned in the Afterword to the reissue of *Stone butch blues* (Feinberg, 2003: 303). Though based heavily on Feinberg's experience of growing up as a working-class Jew in upstate New York after the Second World War, the 'autobiography' is of a fictional character (Jess Goldberg) whose rejection of her femininity (in both sexual and gender terms) is recounted using all the imaginative resources usually associated with creative writing (see Plate 5.2). '[Using] fiction', Feinberg has written,

> gives you the ability to tell a very painful story that's filled with all the shame of growing up differently in [a] . . . society. I felt, by telling it autobiographically, that I would pull back in a lot of places. I really felt that by fictionalizing the story, that I would be able to tell more of the truth . . . .
>
> (Cited in Prosser, 1998: 190)

So much for the genre blending and bending features of *Stone butch blues*. What representation of Feinberg's life, and by implication of other trans people, did the novel seek to convey? At the considerable risk of oversimplifying a complex story, let me make three summary observations. First, in *Stone butch blues*, Jess Goldberg tries, for much of the story, to 'pass' as a male. Her choice of clothes, her double mastectomy, her decision to undergo hormone therapy: these and other things show her conforming to 'normal' transsexual behaviour and to conventional gender roles rooted in 'sex reassignment'. Second, Jess suffers the usual humiliations at the hands of men who discover her transsexual proclivities and gender-bending appearance. In one scene, for example, she's raped by policemen who deliberately remove all her clothes, 'their sexual violence . . . an attempt to enforce on the transgendered butch the indisputable fact of her femaleness against their maleness, her essential and antithetical bodily difference from them' (Prosser, 1998: 191). The rape is a brutal act of heternormativity (see Box 5.4 again) designed to reinforce sex and gender 'purity'.

If these elements of *Stone butch blues* repeat the narrative conventions of earlier transsexual autobiographies, the third most certainly does not. Jess stops her hormone treatment halfway through; she has no further 'corrective' surgery; she takes no additional measures to 'pass' as a man or to hide her female characteristics. She chooses to inhabit the borderlands between male and female: 'I'm a he-she,' Jess declares towards the end of the novel, 'That's different' (Feinberg, 1993: 147). Her journey doesn't have the 'usual' destination of a 'conventional' trans person. Jess decides to be unique.

Plate 5.2 A transgressor represents transgression

This cover of the 'fictional' autobiography *Stone butch blues* captures the representational politics of the book's creation and contents. The image shows the author, who also happens to be the main character in the story (Jess). Feinberg/Jess looks directly at the reader, perhaps challenging them to determine who or what s/he is (man? woman? transvestite?). His/her's is the face of experience: serious and somewhat weathered. 'Hir' story is intended to represent the lives of others in hir situation. The 'novel' is no mere concoction. It is intended to represent the difficult journey towards personal contentment experienced by people like Feinberg/Jess.

By electing to be sex- and gender-ambiguous, Jess Goldberg takes the huge risk of being 'read' as neither a man nor a woman. S/he refuses to make her 'trans' predilections part of a transition to something 'whole' in physical and identity terms (in this case maleness). S/he chooses not to submit to the prevailing subject positions that structure heterosexual, homosexual and many transsexual lives. Goldberg creates what taxonomic historian Harriet Ritvo (1995) once called 'border trouble'. S/he inhabits a psychosocial and somatic location that her society considers off limits. S/he is thus a pioneer and an outlaw at one and the same time. In critic Jay Prosser's view, as a (still) virtually unique presentation of a trans life, *Stone butch blues* 'wields a powerful representational and representative force that's hard to resist' (Prosser, 1998: 197). Certainly in 'hir' reflections on the book's reception 10 years after publication, Feinberg (2003: vii) remarked on how affected by *Stone butch blues* a remarkable diversity of readers had been within and beyond the United States.

However, for all the book's qualities, one suspects that it's a 'niche' text with a socially distinctive readership whose story is too 'challenging' to speak to the 'average' member of Feinberg's (or any other) society. It has helped to question previous prejudices against trans people within sections of the gay and lesbian community, especially in North America, but one doubts that it's had a wider cultural impact. If I'm correct, then this serves to demonstrate the continued power of those borders and boundaries that have been my focus in this chapter, and which Feinberg and others wish to traffic across. Fiction is a potentially powerful genre for challenging these divisions. But in this case, its effects on the mainstream have been limited.

SUMMARY

As the three cases in this chapter show, the process of delimiting and enforcing lines of (supposed) ontological difference can be as contentious as it is consequential for those involved. Despite their manifestly diverse interests and goals, we've seen that legal professionals, biotechnology firms, commercial farmers, zoophiles, lawmakers, queer activists and 'trans' spokespeople have some important things in common. They are, from time to time, engaged in potentially controversial discussions over where the borders and boundaries of nature and its collateral terms lie, as well as how secure these borders and boundaries are. They deploy different discourses and a range of communicative media in order to shape these discussions and draw them to some sort of conclusion. Such discussions are, of course, hardly exclusive to the cases considered in the preceding pages. With a little vigilance, one can identify similar questions, concerns and arguments arising in a range of other social, economic and cultural arenas.

Building on the previous chapter's arguments and examples, I've shown that attempts to **naturalise** boundaries and borders often involve explicit reference to what are taken to be 'real' differences within and beyond the realm of nature. I've shown too that these attempts, when scrutinised closely, reveal the thoroughgoing *non*-naturalness of boundaries and borders. Does this mean that borders and boundaries of the sort I've discussed here can be readily transgressed? The three cases considered suggest *not*, especially the second and third ones. When the law is involved, the nature/not-nature line can be rigorously policed, leaving trangressors to rely on persuasion (as in the cases of *Zoo* and *Stone butch blues*). In this light, we can see why the arguments made by academic luminaries such as Donna Haraway and Bruno Latour don't find a ready audience outside university seminar rooms. However 'hybrid' or 'amodern' the world may appear to these critics, so much of contemporary life remains committed to policing distinctions and cancelling out ambivalence.[22] The concept of nature and its collateral terms, it seems to me, will remain central to attempts to regulate thought and behaviour for many years to come – even as transgressive technologies such as synthetic biology become more commonplace and current material differences between phenomena are seemingly less ontologically secure. The challenge for those on the 'wrong' side of borders and boundaries is to find ways to get their voices heard louder and longer.

ENDNOTES

1 In this chapter, I will not focus on temporal breaks of the sort I mentioned in Chapter 1, but I will look at both semantic boundaries and, to a lesser extent, their spatial expression on the ground.

2 At this point, let me quickly challenge two ideas that suggest it's an ontological dualism (i.e. somehow inherent in the world, regardless of our classificatory schemas). My challenge will hold no surprises at all in light of the past four chapters and recapitulates earlier arguments. The first idea, long discredited in both political science and political geography, is that of 'natural borders' and 'natural boundaries'. It was once believed that the 'real' political geography of a state, both external and internal, was dictated by 'natural barriers' (such as rivers, coasts and mountain ranges). In this light, 'artificial' divisions were those that, from a geographical perspective, appeared quite arbitrary (such as the straight-line borders drawn by European colonial officials to delimit many African states during the late nineteenth century). As many critics have noted, this natural–artificial distinction cannot account for the fact that *all* political boundaries and borders are in some sense constructions: for instance, an ocean strait only becomes a political dividing line if a political authority decides to place a line there and, if need be, to police traffic across it. Analogous to the idea that some political borders and boundaries are natural, there's the much more tenacious idea of 'natural kinds'. This has been much debated by philosophers, and refers to the ontological differences presumed to inhere in natural phenomena (human and non-human). For instance, if one supposes that atoms are natural kinds, can one say the same about larger-scale phenomena like genes (which I discussed in the previous chapter) or mountains (see Smith and Mark, 2003)? These philosophical debates aside, it's probably fair to say that most people believe in natural kinds (even though they don't use this term and even if the particular kinds they recognise in their daily lives necessarily vary according to experience and education). However, as I've argued from the first page of this book, we need to regard natural kinds as semiotic constructs that are every bit as unnatural as the 'natural' borders that political scientists and geographers now rightly view with suspicion. Whatever the case happens to be, the line between 'natural' phenomena and other entities is always contingent on decisions that are, in principle at least, open to contestation and change.

3 This practice of separating an idea or concept from its physical or material manifestation/realisation recalls Tim Mitchell's exploration of the 'reality effect' that, following Roland Barthes, is characteristic of the Western worldview. I discussed Mitchell's work towards the end of Chapter 1.

4 Intellectual property law is one of several areas of the law in which concepts of nature and its collateral terms are used and applied. Legal discourse, and the hundreds of thousands (if not millions) of professionals globally who create and use it, is an especially powerful discourse. As David Delaney argued, 'Law is authority. What law says is, well, the law. What it says about nature is enforced by . . . the state . . . Attention to the law sharpens our awareness that control over the word, over meaning, over the terms of categorical inclusion and exclusion, is strongly conducive to – if not determinative of – control over segments of the material world that are given meaning' (Delaney, 2001: 489).

5 In most patent law, as in wider intellectual property law, there's usually also reference to a fourth criterion: namely, *stability*. This means that the information contained

in any patent claim describes a process, artefact or other phenomenon that is not protean or evanescent.

6 Douglas ruled on a 1948 appeal case (lodged by Kalo Innoculant Co.) against a patent granted to Funk Brothers Seed Company. A summary of the ruling can be found at http://supreme.vlex.com/vid/funk-brothers-seed-v-kalo-inoculant-20015972, accessed 16 December 2012.

7 The summary judgement can be found on the Supreme Court of Canada website: http://scc.lexum.org/en/2004/2004scc34/2004scc34.html, accessed 10 December 2011.

8 Technically, most patents in biotechnological inventions are not like those granted to Ananda Chakrabarty by the US Supreme Court in 1980. Indeed, judges of the Canadian Supreme Court famously denied Harvard University patent rights to a mouse used in cancer research (Oncomouse™) in 2002. Part of their reasoning was that the mice and their progeny were irreducible to the genetic modifications made by Harvard biologists.

9 Strictly speaking, zoophilia is not the same as bestiality, though it may encompass it. It is 'love of animals', and need not have a sexual component, though it does have an emotional and often (non-sexual) physical component.

10 Bestiality is one of several human practices that appear to produce a reaction of disgust in people that's very visceral. Is this reaction natural, traceable to evolutionary processes that conferred an advantage on our historical predecessors? Some would say so but, consistent with the arguments of *Making sense of nature*, others focus on either (1) the social constitution of this reaction; or (2) the social mediation of this evolutionary adaptation. Daniel Kelly's (2011) wonderful book *Yuck!* offers an up-to-date survey that covers the various extant perspectives.

11 Quoted from the online repository of the Revised Code of Washington: http://apps.leg.wa.gov/rcw/default.aspx?cite=16.52.205, accessed 23 December 2011.

12 Previous anti-sodomy laws designed to outlaw homosexual behaviour between males did, in fact, contain a provision against bestiality; however, after these laws were abolished, in response to the arguments of the gay liberation movement from the 1970s onwards, no new anti-bestiality law was created.

13 I should add that my own analysis can be said to be deeply anthropocentric because I pay no attention to animal rights or welfare, nor do I consider seriously the (possibly) ineffable 'otherness' of animals.

14 This distinction between human sex acts overlaps with the distinctions between 'culture' and 'nature', and 'reason' and 'passion'. As Robert Solomon details in his book *The passions*, Western cultures have long sought to contain 'unbridled passion' because it risks the submission of a person to their 'natural desires' (Solomon, 1993). Yet even a moment's reflection reveals how difficult it is to maintain a secure distinction between 'good' sex and 'bad' sex and apply the distinction in practice on a consistent basis. For instance, consider the couple, be they heterosexual or homosexual, who practise 'good sex' most of the time, but also participate in group sex or who enjoy masochistic sex acts. How is such a couple to be regarded? How would they describe themselves in light of the good/bad sex schema?

15 *Zoo* won an award at the internationally prestigious Sundance Film Festival and was showcased at the 2007 Cannes Film Festival. Relatedly, the autobiographical book *The horseman: obsessions of a zoophile* by Mark Mathews (not the author's real name) commanded a lot of attention after its publication in 1994 (Mathews, 1994). This included television coverage by the UK's Channel 4 in a documentary entitled *Hidden loves* (broadcast in 1999).

16 The critic Susan Stryker distinguishes transgender from transsex rather too sharply: for her the former is 'anything that disrupts, denaturalizes, rearticulates, and makes visible the normative lineages we generally assume to exist between the biological specificity of the sexually differentiated human body, the social roles and statuses that a particular form of body is expected to occupy, the subjectively experienced relationship between a gendered sense of self and social expectations of gender-role

performance, and the cultural mechanisms that work to sustain or thwart specific configurations of gendered personhood' (Stryker, 2006: 3). While this is an excellent summary of what a focus on transgender offers over and above a focus on sex *narrowly* defined, it ignores the complex ways in which 'sex' and 'gender' play formative roles in the lives of many or most 'trans' people. This said, Stryker's definition reminds us that not a few transsexuals do place emphasis on what they take to be 'natural' drives that underlie their desired identities, genders and sexual practices. Many transsexuals would thus challenge the philosophical precepts and analytical framings favoured in this book.

17 An early critique of the wedge Raymond sought to drive between transsexuals and feminists is provided by Carol Riddell in her essay/pamphlet 'Divided sisterhood' (Riddell, 1980). Shortly after, Dwight Billings and Thomas Urban criticised the medical view that transsexualism was a technically 'fixable' condition (Billings and Urban, 1982). They argued that transsexuals should explore other ways and means of 'being themselves' beyond assuming they needed medical intervention to 'pass' across the sexual divide.

18 The term 'passing' has conventionally meant two things within the transsexual community. It's referred to as both a desire to 'pass' as a member of the opposite sex-gender (not have one's 'real' sex-gender recognised by non-transsexuals), and the process of crossing the sex-gender threshold to 'the other side' (usually through surgical and pharmacological interventions into the transsexual body).

19 An outstanding collection of original works that, together, foreshadowed and constituted the now vibrant field of 'trans' studies is that compiled and edited by Susan Stryker and Stephen Whittle (Stryker and Whittle, 2006). In the political sphere, activist groups like Transsexual Menace (in North America) and Press for Change (in the United Kingdom) have done much to press the case for trans people in the public domain and in the law.

20 I'm well aware that some very sophisticated interpretations of the *The Silence of the Lambs* have been put forward. My own simple depiction of Jame Gumb is the one that, I suspect, would have made ready sense to viewers of the film at the time of its release. The sort of complex analysis presented by the likes of Judith Halberstam (1992) is precisely *not* the sort of understanding regular filmgoers achieve. This gap no doubt proves the maxim that radical new ideas or values must first be aired among a tolerant minority before they get 'tested out' in more popular representational media and genres.

21 For a short but incisive introduction to the sub-genre see Linda Anderson's book (1991).

22 In one of his major statements on the characteristics of Western 'modernity', the sociologist Zygmunt Bauman argued that 'ambivalence' was the enemy of the sort of 'order' so prized in the Western imagination (Bauman, 1991). This places Haraway's (1992) hopeful essay 'The promises of monsters' in its less-than-hopeful context: most often, the 'monstrosity' she commends metaphorically (and more literally) is seen as a threat to not an enrichment of life.

6 THE USES OF NATURE: SOCIAL POWER AND REPRESENTATION

It's now time to focus more squarely on an important subject that's been left rather (too) implicit in the previous chapters. In Part 1 of this book, I sketched out the links between epistemic communities, epistemic dependence and the many communicative genres in which references to nature and its collateral terms occur. I observed that, while a great many epistemic workers seek to represent everything from human genetics to anthropogenic climate change to the rights of whales, they don't enjoy an equal ability to capture our attention. What's more, I suggested that we can, depending on the circumstance, actively choose to pay little heed to (or ignore) various epistemic communities and their representations (including very socially prominent ones). Even so, I argued that those representations that *do* capture our attention are, over time, involved in a slow, relatively uncoordinated, complex but nonetheless efficacious process of shaping our sense of self and world. Their effects can run deep, solidifying – though sometimes unsettling – our beliefs, values and sentiments. Our 'individuality', 'freedom' and 'liberty' are not *sui generis*. Rather, they're the products of social relationships, institutions and a panoply of associated discourses, signs and references.

These latter, I claimed, are thereby contributory to the achievement of socio-economic and cultural order, but also to the possibility of change (incremental or otherwise). In Chapter 3, I introduced the idea of 'semiotic democracy'. I speculated about what diet of information, knowledge and experience might be offered to us so that we don't become unduly subject to the claims and contentions of those who exist firmly in the mainstream of society, whose values or goals may disadvantage us, who act in bad faith, or who may not welcome rival discourses. I also posed questions about how epistemic communities govern their own practices and how, where necessary, these practices might require outside regulation. I ended Part 1 by insisting that references to nature and its collateral terms are – and will remain – central to the process of 'governing' in the widest sense of this term. These references are not 'apolitical' and need not emanate from the domain of 'politics proper' (namely, the state and political parties) to exert a non-trivial influence on many or most people. Any democracy worth its name must, I contended, recognise that the essence of a democratic society does not simply lie in a written constitution, periodic elections or referenda. The challenge is not to somehow reduce the scale or scope of 'rule' (notwithstanding the aspirations of anarchists). In complex, large-scale, globally interconnected societies characterised by detailed divisions of manual and mental labour, the challenge instead is to resist the attenuation of democracy. Democracy is a form of rule that aspires to empower those who are subject to that rule. It is, self-consciously, self-rule. Freedom and submission are thus two sides of the same coin.

As I intimated in Chapter 3, and as some of the case material considered in the past two chapters suggests, to fully understand how thinking and action are governed we necessarily have to consider the subject of **social power**. I've touched rather lightly upon this important subject so far in this book. Only a Pollyanna could believe that we live in a relatively power-free world where everyone enjoys a roughly equal right – and ability – to influence others meaningfully.[1] Even ostensibly democratic societies are characterised by relationships of power, never mind those we might consider to be autocratic or despotic (like Robert Mugabe's Zimbabwe, Bashar al-Asad's Syria, or the former regimes of Saddam Hussein and Muammar Gaddafi). This is why Jurgen Habermas, the celebrated German critical theorist, famously regards the realisation of 'true' democracy as an 'ideal speech situation' – ideal, because power-free discourse among diverse members of a democratic society is a possibly unrealisable aspiration.

The subject of social power has preoccupied social scientists like Habermas for many decades. It remains a central concern of political scientists, sociologists, political economists, media analysts and cultural studies scholars (to name but a few). It's a subject that conjoins analytical and normative reasoning. What is power? Who (or what) has power? What different forms does it take? Is power always 'bad'? Should we aim to eliminate it? And how, in certain circumstances, are references to nature and its collateral terms important instruments of power? I explore answers to these questions in this chapter, thereby amplifying ideas and key claims presented earlier in the book. Needless to say, because these are very weighty questions – a virtual library of books and essays exists to address the first five, let alone the sixth – my answers will inevitably strike some readers as hopelessly partial or overly succinct. Even so, it seems to me important to examine (however sketchily) the connective imperative between social power and how 'natural' phenomena are made sense of by influential epistemic communities.

I will begin, in the next section, with some general observations about social power. I'll then relate this discussion to representations of biophysical phenomena with reference to one extended case (about deforestation in West Africa). However, a case considered towards the end of the next chapter, whose subject is the major media and genres through which representations of nature are communicated today, will also examine social power explicitly. So, too, will the major cases explored in Chapter 8.

THE NATURE OF SOCIAL POWER

Definitions

'Like love, we experience power in our everyday lives, and it has real effects [on us] despite our inability to measure it precisely' – so writes the political theorist Joseph Nye (2011: 3) in his recent book *The future of power*. Power is tangible but often difficult to quantify, except perhaps in situations where the powerless have so little opportunity to think, speak or act of their own volition that they perceive themselves as (or are considered by others to be) dominated, oppressed or enslaved. *Contra* Nye, some argue that these situations are the only ones where we can legitimately talk of power relationships between people. Here power is equated with such things as intimidation, threats of violence or the use of physical force. According to this view, women coerced into prostitution and drug dependency by criminal gangs are one of the many present-day social groups who are, regrettably, victims of power relationships.

Study Task: Do you think 'social power' is synonymous with the capacity of a person or institution to control others through the exercise of physical force or the threat thereof? Are there examples you can think of where force was not used but which, in your view, appear to involve the exercise of social power? Conversely, think of situations where you feel social power is not operative.

Many analysts of social power regard a focus on 'extreme' cases as unduly limiting. Like Nye, they prefer an ecumenical definition of social power as 'the capacity to produce a change' (Miller, 1992: 241) or 'a relation in which one agent or agency somehow affects the attitudes or actions of another' (Ball, 1992: 14). The latter is more inclusive because it allows for the logical converse of the former: the recognition that social power is also the capacity to *prevent* or *slow down change that might otherwise occur* (see Box 6.1). Interpreted in the broadest sense, this means that power, which exists in a great many forms, is possessed by a wide range of people and institutions, and isn't only synonymous with things like coercion; however, because this definition risks being rather *too* inclusive, we might add that power entails the capacity to produce (or prevent) *significant and/or enduring changes* (rather than trivial or evanescent ones). Typically, this capacity arises because a person, group or institution possesses a large amount of material or symbolic resources compared with most other sections of a society.

BOX 6.1

THE POWER OF INACTION

Theorists of power in the social sciences and humanities have long considered social power as a capacity to *act* – as an ability to influence other people or the material world in non-trivial ways. This understanding, however, necessarily implies that power is also a capacity *not* to act, unwittingly or intentionally. Such *inaction* can legitimately be regarded as a form of social power with potentially significant consequences. It can hinder or even prevent important changes that could otherwise occur, leading us to consider so-called counter-factual situations. Consider what happened (or rather did not) in Friendly Hills, a suburb of Denver, Colorado, in the 1980s. Mothers in the neighbourhood began to notice a higher than average incidence of birth problems and childhood illness. Specifically, several children had died from cancer, had other serious diseases or suffered severe birth defects. Their mothers came to suspect that the root cause was toxic waste discharged from a nearby industrial facility. It was operated by the now-defunct Martin Marietta chemical and armaments/defence company. The facility had opened in 1956 and was suspected of contaminating the public water supply upstream of Friendly Hills, managed by the Denver Water Department.

A health assessment of Friendly Hills residents was begun in 1987 by the Colorado Department of Health. This was three years after residents had identified the potential link to the Martin Marietta plant. With no sign of action by the Department two years later, 14 families in Friendly Hills and nearby Harriman Park filed a lawsuit against Martin Marietta and the Denver Water Department. They sought damages for the deaths of four children and the serious illnesses suffered by several others. The suit was thrown out; however, the judge, based on the voluminous evidence presented in court, acknowledged in her ruling that there was clear and overwhelming evidence of groundwater pollution in a plume stretching from the Martin Marietta plant towards a nearby the public water plant. The doubt was about whether this groundwater pollution had affected the public water supply channels.

Clearly, Martin Marietta and the Denver Water Department felt there was no case to answer. By contrast, many Friendly Hills residents felt there was. They felt especially let down by the public authorities tasked with delivering basic services and upholding their rights as local citizens. These authorities decided not to act to redress a perceived injustice and were accused of colluding with an economically important local employer by withholding important hydrological and health information. By not acting they avoided a potentially expensive legal settlement and prodigious clean-up/remediation costs, and a legal case of their own against Martin Marietta. But, from the perspective of aggrieved local citizens, by not acting they failed to exercise their powers responsibly, with tragic human consequences.

For more detail see Richard Fleming's feature article 'Hard to swallow' in the *Denver Westword News*, 1994: http://www.westword.com/1994-08-31/news/hard-to-swallow/, accessed 1 July 2012.

If these resources aren't mobilised then power is latent – mere potential – and thus, in an important sense, not power at all.[2] The power *not* to act is only meaningful if the converse is realised on more than an occasional basis. This is why several analysts of power have distinguished between 'power to' and 'power over'. The latter involves *actualising* the capacity to 'affect the attitudes or actions of another', though there are, depending on the situation, no guarantees that the desired effects will be achieved. As Nye puts it, 'Having . . . power doesn't always ensure that you will get what you want' (2001: 10). In part, this is because those on the 'receiving end' of a power relationship may have the capacity to resist or ignore whatever inducements, pressures or sanctions come their way. As Amelie Oksenberg Rorty once observed, 'Power is relational . . . : it is [usually] exercised in a field of counter-forces' (1992: 5).

This field metaphor (derived from magnetics) helps us distinguish social power from other capacities and actions that might feasibly be included in the ecumenical definition of power offered above. Having your wallet stolen at knifepoint on the street one unlucky evening is not an example of *social* power in action. Nor is the bullying of one person by another, in (say) a school or a workplace. Social power is operative when those who are empowered inhabit institutions, assume roles or enjoy a status which transcend/s their individual characteristics. It is 'power with' insofar as any empowered individual is drawing their capacity from a wider network of people and a wider pool of symbolic or material resources, knowingly or otherwise.

Power shift: from 'hard' to 'soft'?

Several analysts of power have argued that many societies worldwide have experienced a historical shift. They argue that 'hard' forms of power, such as the public execution of dissidents by a monarch, have given way to 'softer' forms of power based on the creation, dissemination and application of (often esoteric) knowledge, norms, standards, values and information. In Chapter 3, I summarised this as 'power *over*' and 'power *in*' communication, following Luigi Pellizoni (2001). **Soft power** seeks to 'elicit cooperation without commanding it' (ibid.: 2).[3] Primarily, it is not based on material inducements (carrots) or sanctions (sticks), though it may relate to their use in certain situations. It may thus not be perceived by those subject to it as power at all, nor even by those exercising it. Ernest Gellner (1988) suggested as much in his sweeping historical monograph *Plough, sword and book*. More recently, Manuel Castells, discussing mainstream and 'alternative' media, has observed that 'The fundamental battle being fought in ... society [today] is the battle over the minds of people' (2007: 238). Famously, Michel Foucault, whose ideas I discussed briefly in Chapters 1 and 3, argued that 'power-knowledge' began to be operative far beyond the realms of organised religion, sovereign diktat or the national state during Europe's Enlightenment period. It became 'dispersed' among a set of new (or significantly altered) institutions whose members gained authority on the basis of their specialised or esoteric knowledges about some (or all) sections of society. For Foucault, people became subject to power day-in, day-out and not only on those occasions when, for example, somebody breaks the law and is fined or imprisoned. He thus talked of power as 'capillary', as something that permeated the entirety of a society like his native France. Foucault argued that 'power to' was not centred on just a few 'major' organisations but distributed among many. Each created 'domains' in which knowledge was manufactured and put to work (such as psychotherapy, commercial law and 'special needs' children's education). 'Social power', he remarked on one occasion, is 'coextensive with *every* social relationship' (Foucault, 1982: 224, emphasis added). See Box 6.2 for more on soft power.

BOX 6.2

KEY REPRESENTATIONS OF SOFT POWER: 'IDEOLOGY' AND 'HEGEMONY'

Aside from the writings of Michel Foucault, those of Karl Marx (1818–83) and Antonio Gramsci (1891–1937) have arguably done more than most to shape Western academic conceptions of what I've called 'soft power'. Among his many intellectual and political contributions, Marx argued that ideology was a key way in which social inequality was concealed (specifically, that on which capitalism is predicated and which it reproduces). If alive today he'd undoubtedly point too to the concealment of huge environmental damage. Ideologies are a set of beliefs, arguments and norms that cohere into a definite 'way of thinking' that then informs social action (and inaction). Marx (like Antoine Destutt de Tracy, the French aristocrat-turned-revolutionary said to have coined the term 'ideology') was interested in how the means of social communication could be monopolised so that the mindset of a whole society could be influenced. By 'means of social communication' I mean everything from the newspapers to the law to novels to public rituals – what Marx sometimes called the 'superstructure' of society in which people's beliefs and values were conditioned and contained. This superstructure, Marx (and Friedrich Engels) argued in *The German ideology* (1846/1976), misrepresented the realities of class inequality upon which the capitalist economy ('the base') depended. This is because the content of superstructural discourses and signs was controlled by capitalists who sought to legitimise their interests and ensure social order by concealing this inequality. For example, the concept of the entrepreneur that gained currency in nineteenth-century Europe and North America was, in Marx's view, a misrepresentation of the real conditions in which economic agents are obliged to exist in a capitalist society. Far from being 'free' to become successful entrepreneurs through their guile and inventiveness, most people were consigned to a life of selling their capacity to work (their labour power) to others for a wage (often a very low one). This consignment, Marx argued, was based on the 'original sin' of 'primitive accumulation' – an eighteenth-century process of separating people from the 'means of production' that involved 'hard power' (e.g. physical enclosure of land, the dispossession of farmers) and which created a class of asset-less people 'free' to work for capitalists. In summary, for Marx, 'ideology' was a mesh of ideas and representations that sustained ruling class interests by obscuring the inequities those interests depended on. Today he might point to the discourse of 'sustainable consumption' as ideological. It invites consumers to believe – wrongly Marx would doubtless argue – that ceaseless commodity consumption and environmental protection are compatible.

My gloss of Marx is necessarily brief to the point of being inadequate, but I've said enough to introduce Gramsci's particular conception of soft power. Writing in the 1920s and 1930s in Italy, he proposed a subtler and more 'positive' conception of ideology than that associated with the Marxism of his day. He was unsatisfied with the ideas that (1) ruling ideologies operate without much dissent; (2) they are determined by dominant class interests of the bourgeoisie; and (3) ideologies are 'negative' (they always hide or conceal things so as to disadvantage large sections of a society). Gramsci wrote at a time when national states were far more socially powerful and visible to citizens than in Marx's day, when 'civil society' was denser, and when Marxism itself had successfully challenged capitalist ideology (notably in Russia in 1917 when communists led by Lenin had taken control of the country). He was interested in how social power was not so much imposed as *negotiated* by ruling groups and institutions. His notions of 'hegemony' (rule by consent), 'historic bloc' and 'war of position' all spoke to how power was exercised in complex societies marked by a range of often colliding interests (economic, religious, cultural, political) and goals. Gramsci argued that dominant groups had to operate in a 'force field' such that give-and-take was often required, and clever means of persuasion necessary in order for these groups to realise their perceived interests. Ideologies can thus go well beyond 'class' issues and are not monopolised by the bourgeoisie, but, equally, society is not pluralistic in the sense some political scientists later came to represent it – that is, composed of multiple actors who enjoy relatively *similar* rights (through the democratic process) to have their agendas taken seriously by others. For instance, today we might say that the idea that 'government-led regulation + technology = environmental sustainability' is hegemonic in the West. While there are plenty of radical critics of this idea, Gramscians point out that they have not won the war of position with big business, national government or libertarian think tanks.

For a more textured appreciation of Marx and Gramsci on ideology and hegemony, see the early chapters of Terry Eagleton's *Ideology* (Eagleton, 1991).

Apart from advancing a 'decentred' conception of social power ('[L]et us not look for the headquarters,' he once memorably said (Foucault, 1981: 95)), Foucault also famously described modern social power as 'productive'. Consider this much-quoted passage from his book *Discipline and punish*:

> We must cease once and for all to describe the effects of power in negative terms: it 'excludes', it 'represses', it 'censors' ... it 'conceals'. In fact, power produces: it produces reality, it produces domains of objects ...
>
> (Foucault, 1979: 194)

Likewise, in the essay collection *Power/knowledge* he argued that:

> What makes power ... accepted is simply the fact that it doesn't only weigh on us as a force that says 'no', but that it traverses and produces things, it induces pleasure, forms knowledge, produces discourse. It needs to be considered as a productive network which runs through the whole social body ...
>
> (Foucault, 1980: 119)

This conception of **productive power** is useful insofar as it stops us from fixating on overtly 'disciplinary' institutions (like criminal courts or prisons) and the epistemic communities that sustain them. As I intimated in Chapter 3, it also encourages us to see the everyday 'subject positions' comprising all selves as the media of power, rather than already established 'surfaces' that power seeks to 'penetrate' from the outside.[4] To cite Foucault again: 'individuals are the vehicles of power, not [merely] its point of application' (ibid.: 98). What's more, his notion of power obliges us to consider the possibility that information, discourses and representations presented as 'truthful', 'objective', 'in the public interest' and so on are, in fact, constitutive of social power relationships. This is most obvious when a ruling political party in an autocratic state uses public news channels to disseminate propaganda (refer back to Box 3.4). For Foucauldians, there's no 'exogenous location' where we can examine, hide from or counteract social power.[5]

Debates about social power

These plus-points notwithstanding, Foucault's ideas have been subject to searching criticism even before his untimely death in 1984. In many ways these ideas have, for a generation, been *the* reference point for anyone in the Western social sciences and humanities who wants to analyse social power and be taken seriously by their peers. Though many commentators have endorsed his focus on 'epistemic power' as distributed capacity, others have regarded Foucault's conception of power as far too indiscriminate. As the American political philosopher Nancy Fraser noted many years ago, 'The problem is that Foucault calls too many different sorts of things "power" and simply leaves it at that' (1989: 32). Let me itemise some of the reasons why this problem has been regarded by many critics (not just Fraser) as a serious one (these items are summarised in Table 6.1).

- *Arguably soft social power takes significantly different forms that shouldn't be conflated*
  For instance, one might ask if the relationship between producers and consumers of *National Geographic* magazine involves power in the same way (and towards the same ends) as the relationship between an expert on the risks of a new swine flu outbreak and the public s/he communicates with via the news media. Foucault, in much of his writing, did little to distinguish the various forms that soft power takes. From the perspective of those subject to social power (in both its 'hard' and 'soft' forms) this is problematic. Depending on one's social location, the precise mixture of power relations one is subject too will, critics of Foucault have pointed out, vary significantly.
- *Arguably some forms of social power are perceived as socially acceptable, others not*
  One might wish to distinguish between 'legitimate power' (like that which the British Environment Agency exercises) and 'illegitimate power' to account for how social perceptions of power matter to the way it operates. Or one might, relatedly, identify cases of the illegitimate use of legitimate power (or, as Box 6.1 showed, the illegitimate *non-use* of power). Again, Foucault, in much of his work, did not delineate forms of social power according to how they are understood by lay and authoritative actors in any given society. He tended to focus on situations and institutions

Table 6.1 Axes of debate about social power in light of Foucault's germinal analyses: a summary

Analytical problem	*Analytical solution*
Social power takes a range of distinct forms	Identify these forms and study their interactions and non-interactions
Some forms of social power are perceived as legitimate	Distinguish socially legitimate and illegitimate forms of social power, and cases where legitimate power is exercised illegitimately
Different forms of social power interfere with one another, by design and by accident	Identify the contradictions and tensions among different forms of power, not just the complementarities
Some forms of social power are more efficacious than others	Recognise asymmetries between forms of social power
'Soft' social power can work by hiding and silencing as much as by making certain things known	Attend to the absences as much as the presences in socially powerful discourses and knowledges
Social power can sometimes empower those 'subject' to it	Avoid presuming social power is a zero-sum game between social actors
Some forms of social power escape the intentions of those directly involved in exercising it	Look for the unintended and unanticipated effects of social power
Social power does not always achieve the goals of those discharging it	Look for signs of resistance to, or indifference to, social power in the wider society
Social power is not separate from the process of naming and analysing it	Recognise that the analyst of social power is necessarily implicated in that which s/he studies

in which those involved tended not to perceive social power as being operative.

- *Arguably, some forms of social power serve to countermand, stymie and resist other forms*
  At various points in his career, Foucault looked for widespread 'rationalities' of social power. This risked eliding the tensions, even contradictions, between different modalities of (both 'legitimate' and 'illegitimate') forms of social power. For instance, the representational power of the mass media is regulated by powers exercised by press complaints commissions, courts (who enforce privacy and decency laws), and so on. Is this the same as the power of a national government to prevent environmentalists from protesting outside a nuclear fuel enrichment facility?
- *Arguably some institutions, epistemic communities or sections of a society exercise far more soft power than do others*
  To say, as Foucault frequently did, that social power is dispersed is not to say it's evenly spread. For instance, Marxists since the time of Marx have been insistent that capitalists, in various ways, seek to advance

their own particular interests through persuading the wider society that these are shared or common interests. Marx's notion of 'ideology' and Antonio Gramsci's concept of 'hegemony' are just two examples of how Marxists have, historically, highlighted the differential power of some to 'set the agenda' for a wider society. Likewise, radical feminists have long pointed to the powerful discursive dimensions of patriarchy, wherein men, in different ways, define, communicate and enforce particular ideas of 'masculinity', 'femininity' and 'gender roles' that tend to disempower women.[6] Foucault, in part because he took issue with the conventional social science idea that social groups have distinct and coherent 'interests', was mostly unwilling to concede that soft power was a capacity far less 'decentred' than he often insisted it was.

- *Arguably soft power works as much by concealing and constraining knowledge, discourse and information as by creating and mobilising it*
  Though Foucault usefully pointed to the 'productive' aspects of soft power (as I noted), he's been criticised for underplaying the power of some to *withhold knowledge deliberately* or to represent the world in *highly and knowingly selective ways*. Historian of science Peter Galison has called the study of such power 'anti-epistemology' (Galison, 2004). Such study need not rely on the Marxian notion of 'false consciousness', a notion that Foucauldians have understandably been suspicious of (as well as many Marxists).[7] We might also add that power sometimes operates when epistemic actors or communities use uncertainty (e.g. about the extent of future climate change) as a basis for inaction or slowing down reform. It operates too when they rule out certain lines of inquiry, leading to what we might call 'systematic ignorance'. For instance, Daniel Kleinman and Sainath Suryanarayanan (2012) have shown that professional agro-ecologists have favoured a particular hypothesis to explain Colony Collapse Disorder in bee populations. By committing research time and resources to testing this hypothesis, they have excluded alternative explanations advanced by beekeepers, ones that point to harmful manufactured insecticides as the root cause.
- *Arguably some forms of social power empower those subject to them*
  The political theorist Mark Haugaard (2012) has detected an implicit 'zero-sum' conception of social power in much of Foucault's work. This conception, he argues, ignores forms of social power that are 'positive sum'. These forms *increase* the capacities of one or more of the parties involved to act; the relatively *disempowered* can become *empowered* by being subject to the advice, guidance or counsel of others in positions of authority or influence. For Haugaard, this is only a seeming paradox and it calls into question purely *negative* evaluations of social power as something that distorts, corrupts or oppresses. Arguably a 'just society', however 'justice' is defined, *requires* social power relationships in order that injustice be minimised. The case considered in Box 6.1 suggests as much, among many other similar cases.

- *Arguably, some forms of social power have major unintended and uncontrolled effects irreducible to their causes*
'Power over', some critics of Foucault have observed, is a frequently unforeseen and unplanned result of the intentional use of 'power to'. It thus greatly exceeds the intentions and aims of those who exercise it, and may even disrupt their plans and activities. Though Foucault realised this, he paid little attention to the mechanisms whereby unintended effects are produced. The global financial crisis of 2008–9 is a prime example of these mechanisms. Financial services professionals used a tremendous amount of knowledge and information to lend/invest colossal sums of money, yet the result of their collective actions, realised through various investment funds and new financial instruments (like collateralised debt obligations), was ultimately 'irrational' for them and painfully real for hundreds of millions of people. Banks went bust, people's homes were repossessed by mortgage lenders, a 'credit crunch' hurt businesses in need of loans, and so on. The further knock-on effects have, quite unintentionally, been good for the non-human world. A period of economic recession has temporarily reduced the West's ecological footprint.
- *If social power is decentred, as Foucault argued, then there must be the possibility of resistance to it*
Though his later writings began to address the issue, Foucault's early and middle writings said little about opposition to social power. Indeed, some of his followers argue that 'subjects do *not* cease to be governed when they undertake certain practices we can categorize as "resistance" or "dissent"' (Odysseos, 2011: 440). Critics have pointed out that ostensibly powerful groups, institutions or individuals are only powerful in certain arenas and relatively less powerful in others. Likewise, they've noted that those with a relative lack of power in most arenas of life are nonetheless able to question or challenge some of the power relations in which they're embroiled. What's more, if social power was always wholly effective would it permit room for analysts like Foucault to name and anatomise it? Presumably not. Accordingly, John Scott defined power as 'the socially significant affecting of one agent by another *in the face of possible resistance*' (Scott, 2001: 3, emphasis added).

This 'gap' between 'power to' and 'power over' has, to take a notable example, been the consistent focus of political historian James Scott's work since his germinal book *Weapons of the weak: everyday forms of peasant resistance* (Scott, 1985). As Scott shows, even in the most unpropitious situations, people protest, in some cases removing or curtailing the powers of important institutions and actors. Relatedly, cultural studies scholars reaching back to the 1970s have explored how 'sub-cultures' and 'counter-cultures' form, and how (un)successful their members are at resisting the force of dominant social norms and practices. While not all resistance to power is effective, it would be cynical to suggest that even ostensibly successful acts of opposition are, 'really', just incremental accommodations made by the powers that be. Equally, it would be analytically and politically unhelpful to regard such acts as themselves expressions of social power akin to the capacities possessed by those whose power is being opposed. This is because acts of opposition are sometimes only symbolic, of temporary efficacy, or take a 'one-off' form that's hard to reproduce.

We should also note that the discourses of potentially powerful institutions, like climate science (discussed in Chapter 8), may have little or no effect, despite the hopes and intentions of their creators. This is because others (e.g. politicians) are free to *ignore* or not act on the full implications of these discourses.

- *The study of social power and resistance to it are arguably implicated in that which they analyse*
  For all his reflexivity, some critics have detected in Foucault's writings a curious tendency to assume that social power exists 'out there', waiting to be represented by the analyst clever enough to identify its hidden or hitherto unfathomed forms. But what if representations of power in the social sciences and humanities are *themselves* part of the process whereby social power is reproduced? Is it possible to have an 'objective' view of what social power 'really is'? Is research into social power contributory to resistance to it? Should the analyst of social power have a justified normative stance on that which they study, identifying 'good' forms of social power from 'oppressive', 'illegitimate' or 'harmful' ones? If so, should they criticise the powerful and also lay actors who (in the analyst's view) fail to understand how social power affects them?

Foucault gave different answers to these and related questions at different points in his career. He seemed to recognise the tensions inherent in any attempt to analyse and evaluate social power. There appears to be no way of making sense of social power – either what forms it takes or how we should judge them – that does not beg some very large questions about the analyst's role (questions that won't admit of clear-cut answers).

How best to make sense of social power?

What can we learn from these debates about the nature and effects of social power? There's clearly no consensus on what social power is or the means, ends and outcomes of its operation.[8] In this light, we appear to have two analytical options. On the one hand, we can accept that social power may take several different forms whose nature and precise modes of operation are still subject to debate. We can focus on one or more of these and argue that the conception of social power that *we* are advancing is preferable to all the others: in some way it's more 'accurate', 'truthful', 'perceptive' or what-have-you. We thereby enter the fray and seek to shape the debate from the inside. On the other hand, we can infer that different conceptions of social power are *themselves weapons* in the process whereby both hard and soft power are variously obfuscated, reproduced, revealed, resisted and changed. Rather than try to adjudicate between these conceptions, we can regard them as *interventions* that are jostling to shape thought (and practice) about social power. We don't then look for the 'right' conception because ideas about social power, like all ideas, knowledge or information, are being judged according to the effects they seek to engender. From this perspective, academic analysts of social power are using their epistemic credentials to produce representations that they hope will be taken seriously by their peers, students and members of the wider society. We can borrow these representations and put them to work insofar as they're consistent with the general arguments we're trying to advance. We give up any search for the 'essence' of social power and regard claims about its 'true' nature as rhetorical, not only cognitive, as attempts to displace 'rival' conceptions. We need not be bound by these claims nor swayed by these attempts.

As was evident at the end of Chapter 4, where I discussed a pragmatist approach to discourse, representation and knowledge, I prefer the second to the first option. Rather than link my earlier claims about epistemic dependence, communicative genres and epistemic communities to one particular conception of soft power, I prefer to point out that, depending on what available notion of soft power we cleave to, the link can be (and has been) made in several ways. I will shortly detail one of these ways, referring again to empirical cases (an extended case in this chapter and others in the next chapter), and show how representations of nature can be regarded as media for attempts to turn 'power to' into 'power over'. As I've done previously in this book, I'll consider cases analysed by members of the loose epistemic community in which I'm situating myself. The case discussed in this chapter (relating to supposed deforestation), and those to follow, should be read in the context of my various arguments in Chapters 2 and 3 about the uneven effects of epistemic dependence and their implications for the quality of democracy. Though wedded to rather different conceptions of soft power, the cases are consistent with the general claims made in these two chapters. For instance, my generic conception of 'governmentality' (one of Foucault's terms) can accommodate several possible understandings of how representations become vectors of social power relationships. These understandings become resources I can use rather than ideas whose accuracy should be assessed with reference to a putative ontological 'court of appeal' existing 'out there'.

Let's now consider one way in which social power might be said to depend upon representations of nature for its efficacy, recognising, in light of the discussion above, that it's hardly the only way.

DEFORESTATION DISCOURSE

Since the early 1990s, a major component of the discourse of 'environmental crisis', part of the 'end of nature' narrative I discussed in Chapter 1, has been talk of deforestation. The Amazon Basin has become emblematic of tree loss worldwide, along with associated problems of soil erosion, loss of habitat for wildlife, heightened flood risks and reduction in biodiversity. In addition, the huge increase in atmospheric greenhouse gas concentrations since the early nineteenth century means that forests are seen as key to the mitigation of global climate change because they 'lock in' carbon dioxide. Subsequent to the first United Nations Earth Summit, in 1992, heightened efforts have been made worldwide to halt the loss of old growth forest, to encourage natural regeneration of deforested areas and to increase the number of forest plantations. Even so, tree loss is ongoing in many regions according to every major international environmental organisation and not a few environmental education organisations. Consult the FAO's (Food and Agriculture Organisation) biannual *State of the World's Forests* reports, UNEP's *Vital Forest Graphics* (2009), World Wildlife Fund's *Earth Book* and *Living Planet Report* (both 2012), Conservation International's *Climate for Life* (Mittermeier *et al.*, 2008) book and its 'Deforestation, logging and GHG emissions' fact sheet (2009), or the 'eye in the sky' images on the *National Geographic* website and there's a consistent message: a combination of illegal logging, weak regulation, inappropriate regulation, corruption, increased demand for food, increasing population, heightened wood-fuel and charcoal demand, and irresponsible or ignorant farming practices are eating into an already greatly diminished global forest area.[9] Indeed, the problem could be even worse than many believe. Critics claim that FAO statistics, which are usually considered to be the most authoritative, utilise a very weak criterion for what counts as 'forest'. This means that they are probably misclassifying scrub and wooded savannah as much denser vegetation cover.

Though deforestation is occurring globally, and reforestation is seen as a climate change mitigation 'good' wherever it occurs, public perceptions of where the tree-loss problem is most acute are highly selective. Typically, we focus less on temperate locations like British Columbia (discussed early in Chapter 4) and have, instead, learnt to focus on the tropics – not just the Amazon Basin (and Brazil in particular) but also countries like Madagascar and Indonesia. Trees in such countries have become virtual icons of biodiversity loss and climate change. This is not unreasonable: after all, there *is* ample evidence of huge forest ecosystem destruction in many equatorial and sub-equatorial countries. But how accurate is this evidence? How sensitive to variable de- and reforestation dynamics are official reports, news broadcasts, newspaper stories, television documentaries or radio reports? And how are representations of tropical deforestation consumed by publics far distant from the countries where chronic tree loss is said to be occurring?

I will focus on the vital epistemic role of the mass media in Chapter 7, but want to examine the deforestation discourse it draws upon – a discourse advanced by the FAO, among many other governmental and non-state organisations. This is very much an expert discourse, one produced by scientists employed by universities, government agencies, conservation organisations, ENGOs and quasi-governmental bodies. These scientists are typically university trained and many make periodic 'field visits' to (de)forested environments (rather than living in or near them). They 'speak of' and 'speak for' trees simultaneously. Their discourse can exert power in two directions: aside from shaping mass media representations (and thus public perceptions) of de- and reforestation, it also influences governmental policies in myriad countries – and thus the livelihoods of people living in (or dependent on) forest environments. It is thus a **polyvalent discourse**.

When I say discourse I should really say discourses in the plural, ones that are sometimes synthesised into a composite account. For example, the above-mentioned *State of the World's Forests* reports take country-level studies and meld them into a global overview that seeks to identify patterns and trends. This is also true of the FAO's *Global Forest Resources Assessment* (2010), which is its most comprehensive amalgamation of country-level data in the world.[10] These national studies are themselves aggregations of local and regional investigations such that micro-, meso- and macro-scale realities are, in theory at least, captured rather than concealed. Needless to say, proper monitoring of changes in forest area across several continents is no mean feat. To do it properly requires many trained and experienced personnel, a good deal of technology (notably, granular satellite imagery), and much time and money. If countries are to know when and how to arrest deforestation, then they need a reliable evidence base concerning the nature and drivers of land use change within their borders.

Some believe that this evidence base has been absent or wanting – if not today then certainly in the recent past. They argue that local and regional specificities in land use change have been misrepresented and assimilated to a more global discourse of tropical tree loss. In this way, they suggest, an **environmental myth** of generalised equatorial deforestation has been maintained over several decades. It follows, for them, that the current discourses about tropical de- and reforestation have been insufficiently diverse or precise, often with important consequences for land users. We can better understand what they mean by scrutinising a particular case, that of the forest–savannah transition zone that crosscuts the countries of West Africa, and with a specific focus on southeastern Guinea (see Map 6.1).

The crisis of tree loss in West Africa?

The several countries comprising coastal West Africa (Benin, Togo, Ghana, Ivory Coast, Liberia, Sierra Leone and the Republic of Guinea) experience a humid-tropical climate that becomes progressively drier towards the north, where one finds semi-arid countries like Burkina Faso. During the Holocene, or at least the last several hundred years of it leading up to the present, evidence suggests that the coastal zone has been dominated by dense evergreen and semi-deciduous forest, giving way to dry forests, then forest-scrub and eventually treeless savannah the further inland one goes. Since the late

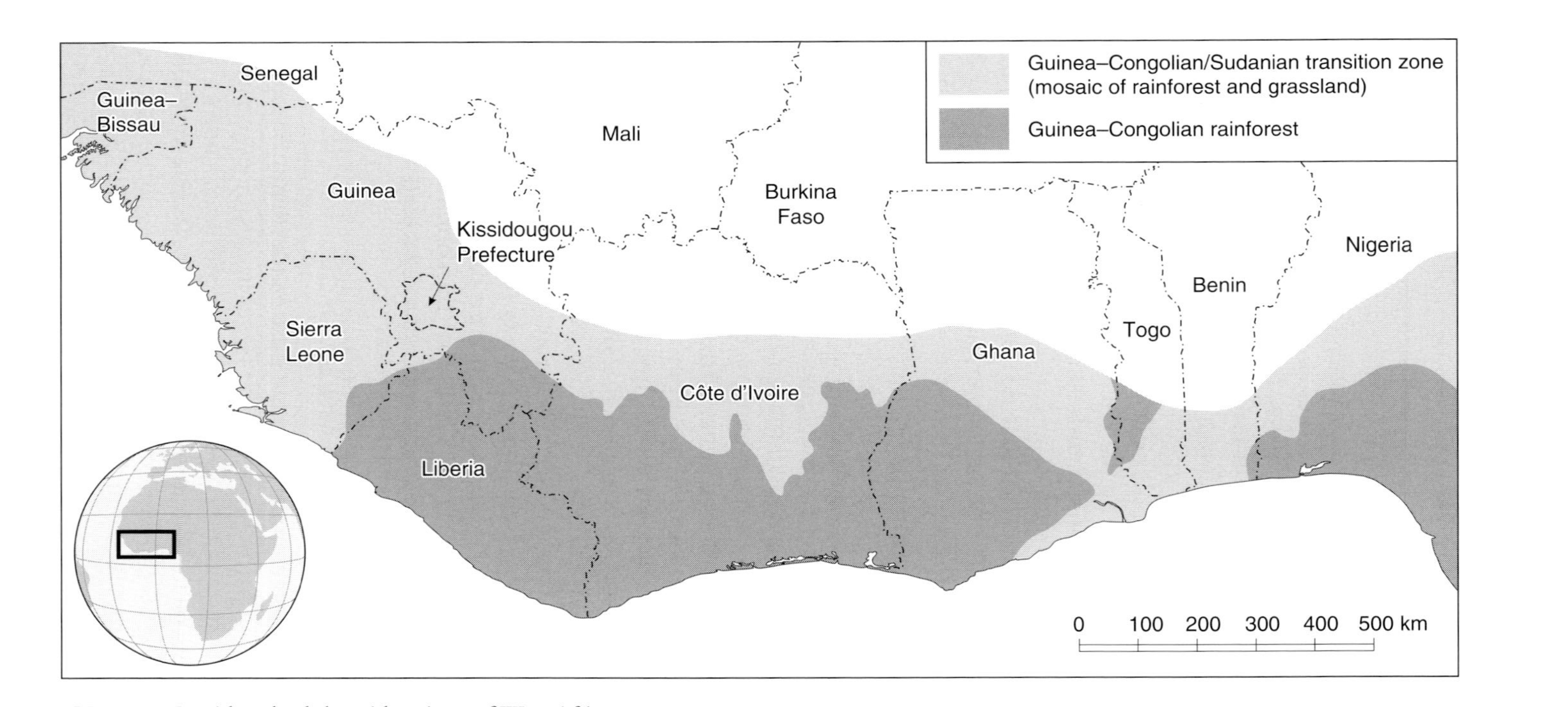

Map 6.1 Humid and sub-humid regions of West Africa

1980s, West Africa has been seen as a deforestation 'hot spot'. For instance, in the year of the first UN Earth Summit, the World Conservation Monitoring Centre reported that only 13 per cent of original West African forest cover remained (Sayer *et al.*, 1992: 74). Likewise, a slightly earlier research paper co-published by NASA (the famous US National Aeronautics and Space Administration) reported up to 96 per cent forest loss during the twentieth century in Benin, Ghana, Liberia and Ivory Coast (Gornitz and NASA, 1985).

Consequently, most West African governments have intensified earlier efforts to conserve and regenerate forest since the early 1990s. Like other governments with deforestation problems, they have received considerable encouragement from the FAO, UNEP, World Bank and a host of international conservation organisations. This encouragement is often financial: development aid from wealthy countries often has 'green conditions' attached to it, while other monies hailing from private, charitable or activist sources have expressly green aims (such as 'carbon offset' funds that pay for forest protection and expansion projects in the tropics).

What does this attempt to prevent West African deforestation mean on the ground? It means a raft of government-led or government-backed policies and schemes, from new forest plantations, restrictions on logging of existing forest, the creation of forest reserves and attempts to encourage natural reforestation of designated deforested areas. Because West African countries have large rural populations, these policies and schemes necessitate working with, or controlling the activities of, many farming communities, including groups of shifting agriculturalists who move locations in the short, medium or long term. In theory, and very often in practice, rural people must work within (or around) measures designed by state officials based in Guinea's cities and interested parties hailing from overseas.

These government-led measures may seem necessary, 'obviously required', in deforested areas within roughly 400 kilometres of the West African coastline (the zone of closed canopy evergreen and semi-deciduous forest). But what about the transition zone to the north where forest gives way to mixed and ultimately very different (grassland) vegetation cover? When and where are pro-forest measures required, given that only some of the zone is (or could be) dense or even open woodland, given the prevailing climate? Unlike political borders and boundaries that divide countries, those separating ecotones are usually understood to be gradual rather than abrupt. This absence of clear demarcatory, let alone *straight*, lines presents an analytical challenge to environmental scientists, policymakers and policy implementers. In the transition zone, what does any given patch of forest, brush or open savannah signify in a wider biogeographical sense? To see how this question was answered, and what its specific implications for local land users have been, I want to consider what happened in Kissidougou prefecture in southeast Guinea. The dynamics of forest cover in this area were the focus of sustained research by the British anthropologists James Fairhead and Melissa Leach in the 1980s and 1990s, and I draw substantially on their findings in the story recounted below.

The roots of epistemic inequity in upland Guinea

Kissidougou covers part of a large plateau punctuated with hills and mountains that extends into countries adjacent to Guinea. If you were to walk from the coast through Liberia and into Kissidougou prefecture you would, during the course of your very long journey, notice that the tropical forest gradually thins out until you see numerous patches of dense forest as seeming outliers in a landscape dominated by grasses, shrubs and small trees (see Plate 6.1). There are over 800 such patches. If you looked closely, you'd see that some exist near watercourses and swamps, but most do not. They are small relative to the area of surrounding savannah, metaphorical islands in a sea of mostly non-forest vegetation. They contain many indigenous tree species found further south in the moister coastal zone. If you walked into some forest patches you'd discover (probably to your surprise) settlements at their heart, perhaps dominated by members of the Kissia ethno-cultural group (sometimes known as 'forest people'). If you talked to these Kissia farmers, who live in and largely off the forest, you might hear older members mention the hunter-gatherers of Mandinka origin who inhabited the grasslands and whose local relations (the Kuranko) migrated south during the twentieth century. As well as being an ecological transition zone, you'd then realise you're also in an ethno-cultural one where, over many years, Kissia and Kuranko have coexisted, sometimes mingling and mixing too.

Study Task: Image yourself having really made this journey as recounted above. You're now standing in the transition zone. What might you reasonably conclude about vegetation dynamics in Kissidougou from observing stands of dense forest in an otherwise savannah landscape?

The existence of forest patches suggests that the regional climate (which supplies some 1,600 millimetres of rainfall per annum) can support far more than open woodland, shrubs and grasses. But their isolation, often existing some considerable distance from continuous forest further south, might cause you to make the following deduction: that the forest islands are relics of an earlier period when the forest–savannah frontier was further north than it is today (refer to Box 1.2 and the discussion of metonymy and the 'nature effect'). It was precisely this deduction that was made over a century ago by the first European botanists to survey the prefecture and others like it in West Africa. Processes of vegetation change were presumed to have occurred based on logical reasoning about what observed forest islands represented.

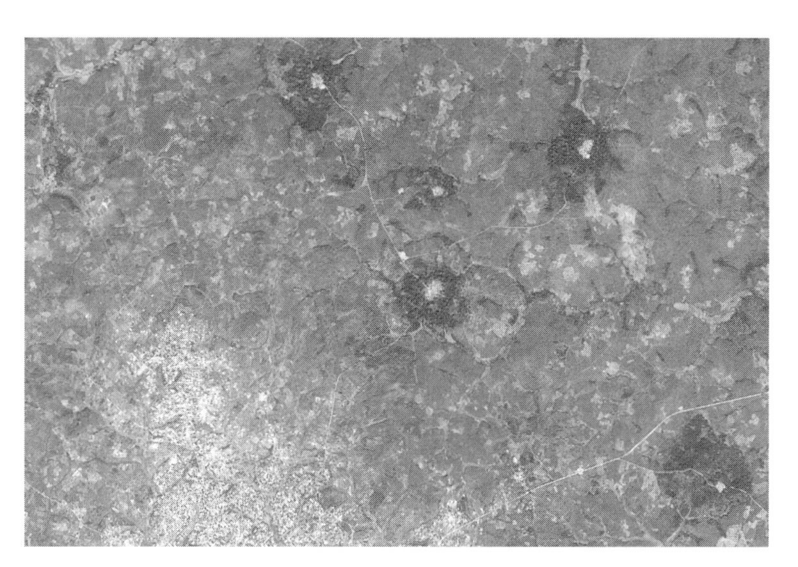

Plate 6.1 Forest islands in Kissidougou prefecture, Guinea: markers of historic tree loss?

Image reproduced from Google Earth. Aside from Kissidougou town to the south, most of what you see here is grassland, shrubland and farmed land. The tree islands are the small dark circular areas.

Because more forest *could*, seemingly, exist in the present it was presumed that it *had* existed in the past. This meant that the 'natural' zone of transition to 'true' shrub- and grassland was presumed to be many kilometres further north. The image that an advancing savannah evoked became a key aspect of Guinea's forest policy until the late twentieth century. Let me explain how and why, and what (until fairly recently) the practical consequences have been.

The reasoning underpinning deforestation discourse

Guinea became part of French West Africa (FWA) in the 1890s. Like Britain and Germany, France sought to extend its sphere of political and economic control by colonising large parts of Africa during the late nineteenth century. It appointed military governors to rule the territories of FWA, who were soon followed by the apparatus of a modern national state, namely administrators, technicians and scientists responsible for managing separate domains (e.g. farming, mining, taxation). Early observers and surveyors of Kissidougou interpreted its forest islands as evidence of advancing savannah, just as we might today if we passed through the prefecture on our imagined coast-to-desert transect. Kissidougou's first administrator reported in 1893 that 'immense forests ... cover a large part of the soil, and ... covered it entirely at a period relatively little distant from our own' (Valentin, 1893: 1G188). Twenty-one years later, the director of the prefecture's agricultural research station lamented that 'A region so fertile [has] become a complete desert. Now there exists no more than a little belt of trees around each village . . . ' (Nicholas, 1914: 1R12). More than a generation after this observation, Kissidougou's chief botanist wrote of 'oases of equatorial vegetation in the middle of savannah burned by sun and fire . . . all in regression' (Adam, 1948: 22).[11] On what basis were these and other confident claims by European settlers made and sustained?

Valentin and others did not have the benefit of modern time-series aerial photography or satellite imagery. What's more, though they could have consulted Kissia and Kuranko farmers about their experience of decadal vegetation change, they chose not to – at least not in any sustained or deep way. This no doubt reflected the attitudes of superiority that underpinned the everyday racism, be it malign or 'paternalistic', that structured everyday life in Europe's African colonies up until the 1950s and 1960s when they became independent nation states. In addition, West African botanists before Valentin's time had observed that palm oil trees grow quickly in 'secondary forest' (i.e. that which replaces original forest after, say, logging) but do not thrive in savannah. Thus, when they noted several palm oil trees in most of Kissidougou's forest islands, it was taken to indicate the loss of mature semi-deciduous forest to land clearance for farming or grassland-scrub fire.

Moreover, the absence of settlements in many forest islands reinforced the idea that they were natural, vestiges of once continuous tree cover. As important, if not more, than all these reasons for Kissidougou's deforestation discourse were influential new ideas in ecological theory. In the early twentieth century, American botanist Fredric Clements (1928) published what became a germinal book, *Plant succession*. Based on field observation and experimental data, Clements proposed that each climatic zone gave rise to what he called a 'climax community' – an ensemble of vegetative species that are well adapted to prevailing patterns of temperature and precipitation, varying according to local relief, soils and hydrology. When 'disturbed', a vegetative community will, Clements argued, return to its climax condition (given sufficient time), with 'secondary succession' species colonising the soil only to give way to larger and denser vegetation (e.g. trees) later on.

Clements's arguments about stable climate–vegetation equilibria were, intentionally or not, normative. They implicitly proposed a biogeographical 'baseline' against which current vegetation patterns in a given region could be compared. His work also popularised the idea that sub-climax vegetation was an aberration, either because of extreme natural events (like a prolonged drought) or human interference with natural processes. *Plant succession* positioned people as exogenous variables, not constructors or co-producers of vegetative landscapes.[12] Coincidently, Clements's book was published the same year Woodrow Wilson created America's first national parks, one of several federal responses to the large-scale destruction of plant and animal species (like the bison and passenger pigeon). His work soon proved influential in the then relatively small world of botanical education and agro-forestry practice. In West Africa, its spirit came through strongly in the monumental works of French scientist and administrator André Aubréville (1897–1982), who is credited with coining the term 'desertification' in 1949.[13] Aubréville's (1949) book *Climate, forests and desertification in tropical Africa* took a macro-perspective, generalising about regional biogeography from local observations made by him and his (small) peer group in French and British West Africa. He rose to become Inspector General of Water and Forests in the whole of FWA.

A key epistemic worker, Aubréville was confident that the retreat of mature forest in West Africa was anthropogenic. It scarcely occurred to him and his peers that forest loss, if indeed it *was* occurring en masse, might be due to long-run climatic variability. Still less did it occur to them that forest areas might be *advancing* in regions like Kissidougou, and advancing in the *absence* of colonial state-controlled afforestation projects. These possibilities were considered fleetingly, if at all. The result was that by the 1950s, Guinea's forest policy was premised on the idea that ordinary people in rural areas needed to have their land use decisions monitored and regulated robustly. Several purported anthropogenic causes of deforestation underpinned this idea – often focused on Kuranko 'savannah people' migrating south.

First, it was suggested that the political stability created by French and British rule brought an end to internecine conflict in the transition zone. This, some believed, allowed population numbers to grow and encouraged Mandinka peoples to slowly move towards the forest–savannah boundary. Second, these people's subsistence practice was, among other things, to hunt meat and collect honey. Neither is possible in dense forest, and fire setting in shrub- and grassland was often employed to flush out prey. This was seen as one cause of forest loss. Third, by the 1910s, if not sooner, both Kissia and Kuranko peoples were farming on open or shaded savannah, in part to meet their own dietary needs but also to sell any surplus in order to acquire other foodstuffs or goods through local trade. Felling trees or burning wooded savannah to suppress shrub and tree growth was seen as leading to a greater area of upland rice and coffee farming, among other staple crops. Many of these crops found their way to householders in Kissidougou's increasingly populous towns, consistent with twentieth-century urbanisation in Guinea as a whole. New demand for food, apparently, resulted in new supply – at the expense of tree cover. Fourth, the second and third supposed causes of forest loss in the Guinea transition zone were sometimes linked to the open access character of the land. The lack of clear property rights, it was suggested, permitted land users to deforest the terrain because they faced few restrictions – except perhaps from other local residents who happened to value trees more than savannah or farmland.

The policy implications of presumed tree loss in the transition zone

In light of this reasoning about tree loss in Kissidougou, a number of regulations were introduced and enforced by the national forest service before and after Guinea achieved independence from France in 1958. These regulations were intended to conserve natural forest and facilitate new forest growth in what was taken to be 'derived savannah' around Kissidougou's islands of trees. Nature, it was presumed, needed protecting from people.

With varying degrees of success in their enforcement, the following measures became state policy in Kissidougou prefecture and elsewhere. First, the felling of most trees – young or mature – became illegal, unless a villager had applied and paid for a licence to log selectively from the local branch of the Guinea forest service. This policy rested on the idea that most trees did not belong to any individual, family or village community in Kissidougou and similar prefectures. Second, fire setting also became illegal, unless it was undertaken or overseen by state forestry officials. Third, some areas of derived savannah were closed off and made into government farms, in part to meet heightened food demand in Guinea post-1950 and in part to ensure 'rational' land management. Fourth, shifting cultivation was deemed in need of monitoring so that farmers used fewer sites overall. Finally, new tree plantations were created during the 1980s – artificial twins of the Ziama forest reserve created elsewhere in the transition zone in 1932 (and designated a UNESCO biosphere reserve in 1981).

Inevitably, indeed intentionally, these various measures impinged upon the ability of the Kissia and Kuranko to maintain their established practices of living off the land. When tree felling or fire setting occurred, and was noticed by local forest officials, fines were imposed and local people criminalised. Indeed

> agencies ... generally t[ook] ... such [actions] as further evidence of local ignorance and wanton destructiveness, and hence the need to implement policy with greater force, without questioning the appropriateness of the policies or their underlying analysis.
>
> (Fairhead and Leach, 1996: 116)

Together, these measures instantiated a division between state officials and local people, with the latter's agro-ecological knowledge and practice regarded as something to be managed and disciplined, not a resource to be studied or utilised. Natural forest islands and afforestation projects were imagined as 'buffers' protecting continuous forest further south from conversion to shrub- and grassland.

The occlusion of forest history in Kissidougou

A century after the first French colonialists arrived in Guinea, two English anthropologists undertook extensive field and archival work in Kissidougou in order to understand how effective state forestry policy had been in tackling land use problems. They interviewed local farmers living in and near many forest islands, examined closely aerial photographs and satellite imagery covering several decades, and scrutinised the written claims and field observations of travellers, scientists and government officials dating from the 1890s onwards. Their research suggested a very different interpretation of what Kissidougou's 'relict forest' signified: *the advance of trees into the savannah zone not their retreat.*

More specifically, James Fairhead and Melissa Leach supplied considerable evidence that it was local people, descendants of Kissia and Kuranko settlers who had intermixed over time, who were directly responsible for forest expansion. 'Paradoxically', they argued in their book *Reframing deforestation*, 'what appeared to be ... *undisturbed* vegetation ... turn[ed] ... out to be the *most disturbed*' (Fairhead and Leach, 1998a: 21, emphasis added). This implies that if there *has* been forest loss in Kissidougou and similar prefectures over the past century, it has been loss of *anthropogenic* (not 'natural') *forest*, and may have been *temporary* loss before farming practices created new forest islands over a period of many years. It also implies that government policies enacted in the interests of the 'natural environment' had, in fact, been protecting forest islands that were not, according to the conventional definitions, natural at all!

Study Task: Let's take Fairhead and Leach's thesis seriously. Assume their proposition about forest islands is correct. Now try to think of causes: how and why – intentionally or perhaps as a by-product of other practices – did local land users in Kissidougou create tree stands?

How and why did Kissidougou's rural inhabitants produce forest islands from as far back as the nineteenth century (and possibly earlier)? Fairhead and Leach's research suggested the following answers. First, the long-established practice of shifting cultivation in the transition zone enriched the soil rather than depleting it. By growing (for example) peanuts, cassava and various other root crops in open woodland or grassland, farmers actively worked organic matter into the upper soil layer – often supplemented by the use of household waste and cattle dung as manure. Fertility was thus maintained, while soil moisture retention was enhanced. Once a farmed site was left fallow, tree species seeded naturally because of the conducive soil conditions.

Second, herders used natural bush and tree species to create temporary stockades for cattle, but also to create a degree of community protection from intruders. By both accident and design, transplanted stakes and seeds grew into mature trees – especially fast-growing species like the African whitewood and *Ceiba pentandra*. Because of the persistence of inter-group warfare through the nineteenth century, tall trees were often planted deliberately to afford a view of the surrounding savannah, with silk-cotton trees being particularly favoured.

Third, Kissia people from further south inhabited 'forest fallows' (afforested patches growing on previously cultivated land) for a variety of reasons. They afforded families a degree of independence from their home communities, while offering the chance to pursue traditional forest-based agriculture. At the heart of this agriculture were (and remain) crops that required a humid microclimate and shelter from direct heat and strong wind that forest patches offered. At various points in Kissidougou's history, this involved planting kola trees, whose caffeinated fruit is edible and was traded with Kuranko peoples to the north. Similarly, coffee was grown in forest islands from the 1930s for trade purposes, later still bananas. What's more, cattle could be grazed on the island edges, and benefit from tree shelter during the hot season, along with the residents of island settlements themselves.

Fourth, to protect cultivated plots in the savannah and forest islands, the deliberate burning of grass early in the dry season became common practice among forest island residents. This practice prevented the build up of stored energy in late season grasses that often resulted in very intense savannah fires. It thereby permitted deciduous tree species to increase in number and density at forest island edges because these were normally suppressed by such late season fires. In addition, the harvesting of grasses to make and replenish the roofs of family homes inside the forests reduced the risk of intense fires in edge areas. Relatedly, many island communities created firebreaks *within* islands to protect older trees, homes, cattle, crops and people from especially dangerous savannah fire events.

Finally, because of changing gender relations within Kissidougou's forest island communities, more women had been able to cultivate their own cash crops from the 1950s onwards. This produced more 'gardening' around and in between islands and, once fallowed, these cultivated sites tended to be colonised by trees and shrubs rather than grasses (for reasons mentioned previously).

I have necessarily simplified a complex story of anthropogenic environmental change in Kissidougou. This story is derived from local farmers' own accounts of their land use practices and cultural habits, so too those of their forebears. That Fairhead and Leach took these accounts very seriously differentiates their representation of the forest–savannah mosaic from that of Guinea's forestry officials going back to Valentin's time. The accounts suggest that ordinary land users in Kissidougou have helped to bring the forest–savannah transition zone further north (not south) as a result of their ordinary subsistence and cash-cropping practices. These land users can be seen as a *component* of ecosystems (not exogenous forces), switching biomass characteristics between two or more different states in a transitional zone where the 'climax' vegetation is uncertain. They suggest too, though I haven't detailed it in the previous paragraphs, that forests and trees have long played an important cultural role in community life. Individual trees, and particular sites within forest islands, assume important roles in family histories, hierarchies of status within communities, rituals, rites-of-passage and the identities of males and females. Thus, when government-licensed loggers entered Kissidougou to cut down selected mature trees, it would often cause cultural offence because the trees in question possessed symbolic value. The question then arises: how, for so many decades, did state forestry scientists, managers and field agents manage to miss what Fairhead and Leach uncovered? Why did a generalised discourse about deforestation at the hands of errant farmers come to be applied in areas like Kissidougou where it ought not to have been?

The persistence of deforestation orthodoxy

We will never have a full answer to these questions. The historical record is incomplete and the various epistemic workers in question, such as André Aubréville, are long deceased. What the multi-decade persistence of Guinea's deforestation orthodoxy demonstrates was well summarised by Leach, writing with geographer Robin Mearns:

> It is not merely that 'knowledge itself is power' ... but that 'what *constitutes* knowledge, what is to be excluded and who is designated as qualified to know involves acts of power'.
>
> (Leach and Mearns, 1996: 16 [quoting Foucault, 1971], emphasis added)

Leach and Fairhead have offered some informed speculation about how Guinea's forest service was able to able to represent the forest–savannah mosaic as a zone of persistent tree loss for so long. I now list their suggestions in no particular order of priority.[14]

First, a pattern of non- or limited communication with local farmers established by French colonists was subsequently hard to overturn, even after Guinea became independent. Though some travellers and botanists did recognise that tree islands in Kissidougou and elsewhere may not have been historical relics, others (like Aubréville) disagreed or classified them as exceptions that proved the rule. Because the epistemic community of botanists and agro-foresters was so small in West Africa until the 1990s, it arguably required considerable courage (and compelling evidence) to be the person who would challenge – if only at the level of a few localities – the prevailing deforestation discourse. Second, ideas about 'the rule' (deforestation) came from many observed instances of tree loss elsewhere in Guinea and West Africa; they were bolstered by Fredric Clements's idea of spatially continuous 'climax communities' of vegetation, which, as I've previously noted, created an expectation of biogeographical consistency at the regional level.

Third, national and local forestry department officials, like most state employees, were apt to define themselves as participants in Guinea's 'modernisation' and 'development'. This perhaps disposed many to see peasants and smallholders as hide-bound to 'tradition' and embodiments of a past that Guinea's post-independence president (Sékou Touré, who ruled autocratically from 1958 until 1984) wanted to leave behind. Fourth, local forest departments acquired assets and revenues courtesy of the deforestation discourse. As the decades wore on it was, perhaps, rational to avoid challenging this discourse: for instance, jobs monitoring illegal logging and fire setting might be lost, so too monies from fines and logging permits. Fifth, the late twentieth-century appearance of more systematic remote sensing (by planes and satellites) did not, in fact, increase certainty about land cover changes. This is counter-intuitive and requires explanation. Complete the following task before reading on.

Study Task: One would think that the availability of aerial images of land cover in rural Guinea would help to dispel any myth of deforestation. Can you think of any reason why this did not occur?

Close examination of aerial images of Kissidougou taken in 1952, 1979 and 1982 by Fairhead and Leach confirms what Kissia and Kuranko farm communities told them. The images revealed remarkable consistency in forest cover and, where it was lost, was usually compensated for by increases nearby or further afield; yet Guinea government officials and overseas researchers, in the 1970s and 1980s, either missed or ignored these manifest facts. Why? Feasibly, they made incorrect inferences and guesses from strictly incomparable images. The resolution of photographic and satellite images differed (the latter are grainier); the time of year images were taken often varied (meaning dry and wet season vegetation patterns were being compared); and there was a lack of personnel or will to ground truth image interpretations by visiting enough field sites. In light of this, it was unlikely that Guinea's forest managers would 'see' afforestation signatures in the images available to them. Finally, by the 1990s, the considerable international development aid and nature conservation revenues mentioned earlier became tied, in West Africa and elsewhere, to pro-forest projects in many cases. To challenge deforestation discourse might have reduced the flow of such monies into a still poor Guinea after the first Earth Summit.

Since Fairhead and Leach published their research, there has, by all accounts, been a much greater attempt by governmental agencies in West Africa and beyond to engage more closely and sympathetically with local land users living in or near major forest zones. For example, an important new trend in international natural resource management since the mid-1990s has been to facilitate 'community-based' and 'participatory' approaches. The number of remotely sensed images has also increased and there's greater care taken when comparing and ground truthing photographic and satellite representations of land cover.

The power of discourse

Let me conclude this extended discussion of one case (of deforestation discourse in Guinea) by linking back to my earlier examination of social power. Courtesy of Fairhead and Leach's research, we've considered a situation where a particular representation of the forest–savannah transition zone advanced by a scientific–bureaucratic elite in Guinea had significant long-run effects. This elite stood for the forest (they were its spokespersons politically) by claiming to represent it epistemically (as scientists or experts). Their depictions underpinned new land management policies, created government jobs, curtailed ordinary farmers' freedom, and – as the twentieth century drew to a close – channelled overseas money into Kissidougou and similar districts. Its power derived not so much from its *content* as from *who produced it*. The predecessors and successors of Aubréville authorised themselves to create and mobilise knowledge of whole landscapes as part of their professional modus operandi as state employees. They viewed these landscapes as largely natural – in actuality or in their ideal state. Though they had the means, they had neither the will nor the wisdom to challenge a deforestation discourse entirely of their own making, yet which they presumed reflected Guinea's biophysical 'realities'. Once this discourse became institutionalised, it created inertia and made it hard to see the 'realities' of land use change on the ground.

Ironically, Fairhead and Leach found that Kissidougou's farmers often behaved as if the deforestation discourse was correct. They suggested two reasons why. First, it helped farmers avoid conflict with local forest officials. Second, it often yielded collateral benefits. For example, European Union-funded forest plantation projects in the 1990s brought new schools (among other things) as a reward for community support and compliance. To pretend deforestation discourse was true may perhaps be regarded as a 'weapon of the weak' in this case. There's little or no evidence that the discourse successfully interpellated Kissidougou land users in the way I discussed in Chapter 3. Its power over them stopped short of affecting their self-understanding (cf. Leslie Feinberg's struggles to establish 'hir' identity, recounted in Chapter 5). In fact, despite its efficacy, the embrace of the discourse by forestry officials, Fairhead and Leach showed, produced a grassroots lack of faith in the competence and integrity of these officials.[15] To be effective, social power need not persuade those subject to it that its discursive claims are valid. Conversely, nor should the lack of persistent or periodic protest by Kissidougou's farming communities be taken to signify their contentment with prevailing forest policies through much of the twentieth century.

In recounting this story of epistemic inequity between Kissidougou's land users and forestry managers I have, necessarily, given credence to Fairhead and Leach's particular representation of forest–savannah dynamics.

Their account of soft power in Guinea (and West Africa more widely) suggested that influential individuals – influential because of the institutional positions they occupy – can, as a group, deceive themselves unwittingly. Fairhead and Leach's conviction, it seems to me, was that it's possible to speak truth to power. By gathering new or overlooked evidence, they sought to refine an, only partly, accurate representation of forests. While 'false' representations might prove efficacious, Fairhead and Leach believed that more 'truthful' ones would (or should) win any epistemological battle. Their research was, in effect, a plea for more accurate representations based on better evidence and sensitive to geographical differences within and between nation states.

In light of the discussion of social power early in this chapter, we might want to question these epistemic workers' self-representation. Their depiction of the history and effects of soft power in Kissidougou was intended to exert a soft power of its own, using the rhetoric of 'truth versus falsity' as a tool of persuasion. Or, if you prefer, it employed this rhetoric to challenge and resist soft power on behalf of all those West African land users whose voices went unheard for so many years. It was a political intervention, one designed to alter a reality it was ostensibly examining in a 'reasoned and objective' fashion. Accordingly, my summary of Fairhead and Leach's research in the previous pages should be viewed in the same way. I'm not suggesting that their exposé of the false, but consequential, representations upon which soft power was built in Guinea spoke the language of unadulterated truth and is to be trusted by you the reader. Instead, I'm arguing that their representation of trees, both 'natural' and anthropogenic, was intended to empower those disempowered by deforestation discourse, and to make the authors of this discourse aware of its limitations. Fairhead and Leach's normative stance was that local land users should have the opportunity – even, perhaps, the right – to shape the ways in which their livelihoods and environments were (are) depicted by authoritative actors. They were not, it seems to me, arguing against state-led forest management. Their point, instead, was that 'power to' was operating on a flawed epistemic basis and was thus being misapplied. We should therefore treat with 'positive scepticism' their claim to correctly represent what others had long 'mis-represented'. As I said earlier in this chapter, exposés of social power can themselves be seen as power plays.

SUMMARY

In the previous pages, I've focussed on soft social power, namely power that depends not on force or violence but on the production and deployment of information and knowledge. I've explored the principal issues and fault lines that have preoccupied analysts of social power in the social sciences and humanities in recent years. I suggested that these analysts' representations are plural – there's no consensus on what, precisely, social power is, who (or what) exerts it, what its effects are, or whether it has an 'outside' (a place where it can be ignored or resisted). It's an open question, whether this reflects the plural character of social power, academic confusion or something else. Side-stepping this question, I focussed on one way in which soft social power has been said to work by way of influential epistemic communities (in this case botanists and forest managers) who rarely consult with those subject to knowledge they are authorised to produce and act upon. I showed, taking one extended example, how representations of what is taken to be natural (in this case 'climax vegetation') can be pivotal to the exercise of soft social power. 'Nature' has its uses and the users' tools are linguistic, epistemic and representational ones as much as physical ones. In situations, like Kissidougou, where the users have real power over others it is important to inquire into the quality of their epistemic practices (in Chapter 8 I will talk about governance issues in some detail).

Of course, readers of this book aren't subject to the discourse of tropical deforestation in the same way that farmers in Kissidougou were. In highly urbanised Western countries, most of us, most of the time, don't confront regulations and sanctions designed to protect 'nature' that directly curtail our actions or affect our livelihoods. However, in other ways we *are* subject to discourses about tree loss, climate change, our own genes and much more besides. These 'other ways' speak to the power of the mass media – notably, its non-fictional components (news, current affairs programmes, documentaries, etc.). As I mentioned in Chapter 3, whatever the origins of any given representation of nature and its collateral terms might be, it must usually be noticed and re-presented by the mass media in order to have any wider purchase. As I noted earlier in this chapter, while Foucault's 'dispersed' notion of social power is valuable, analytically and politically, the power to represent reality remains uneven nonetheless. In the next chapter I want to focus on the mass media since, for most us, it's arguably one of *the* media through which our understanding of self and world is shaped and reshaped.

Those working in large institutions that supply television, radio and internet content, like Britain's BBC or Rupert Murdoch's News Corporation, are among the most powerful epistemic communities in the world. In part, their modus operandi is to assemble, interpret and reframe the representations of others (e.g. scientists, politicians, entertainers and activists). In part, their role is to produce their own distinctive genres of representation, such as wildlife documentaries. Without the mass media we'd not only be less well informed (and less entertained) but, arguably, different people. The quality of the mass media is a key determinant of the quality of any democracy. As the next chapter reveals, mass mediated representations of biophysical phenomena are very important ingredients. The following chapter introduces the final part of *Making sense of nature*, where I look closely at the epistemic communities whose representations of nature and its collateral referents, arguably, overshadow most others.

ENDNOTES

1 Though he's not exactly a Pollyanna, British climate scientist Mike Hulme (2009) offers us a good example of this belief that social power does not significantly restrict thinking or action. In his book *Why we disagree about climate change*, he rightly argues that it's not disputes about the quality of climate science that are the root cause of today's unresolved debates about the severity of the 'problem' and what to do about it. Despite the efforts of 'climate change sceptics' to discredit the science, Hulme argues that we disagree about climate change because people possess a wide variety of values, norms and goals. Because climate change is perceived as a 'global problem' that may be 'dangerous', Hulme argues that the idea of it precipitates debates over how we should live, debates that often get mistaken for debates about what the 'facts' of climate change 'really are'. I entirely agree with his diagnosis of what the 'real' debate over climate change is and should be about (i.e. *not* about how best to 'solve' climate change as if it's a technical problem amenable to single 'correct' solution in light of the available scientific evidence). However, Hulme's book is marred by the assumption, which is rarely problematised in its pages, that the various extant viewpoints on climate change are equally visible. Hulme appears to have a 'pluralist' view of debate, in which interlocutors enjoy roughly equal rights to join and shape the conversation.

2 There are exceptions to this, depending on one's preferred conception of social power. For instance, national governments control armed police, soldiers and other military personnel who are rarely asked to use their weapons at home or abroad. Even so, this 'latent power' might be seen to be a highly effective way in which governments ensure social order domestically, and ensure peaceful relations with other countries.

3 Joseph Nye popularised the term 'soft power' is his analysis of intergovernment relationships and the powers deployed by leading states like the United States. I use the term here in a more generic sense, extending beyond the field of international relations to domestic affairs, non-governmental arenas and everyday life. Because Nye equates soft power with language, information, persuasion, representation and negotiation, this extension is not unreasonable. It has filiations with the idea critical analysts of culture (such as Raymond Williams) proposed from the 1950s: that language and 'culture' in the broadest senses are arenas in which power is exercised and opposed. Note that I'm not at all suggesting that forms of 'hard power' have entirely disappeared. To different degrees and in different ways, they are still visible in most countries worldwide, for instance, the death penalty is legally enforceable (in customary or now customary law) everywhere from Iran to the United States.

4 And, as I showed in the discussion of Clayoquot Sound in Chapter 4, these subject-positions can crosscut otherwise different representations of 'nature', meaning they are implicated in what we can call 'cultural power' – that is, the influence that deep-seated habits of thought specific to one culture exert on those not from that culture.

5 In an endnote in Chapter 3, I listed some further reading for those who know little about Foucault's writings on government and subjectivity. In the present context, I'd also recommend Barry Hindess's (2004) short but dense presentation.

6 These arguments by left-wing social scientists were presaged by several post-war political scientists writing in the United States and United Kingdom. The so-called 'three faces' of power were identified over time by Robert Dahl; Peter Bachrach and Morton Baratz; and Steven Lukes. Dahl argued that power is observable: for him, its 'face' can be seen in the interactions among competing actors and institutions. Bachrach and Baratz argued that this empiricist approach is inadequate because power has a second, *hidden* face. Some actors and institutions, they argued, pull their punches to avoid conflict over the prevailing 'rules of the power game'. They thus self-censor, knowingly or otherwise. Lukes built on this to suggest that power's third face resides in determination of these rules: who can bend or amend the rules has the greatest power of all, he argued.

7 Though the term has several shades of meaning, at root, 'false consciousness' refers to dominated people who are unable to recognise their 'real' interests because powerful groups actively occlude these interests. The idea has been criticised because it conflates the positive and the normative: how can an analyst presume to ascribe 'real' interests to people when their current ('false') interests are very obviously also 'real'? The false consciousness idea rests on the contentious conviction that people would be better off if they discovered, and acted in accordance with, the 'truth' belied by their false beliefs. However, if 'better off' is based less on the presumption of a revealed truth and more performatively as a possible alternative future state that can be argued to be 'better', then the false consciousness idea can be as a *political intervention* rather than an analytical concept. To my mind, this is more plausible and defensible: the critic of 'falsity' is seen to be trying to *persuade* people that a different way of living is possible, which the critic regards as preferable. The idea of 'real interests' can then be seen as part of the rhetoric of persuasion.

8 This much is obvious in the recent *Handbook of power*, edited by Stewart Clegg and Mark Haugaard (2009). This collection of commissioned essays showcases sometimes conflicting, sometimes complementary conceptions of social power current in the Western social sciences and humanities (including Foucault's). But it doesn't aim to achieve a synthesis of perspectives; nor does it stage a discursive battle in which pretenders to the analytical throne are eventually vanquished by a single victor. Though some, such as the historical sociologist Michael Mann (1986, 1993), have sought to bring different ideas about social power together in a single overarching account, such grand analyses have not significantly reduced disagreement in the wider community of interlocutors.

9 These publications [all accessed 15 October 2012] can be viewed at: http://www.fao.org/forestry/sofo/en/;
http://www.unep.org/vitalforest/;
http://www.earthbook2012.org/earth-book/forests/;
http://assets.wwf.org.uk/downloads/lpr2012_online_single_pages_11may2012.pdf;
http://www.conservation.org/Documents/CI_Climate_Deforestation_Logging_Greenhouse_Gas_Emissions_Facts-12-2009.pdf; and
http://www.nationalgeographic.com/eye/deforestation/deforestationintro.html [link has subsequently been removed]

10 Details of this can be found at http://www.fao.org/forestry/fra/fra2010/en/ [accessed 16 October 2012]

11 All the historical sources listed in this paragraph are cited in Fairhead and Leach (1996).

12 After the Second World War, Clements's thinking gave way to a more holistic and inclusive understanding called 'systems ecology', with roots in the work of Clements's British contemporary Arthur Tansley but developed later by Eugene Odum and others in the United States. Despite the important differences of emphasis, ecosystems ecology tended, like Clements, to emphasise ecological stability.

13 It's highly likely that Aubréville was influenced both by the 'dust bowl' phenomenon of the United States in the late 1930s and early 1940s, and by the Goma conference on soil erosion in 1948 (held in the Belgian Congo). Both probably disposed him and many others to believe that unregulated human use of natural landscapes can produce dramatic and undesirable results.

14 I take most of the arguments below from pages 43–8 of Leach and Fairhead (2000). Many are repeated in a more general discussion of deforestation discourse across the whole of West Africa in Chapter 8 of *Reframing deforestation* (Fairhead and Leach, 1998a).

15 And Fairhead and Leach reported local farmers' complaints that local forest officers set fires in order to then blame them and impose fines, which they pocketed.

PART III

KEY EPISTEMIC COMMUNITIES: THE MAKING, MOBILISATION AND REGULATION OF NATURE-KNOWLEDGE TODAY

7 NATURE'S PRINCIPAL PUBLIC REPRESENTATIVE: THE MASS MEDIA

It's not too contentious to assert that we know most of what we know about the world beyond our doorstep courtesy of the mass media. The mass media are those in and through which largely one-way few-to-many communications are achieved on a daily basis. Today, radio, television, newspapers, magazines, cinemas, billboards, games consoles and (increasingly) the Internet are the most important mass media – though we might also include such things as zoos, museums, theme parks and galleries because the most successful of these attract very large numbers of visitors.[1] These media communicate content in a wide range of genres, from television soap operas to popular music to video games to investigative journalism. The mass media does today what it's always done: it aims to *inform and edify* millions of people, to *entertain* them and (as product advertising makes abundantly clear) to actively *shape their values, tastes and preferences.* Until relatively recently, it was largely domestic in scale, with each country having a relatively small number of national newspapers and broadcasters. This ensured that the mass media was largely 'mainstream', its format and content remaining within the perceived bounds of what most members of a country would regard as 'acceptable'. Despite some recent diversification and a degree of geographical disembedding – exemplified by the proliferation of pay-TV channels, the rise of 'reality television' and the creation of 'global' news channels like CNN's – the mainstream mass media rarely strays beyond the perimeters of 'decency'. In large part, this is because it's been able to define and over time slowly redefine where those perimeters lie. Whether it simply reflects or sets the pace for changes in its wider sociocultural milieu has been a perennial question for mass media analysts.

This said, it's been suggested that mainstream media is now being displaced by 'alternative media' in many countries. These are not alternatives in the technological sense because they utilise the established communicative channels of the Internet, print, radio, television, photography, etc. Instead, they're alternative in an informational/content sense. It's become technically easy and relatively cheap for all manner of organisations and people to disseminate their own ideas, knowledge and images far and wide – so-called 'many-to-many' communication. The information/content created is often accessible well beyond the bounds of any one nation state. For instance, Wikileaks, YouTube and the website of Earth First!, in different ways, offer Internet users worldwide written or videographic representations that the mainstream mass media might otherwise ignore. Such alternative media promise to democratise and diversify the diet of information, argument and experience available to people living in all parts of the globe. They often push the envelope in a cognitive, moral or aesthetic sense.

However, various national surveys reveal the tenacity of the mainstream media. When asked, even people living in very 'media-rich' environments report that they derive most of their media content from a small number of established newspapers, television channels, radio stations and websites. Relatedly, Paul Watson, the famously radical leader of direct-action environmental organisation Sea Shepherd, observes that:

> New media is increasingly important, but the mainstream media is still very much the main game of the environmental campaigns. You do the ground work; you build the community of opposition. That's the foundation, but then the icing on the cake is still mainstream media opinion. That's just how it is and that's okay.
>
> (Quoted in Lester and Hutchins, 2009: 588)

Generalising, Watson's point is that messages, performances or events that don't get channelled through the mainstream mass media are unlikely to be noticed by more than a niche audience, and a potentially small one at that. To be *too* 'alternative' is to risk being cut off entirely from the social mainstream. In this light, we might say that alternative media have enriched the epistemic ecology in which mainstream media function, without yet ending their ecological dominance. Large media organisations, most of which are private profit-seeking ones with a minority being publicly owned or otherwise not-for-profit, continue to supply the overwhelming volume of mass mediated words, sounds and sights that people the world-over consume.[2]

Accordingly, in this chapter I want to focus on how the mainstream media represents the world to its (very large) audiences. I'll concentrate on the 'serious' (non-fictional) parts of the mainstream, which is to say those engaged largely in the business of 'informing' rather than 'entertaining' or 'persuading'.[3] I will look specifically at newspapers, which remain a crucially important genre of representation, taking their reporting of global climate change as an instructive illustration of their working practices. Before I do so, however, I need to make some general observations about the role the mass media play in contemporary societies. This will be the focus of the first two sections of this chapter.

LIVING IN A MASS MEDIATED WORLD

'Much everyday interaction', writes media theorist John Thompson, 'is "face-to-face", in the sense that it takes place in a localized setting in which … individuals confront one another directly.' However, 'mediated interaction', he continues, 'is "stretched" across space and perhaps also time; the participants ... are unlikely to share a common spatio-temporal framework' (Thompson, 2004: 174). Such mediated interaction became *mass* mediated from the early twentieth century, notably in western Europe and the United States. Rail transportation and faster shipping permitted newspapers to circulate more widely, while the telegraph, phonograph and radio also allowed otherwise disparate populations to consider themselves part of a wider, multi-scalar ecumene. Lives defined largely by local experiences and routines became increasingly defined by access to information, ideas and stories hailing from far afield. As another media theorist (Brian McNair) puts it, commenting on the news:

> when we receive it [the news], and we extend to it our trust in its authority as a representation of the real, [it] transports us from the relative isolation of our domestic environments, the parochialism of our streets and small towns, the crowded bustle of our big cities, to membership of virtual [national and] global communities united in their access to *these* events, communally experienced at *this* moment, through [far-flung] ... communication networks.
>
> (McNair, 2006: 6)

As Thompson's and McNair's words imply, the mass media arose – and continue to prosper – because of technological innovations that facilitate 'time-space compression'. Such innovations bring the world 'out there' close to home, and quickly too. However, lest this sounds like technological determinism, we need to remember that such innovations have long been the product and medium of capitalist market expansion. The geographically extended traffic in goods, services and information has, of course, been hardwired not only to facilitating commodity exchange but also to the realisation of profit. Surpluses of capital are invested in the hope of selling commodities to more people in more places more of the time.

While the spatial spread of capitalism has been key to the growth of mass mediated communication, we must also acknowledge the rise of the nation state as the world's dominant political unit. While capitalism transcends political borders, the formal structures of government largely remain national (even in the European Union). Governments, and the populations they govern, have long needed means of communication that can encompass all the cities, towns and smaller settlements that together comprise the polity.[4] This does not, of course, mean that politically salient communications are exclusively national in focus. On the contrary, it's long been the case that such communications make substantial reference to, or else hail directly from, the wider world. This connects to issues of culture and personal identity. The mass media has played an important role in up-scaling people's loyalties and affiliations beyond the local arena, encouraging (or allowing) them to identify with a range of norms, values and goals held by others at home or abroad. It's helped to de-parochialise everyday life. In both a political and cultural sense, most citizens of wealthy countries now enjoy a more global mass mediated experience than any previous generation, especially when one takes alternative media into account. The 'imagined communities' to which (they feel) they belong can be multiple and arrayed at more than one geographical scale.

In summary, well over a century after the mass media came into existence their importance remains undiminished. Though we all lead ineluctably local lives rooted in place, our actions, and those of others far distant from us, are highly 'distantiated' both spatially and temporally. We affect, and stand to be affected by, what others are saying and doing on the other side of the world, not only in our home country. 'Deep integration' – *interdependency*, not mere *interrelatedness* – defines the existential condition of most people living today. The three functions of the mass media (distinguished previously) allow us to benefit from, (often) recognise and (sometimes) think critically about this interdependency and the effects it gives rise to. The mass media allows Marshall McLuhan's famous image of 'the global village' to be far more than a catchy slogan. It influences us in our roles as workers, consumers, citizens, members of faith communities, immigrants, voters and much more besides. It thus shapes our actions in the economic, political and sociocultural arenas, with implications for what we call 'nature' and its filial concepts. As Steve Fuller (2007a) aptly puts it in *The knowledge book*, 'The growth of the mass media in the modern era is most directly tied to the value placed on the sheer quantity of people – voters, consumers, citizens – whose opinions matter to achieve some desired social effect.'

Study Task: How important is the mass media in your own life? To answer this question you'll need to do several things. First, ensure you include the Internet, television, radio, newspapers and magazines. Second, include situations where you consume mass mediated representations with others (e.g. your immediate family), not just by yourself. Third, beyond simply listing things you routinely watch/read/listen to or how many hours you interface with the mass media, think too about their importance in your life. For instance, if you read a favourite newspaper each day is this as important for you as, say, watching a soap opera twice a week (and if so why)?

The effects of the mass media – more on social power

Given the huge prominence of the mass media in contemporary life, it's no wonder that a large cadre of analysts have been attracted to it. For instance, a great many universities now have departments of 'media and communications' close to celebrating their centenaries, while the interdisciplinary field of cultural studies has devoted a lot of attention to the mass media. An intellectually diverse epistemic community with some considerable pedigree now exists, possessed of its own specialist discourse about 'media effects', 'condensation symbols', 'framing' and so on. A perennial question has, unsurprisingly, been: to what *degrees* and in what *ways* does the mass media influence its mass audiences? Initially, profound interest in this question arose in the 1930s because of the highly effective use of propaganda by the Nazi Party, which enabled it to replace the Weimar Republic (a representative democracy) with the Third Reich (a one-party state that instigated the Second World War). Propaganda, as Garth Jowett and Victoria O'Donnell say in their book on the topic,

> is the deliberate, systematic attempt to shape perceptions, manipulate cognitions, and direct behaviour to achieve a response that furthers the desired intent of the propagandist.
>
> (Jowett and O'Donnell, 2005: 7)

They forgot to add that it appeals largely to people's emotions and sentiments, rather than to logic or 'reason'. Before, during and after the Second World War, propaganda was state-led, occurring during an era where people had relatively little choice on the news stands, in cinemas and on the radio.

The success of propaganda in Nazi Germany and elsewhere encouraged a very strong, one-way conception of the relations between 'sender', 'message' and 'recipient'. But the unusual conditions of the 1930s – international economic recession, a gifted demagogue (Adolph Hitler), and the festering legacy of the Great War (1914–18) – were hardly representative of all times and places. This is why the 'hypodermic theory' of mass media effects soon gave way to less muscular ones that have, with few exceptions, prevailed ever since. Starting with research into people's voting decisions in the 1940 US Presidential election, Paul F. Lazarsfeld and his students at Columbia University suggested that

> mass communication could be persuasive only under special conditions, such as the absence of counter-propaganda, the reinforcement of media messages by face-to-face discussion, and the strategic exploitation of [people's] well-established behaviours.
>
> (Peters, 2004: 19)

In their monograph *Personal influence* (Lazarsfeld and Katz, 1955) and elsewhere, the Columbia researchers argued, with reference to considerable empirical data, that mass media messages often ran up against people's existing beliefs and needed to be relayed through 'opinion leaders' in interpersonal networks (e.g. families and churches) to be effective.

Subsequent research at Columbia and elsewhere in the United States appeared to confirm that mass media audiences enjoyed a degree of autonomy from the communications thrown at them. It suggested too that these audiences were differentiated along significant axes such as gender, age, ethnicity and level of formal education. It also attended to the differences between short- and long-term media effects and tried to distinguish between audience 'attitudes', 'preferences', 'beliefs', etc. so as to unpack the blackbox of 'reception'. This kind of mass media research arguably conformed to America's self-conception as a liberal democracy in which no one actor – be it a government or a private company – should seek to homogenise socio-cultural values or beliefs. By the 1970s, it bled into 'gratifications research', in which 'mass audiences' were seen to be fractionated (*de*-massified) according to which media content and forms they decided to pay attention to and how they tailored their reactions to them to satisfy their own personal needs.

Coincident with this (very American) body of inquiry into the effects of the mass media was the rise to prominence of a more critical tradition originating in Europe. The Frankfurt School of critical theory (in)famously emphasised the mass media's role in telling people what *not* to think, believe or consider. Unlike the state-led propaganda of the 1930s and 1940s, they suggested that the post-1945 'culture industry' in the West operated to bind ordinary people to capitalist social democracy by making it their 'common sense'. With the communist bloc depicted as enemy and threat, the mass media, according to the Frankfurt School, worked to instil non-revolutionary sentiments in Western populations via commodity advertising, televised light entertainment shows, Hollywood films, popular music, high street fashion and selective news reporting. In tandem with their rising wages and improved living standards, working people, so argued the Frankfurt School, were pacified by the mass media at the symbolic and semantic levels. The political radicalism throughout pre-1939 Europe was thus neutered, having barely been allowed to develop in the first place in North America.

Related research of a less resolutely theoretical kind later followed at Birmingham University in the United Kingdom. Inspired, in part, by the writings of Italian Marxist Antonio Gramsci (see Box 6.2), Stuart Hall and colleagues explored the complex relationships between 'encoded' (media/producer) and 'decoded' (audience/receiver) meanings. They showed, through a combination of discourse analysis, ethnography and interview data, that the mass media tends to naturalise a very particular set of norms and values at any one moment in time. These norms and values aren't created by the mass media *sui generis*; rather, they are ether in the slow-changing cultural atmosphere. The mass media, so argued the Birmingham School, serves to make them visible by reiterating them day-in, day-out. This notwithstanding, Hall and others have shown that audiences could subvert, recontextualise or appropriate encoded meanings in complex acts of decoding (e.g. see Hall, 1973). In part, this is because encoded meanings were not always simple or internally consistent. Encoding was thus shown not to guarantee the 'right' interpretations and emotions in recipients, even when they were 'paying attention' (see Box 3.2).

Unlike the pluralist-cum-voluntarist ethos of the research inspired by Lazarsfeld, the Birmingham School highlighted the structural inequalities that conditioned whether and how any given audience would accept encoded meanings uncritically. Mass mediated messages, they contended, were sent and received in conditions of 'hegemonic rule', in which there was give-and-take between the dominant and subordinate sections of any given society. This allowed room for sub- and counter-cultures to form at the grassroots whose signs, symbols and practices consciously departed from dominant habits of thought and action. Sometimes, as with punk rock in the United Kingdom, these became highly visible in the sociocultural mainstream.

Some four decades on and the debates about the role and power of the mass media remain unresolved. Though there's broad agreement that it rarely propagandises – with the exception of autocratic states, and notwithstanding Edward Herman and Noam Chomsky's trenchant claims in 1988 to the contrary (see Box 7.1) – there remains a fault line between those analysts disposed to look for evidence of how the media sets the cognitive, moral and aesthetic agenda for a society, and those disposed to look for evidence of audience independence. Some focus on the effects of resurgent oligopolistic tendencies in media ownership (the likes of Rupert Murdoch and Silvio Berlusconi being metonymic here), while others focus on alternative media, 'cyber-democracy' and the blurring of the producer–user distinction. Some point to the 'dumbing down' of the mass media (e.g. the rise of 'infotainment', 'glossy magazines' and 'lifestyle shows' – all 'weapons of mass distraction'), while others point to the enhanced diversity of media content now available at the click of a button.

In significant measure, the analytical differences resolve to ontological ones: what is evidence of 'audience freedom' from 'media effects' for one investigator is seen by another as evidence of 'audience control' based on a skewed menu of media offerings. This said, there's some consensus that the mass (and alternative) media must be analysed contextually: their operations and influences do not occur in a vacuum, but reflect and shape a wider and increasingly dynamic set of economic, political, cultural and biophysical arenas. The resulting challenge for research is as follows: how can one isolate – in the short, medium and long term – the distinctive content of mass mediated communications and their effects in highly complex, 'over-determined' situations?[5]

BOX 7.1

THE MASS MEDIATED MANUFACTURE OF CONSENT?

Focussing on the news (in both its broadcast and paper formats), Edward Herman and Noam Chomsky provocatively suggested that it's not only autocratic states that, today, channel propaganda through the mass media. In their book *Manufacturing consent: the political economy of the mass media*, they argued, with a focus on the United States, that even ostensibly democratic countries have seen a misappropriation of mass media channels for propagandistic ends (Herman and Chomsky, 1988). They famously posited five reasons why the contemporary news media has, in their view, narrowed and homogenised its field of vision:

1. Public and not-for-profit news sources have given way more and more to a few large, private, profit-seeking media corporations with shareholders – this puts a premium on news that 'sells' and reduces the likelihood of news reporting that compromises the corporate interests and alliances of the owning companies.
2. Advertising is a key revenue source for many news providers – these providers are thus more likely to cater to the political prejudices and economic desires of their advertisers.
3. A relatively small number of news sources (e.g. national governments, large firms, scientific institutions) have become adept at providing 'newsworthy' plans, evidence, findings, etc. – these sources thus become 'primary definers' of the news and reduce the costs that news providers incur in 'making' the news.
4. Threats and recent cases of 'flak' have a chilling effect on news providers – organisations unhappy with news content can take (and have taken) punitive action against news providers (e.g. lawsuits). This kind of organised opposition to 'unwelcome and unwanted news' can make news providers avoid risk in their choices about what is newsworthy and how it's presented to readers/listeners/viewers.
5. Putative 'enemies' and 'threats' help define the terms of news discourse – partly, but not entirely because of points (1)–(4), news providers will rarely question taken-for-granted beliefs that are used strategically by economic and political elites, e.g. terrorists are 'irrational' and 'inhumane', or the solution to a recession is more of the economic growth that preceded the recession.

Herman and Chomsky's book was, intellectually speaking, located somewhere between the pessimism of the Frankfurt School and the determinism of the 'hypodermic theory' of media effects. It showed propaganda to be the inevitable result of a creeping process of self-censorship by wary, co-opted or compromised news providers. Needless to say, *Manufacturing consent* came in for a lot of criticism, even from erstwhile sympathisers. Quite aside from overstating the power of the news media to create and sustain social consensus, the book was said to underplay the enduring liberalism of many news media organisations. More stringent criticism came from those who didn't share Herman and Chomsky's left-wing values (see, for instance, *The anti-Chomsky reader*, edited by Peter Collier and David Horowitz (Collier and Horowitz, 2005).

Conclusions?

We can regard the long-running academic debates about the mass media's 'agendas' and effects as one part of the broader discourse about social power summarised in Chapter 6. As I've intimated previously, some critics see the mass media as 'powerful' in a largely pejorative sense, while others see its 'power over' as 'legitimate' but uneven demographically given the sociocultural diversity of many modern societies. Still others point to the new power (what some prefer to call 'resistance') that dissidents, sub-cultures and 'outsiders' now have to get themselves heard in the mass mediated mainstream by way of their media-savvy activities. My own view is that the mass media, in its diverse forms, clearly possesses distinctive modalities of representation (including a suite of habitual discourses). Though its audience effects are difficult to measure robustly, it's undoubtedly visible in a social sense – as much, if not more, than ever before. Broadening Bernard Cohen's famous observation about the press (Cohen, 1963), we might say that the mass media doesn't necessarily tell people *what to think or feel*, but it does tell them what to think and feel *about*. In addition, one of its most distinctive, and important, functions is to package all manner of information, argument, imagery, etc. that's generated *outside it* so as to inform, entertain and instruct its sizeable audiences. It's at the heart of what I called 'circulating and mutating representation' in Chapter 3. Geographer Max Boykoff phrases it nicely:

> Media representations are confluences of competing knowledge, framing [topics and issues] ... for policy, politics and the public, and drawing attention to how to make sense of, as well as value, the changing world.
>
> (Boykoff, 2011: 3)

As I asserted in this chapter's introduction, the *mainstream* mass media remains the most important set of communicative channels and genres that *non*-mass media (and alternative media) representations *must* be routed through if they're to capture our collective attention. Paul Watson was absolutely right.

The cognitive, ethical and aesthetic 'bandwidth' permitted by the mainstream mass media is both a reflection of, and contributory to, the levels of open-mindedness (or tolerance) of the societies it serves. Its considerable power to represent the world can be abused, certainly, but it remains the case that the mass media can also serve to hold powerful institutions and actors to account. While the fine details of how remain elusive after decades of research, the mainstream mass media is thus necessarily central to determining the speed and character of social stability and social change. In part, this is because it has a connective imperative with the existence of the public sphere and civil society, both of which, as I'll now detail, are (like the mass media itself) key contributors to the political life of nations.

MASS MEDIATED PUBLICS

If 'Democracy [is] ... the idea that political rule should, in some sense, be in the hands of ordinary people' (Barnett and Low, 2004: 1), then the existence of the public sphere is a *sine qua non*. The public sphere is the arena in which individuals can come together freely to discuss societal challenges and goals and, through that discussion, influence the political actions of elected representatives. Despite my use of the word 'arena', the public sphere is not a physical place. Instead, it's virtual (nowhere and everywhere), though no less real than any site of face-to-face assembly for all that. This is well illustrated by the websites of 'serious' newspapers. Most offer readers the chance to post comments on the arguments made by columnists.

Study Task: Try to identify the main reasons why modern public spheres do not resemble the face-to-face ones of the Ancient Greek city states where democracy was first invented. It may help to refer back to the start of Chapter 2.

The reason why modern public spheres are necessarily 'stretched out' is obvious: most democracies are physically large and well populated, meaning that people simply lack the ability to have direct encounters with even a tiny fraction of their fellow citizens. Furthermore, given how things like climate change and the 2008–9 financial crisis stand to affect everyone on the planet, the public sphere is now in some sense global. 'The public' is, today, thoroughly multi-scalar – a dynamic *composite* of overlapping publics with no fixed geographical address.

The public sphere, publics and publicity

In his influential history of the public sphere in Europe, Jurgen Habermas (1962/1989) located it functionally between what he later called 'lifeworld' and 'system' (see Box 3.4). People, he argued, inhabit a *private lifeworld* (the family and community), a *private system-world* (work and commodity consumption, linking people to the 'invisible hand' of the capitalist market), and a *public system-world* (their submission to the law and to the regulations of elected governments, both local/regional and national). In each case, different sets of interpersonal relations are operative, spanning the intimate and the impersonal. However, people also connect lifeworld and system by participation in the public sphere, which is that 'intermediating zone between these two realms' (Barnett, 2004: 186). For Habermas, the public sphere is a domain of discourse, debate and discussion among equals (citizens) pertaining to collective affairs. Through such dialogue, people help to make 'the public' something real, more than just an idea or empty signifier (see Figure 7.1, noting the question posed immediately below it, which I will address shortly).

What does 'public' mean? In the first instance, it's a noun, describing a collective subject, 'the public'. It also refers to the things that concern it, as in the phrase 'public affairs'. As Clive Barnett notes, 'some issues gain their importance both from affecting and being addressed by the people acting together in concert' (Barnett, 2008: 404). However, he continues, 'there is another sense of "public", one that refers to the idea that some things are carried out in the open ... involving free and unrestricted discussions and debates ...' (ibid.). So, in democracies, 'the public' is (ideally) a collective subject that addresses issues of common concern in an inclusive manner so as to influence its elected representatives.

Habermas's concept of the public sphere as a key intermediating zone where the needs and wants of individuals can periodically alter the systemworld through collective decision-making includes what political analysts call **civil society**. This, as sociologist Larry Ray put it, is a

> network of civil associations, churches, sports clubs and the like that generate 'social capital' ... Active, voluntary and informal groups and networks make for more stable democracy and protect against incursions by the state.
>
> (Ray, 2004: 223–4)

Civil society exists at the 'lifeworld' level for individuals and is much less impersonal than their membership of a public (or publics). Individuals voluntarily leave the private realm of the home and participate in organisations and activities that bring material or emotional benefits to others, as well as to themselves. The greater the number of organisations and activities in a locality or country, the stronger the bonds between citizens – people are likely to feel less atomised if they have the opportunity, and desire, to invest in other people beyond their immediate family.[6] In turn, this increases the chances that they'll take their membership of larger publics seriously, beyond simply casting a vote at elections every few years. Indeed, a vibrant civil society is the source of arguments, claims and projects that seek to make direct appeals *to* 'the public' and to sway elected governments. Critics have argued that a weak civil society correlates closely with publics that are a mere aggregation of largely personal, private 'preferences' – and thus scarcely publics at all. The same has been said of countries where 'public goods' are few

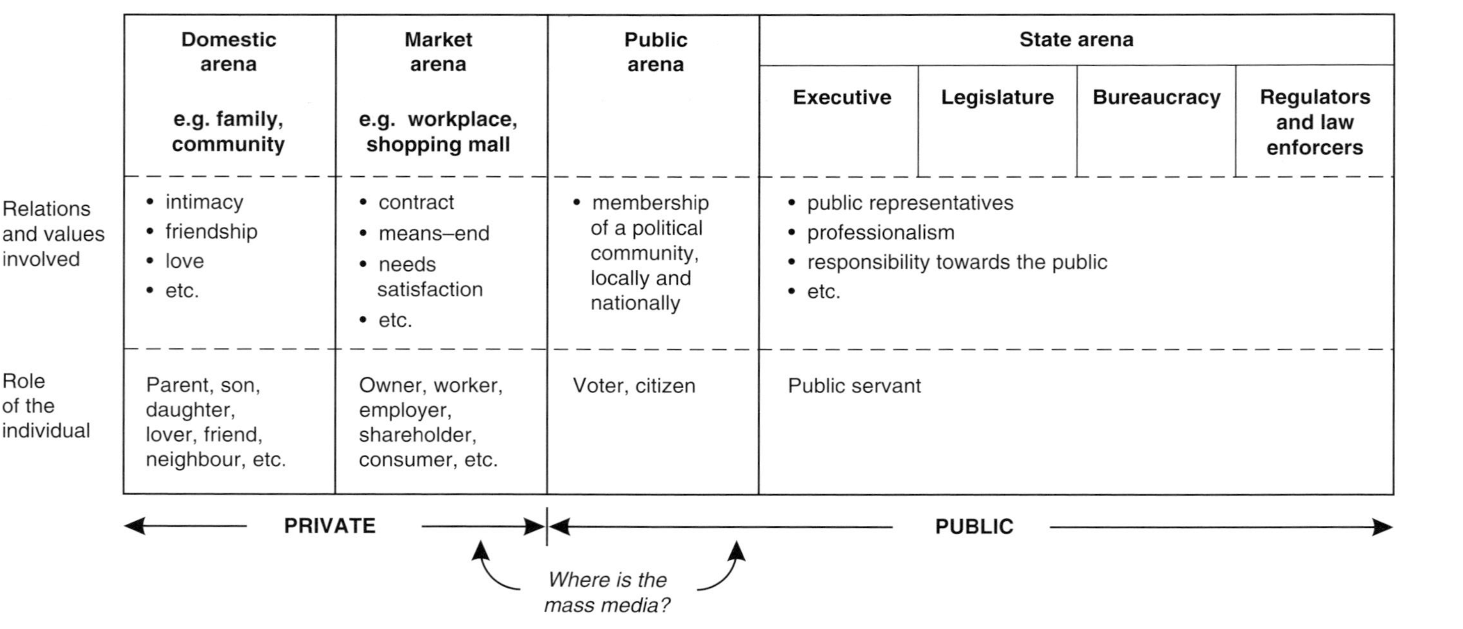

Figure 7.1 The structural characteristics of advanced capitalist democracies (after Habermas, 1989)

in number, poorly resourced or invisible – goods such as public schools, toll-free highways or universal healthcare.

Mediated publicness: the mass media as public resource and public threat

What's all this got to do with the mass media? Whereas involvement in civil society organisations is typically local and face-to-face, membership of publics is, for reasons already mentioned, virtual. Quite simply, a public cannot exist without the technologies of mass communication (and it's worth noting here the etymological connection between 'communication' and 'community'). In the words of media analyst Mike Schafer,

> The mass media constitute *the* most important forum for the public sphere in modern societies, providing an organising framework for societal self-observation, allowing a large number of citizens to inform themselves about political, economic and other developments.
>
> (Schafer, 2010: 1, emphasis added)

For a strong public sphere to exist, it's necessary, but not sufficient, to have an active civil society. For citizens to have sufficient knowledge of and be able to make informed judgements about issues of public concern, the mass media has to perform the first of its three functions to a high standard. Representations disseminated as 'news', 'current affairs', journalism and documentaries are especially important here. This is why one commentator talks of the 'mass media public sphere' (Downey, 2007: 118) in order to signify the indissociable link between mass mediation and the very existence of 'the public', 'public affairs' and 'public debates'. Echoing this, media analyst John Hartley once opined that 'the public can nowadays *only* be encountered in mediated form . . .' (Hartley, 1992: 1, emphasis added). This means that a lot of 'politics' today, in both the narrow and broader senses of the word, is played out in and through the mass media. To quote John Thompson once again, echoing Paul Watson,

> To achieve visibility through the [mass] media is to gain a kind of presence or recognition in the public [sphere] . . . which can help to call attention to one's situation or to advance one's cause. But equally, the inability to achieve visibility through the media can confine one to obscurity – and in the worst cases can lead to . . . death by neglect . . . Mediated visibility is not just a vehicle through which aspects of [contemporary] life are brought to the attention of others: *it has become a principal means by which . . . political struggles are articulated and carried out.*
>
> (Thompson, 1995: 49, emphasis added)

Those parts of the mass media devoted to 'informing' people, such as newspapers, typically make frequent and explicit mention of the public.

Likewise, those hankering for visibility in the media, such as the leaders of political parties or oppositional groups like Sea Shepherds, often phrase their discourses in terms of 'the public interest'. As Barnett rightly argues,

> a public...appear[s] through representative acts of being spoken for and spoken to...Any public utterance does what it says, it brings into being what it presupposes to exist already as the condition of getting off the ground.
>
> (Barnett, 2008: 411–12)

This doesn't mean that 'the public' is a *fiction* or that there's a *single* public within any democratic political unit (or more broadly); instead, it means that the mass media is key to constituting and reconstituting publics and public issues on a continuous basis. 'Publics are always in the making,' as the editors of *Rethinking the public* correctly observe (Mahony *et al.*, 2010: 8), because there's a *variety* of ways in which members of any public can be interpellated by mass mediated representations of what 'the public' should care about, who is implicated and what should be done:

> a public is not best thought of as a pre-existing collective subject that straightforwardly expresses itself or offers itself up to be represented. Rather, ... publics, in the plural, are called into existence, or summoned.
>
> (ibid.: 2)

From the perspective of citizens the mass media is thus both an invaluable *resource* and a potential *threat*. The threat arises because, without proper internal and external regulation, the mass media can hollow out (or debase) the public sphere.[7] Publics are heavily reliant on the mass media, but very few citizens ever get to involve themselves in its internal affairs. Indeed, in a real sense, citizens don't know what their 'public' needs and wants *are* without the mass media, which gives the latter considerable 'power to'. For instance, opinion polls organised by television news programmes can ventriloquise 'public attitudes' towards an issue or event, rather than giving people the opportunity to share their more complex and considered assessments. Relatedly, political parties and private companies utilise 'spin' and 'public relations' professionals in order to engage in what critics, after George Orwell, call mass mediated 'doublespeak'.[8] Speaking for and about the public is a weighty epistemic responsibility and one that can be discharged for the wrong reasons and towards highly particularistic ends. This is why many countries decided many decades ago to create public media organisations existing at arm's length from the national state and free from the profit motive of private companies. Such organisations are intended to give citizens a nutritious and varied menu of materials that inform, edify and entertain, often by challenging the norms of their own society. Increasingly, however, they exist in a global media industry that's dominated by private corporations with shareholders and possessed of a wide portfolio of business interests (spanning news, entertainment and leisure).

GLOBAL CLIMATE CHANGE AS 'NEWS'

Let me turn now, at last, to newspapers and the reporting of anthropogenic climate change in recent years – a subject of profound importance for all of us.[9] I'll begin with some observations about the role of the press and how news gets made.

Newspapers, journalists and news as a communicative genre

As noted previously, the 'serious' parts of the mass media, of which 'quality' (or broadsheet) newspapers are a prime example, have an essential connection to the existence and quality of public literacy and public debate about all manner of issues, topics and questions. Their independence from non-media organisations and sources is important. As political scientist Doris Graber (2003: 143) has noted, in theory such independence permits them to do three things, to which I'd add a fourth:

- Provide a forum for presenting and evaluating diverse, often conflicting, claims, beliefs, goals and values;
- To give voice to, and make claims on, 'public opinion' and 'the public interest';
- To hold elected governments to account on behalf of the citizenry;
- To hold important quasi- or non-governmental institutions and individuals to account (such as senior scientists).

Rephrased, the serious parts of the mass media ideally serve to (1) be a *representational commons* that shares and disseminates a wide variety of cognitive, moral and aesthetic messages, and (2) be a *watchdog* on behalf of the various 'publics' they summon in their discourses.

Anyone who reads a newspaper, at least in the West, usually understands the particularities of the communicative genre well enough. Each day, what's considered to be new or novel is reported ('If it ain't new it ain't news!'), with lead stories garnering more column inches than the many smaller reports that follow them. These stories, many accompanied by relevant photographic images, are complemented by invited columns, by the paper's own editorial view on a selection of live issues and by such things as readers' letters. In these other sections, there's space for debate and criticism rather than just for reporting recent events and 'facts'. In these sections, a newspaper's own political leanings become very evident to readers – indeed part of their *appeal* to readers. Increasingly, newspapers have moved online, with their websites both reproducing and extending what's contained in each day's hard copy. Increasingly, all mainstream newspapers (not just the 'popular press' or tabloids) are blending 'news' with other content, such as 'lifestyle' sections about holidays and 'celebrity gossip'; however, news stories and associated content (e.g. editorials) still rightly tend to dominate.

The journalists who research and write these stories, and the editors who select and edit them for publication, have long operated with a code of ethics intended to structure their professional activities. As communications scholar Robert Cox notes,

> The values of *objectivity* and *balance* have been bedrock norms of journalism for almost a century. In principle, these are the commitments by the news media to provide information [to the public] that is accurate and without reporter bias and, where there is uncertainty, to balance news stories with statements from all sides of the issue.
>
> (Cox, 2010: 165)

'Objectivity' here means that journalists commit to reporting events, or the words and deeds of their sources, free from any personal prejudice or any perceived 'agenda' pursued by their employers. 'Balance' means ensuring more than one side of any story is reported if there are doubts or debates about an occurrence, project or discovery. Together, these norms imply that news journalists and editors aim, however imperfectly in practice, to act as 'honest brokers' between the events and sources that are deemed newsworthy and various publics. To the extent that people trust the news media, it's arguably because these norms are seen to be adhered to most of the time.[10] It is also, more generally, because the news media are understood to be operating within the meta-genre of 'realist' representation. They're perceived by their audiences to be in the business of reporting *actualities* (and future *probabilities* – not mere possibilities). The views and values contained in editorials and columns are understood to be somehow distinct from the 'truth-orientated' content of daily news stories. However, journalist norms and conventional perceptions of newspaper practices must reckon with a range of pressures and demands that are typically hidden from public view.

Study Task: Think of a newspaper, maybe one you read regularly. Aside from the norms just described, what do you think influences the content of the reported 'news'?

News is a competitive industry: newspapers have to sell, which raises questions about how they maintain an edge over rival publications. Space is also limited. For instance, a news broadcast on television is rarely more than 30 minutes long. Additionally, what *counts* as 'news' is, as media analysts have known for decades, highly partial. For instance, nearly 50 years ago, Johan Galtung and Mari Ruge (1965) famously suggested that the following criteria were used by journalists and editors to judge 'newsworthiness':

- *Speed of development:* an event or activity that unfolds over days or weeks is more likely to become news than one that is slow to develop.
- *Threshold:* large events or actions that either affect 'significant people' or large sections of a population are more likely to become news than small events.
- *Unambiguity:* events or actions (while they may be complex) that invite clear interpretation and analysis are favoured over ones that create interpretive uncertainty.
- *Meaningfulness:* events or actions that can be storied in ways directly relevant to the daily lives of readers are more likely to be noticed than those in far-off places, the distant past or likely to occur in the far future.
- *Consonance:* events or actions that can be storied within the broad range of norms and expectations of readers are more newsworthy than ones that might significantly challenge or even alarm readers.
- *Unexpectedness:* events, findings or actions that contain an element of surprise and novelty, when compared with prevailing expectations, will tend to find favour – albeit within the limits of the above five points.
- *Continuity and inertia:* once certain things have become news they are likely to remain so until and unless they have run their course. This 'crowds out' other potential stories because the size of the 'news hole' is finite (most daily papers are no more than 30–50 pages long).
- *The news mix:* if a particular section of a newspaper is already large or 'busy', such as 'domestic affairs' or 'economics', then it's unlikely that additional stories in the same area will be included, unless they meet many of the criteria above.

More recently, Max and Jules Boykoff (2007) emphasise the sixth point above, but also argue that many Western newspapers, including the 'quality' press, now fixate on news that can be *personalised* and contains *drama*. Personalisation involves telling stories by focussing on the actions and reactions of individuals, be they politicians, celebrities, scientists or what-have-you. Dramatisation involves favouring news that can be storied as 'excitement', 'conflict' or 'the extraordinary'. Though the Boykoffs don't explain why these two things have become 'first order' journalistic norms, it's presumably bound up with the wider focus on celebrity and drama in the non-news sections of the mainstream mass media. Newspapers have, it seems, had to (or chosen to?) appropriate representational devices utilised in the broader 'media-sphere'.

These factors conditioning 'newsworthiness' aside, research suggests that most journalists (paper and non) also gravitate towards just a few information sources, with even fewer gaining continued and positive access, leading to sustained reporting of their words and deeds in news stories (rather than those of other potential sources; see Lester, 2010: chapter 4). Additionally, most Western newspapers have grown in size over the past 20 years, while the number of journalists they employ has flatlined or even decreased. This has meant there's (1) less space for long-term, in-depth investigative reporting, (2) more demand for journalists to cover several 'beats' (economics, industry, law and order, etc.), often forcing them to take what authoritative or very vocal sources say at face value, and (3) expectations that journalists write two to three times more stories than their forebears of the 1980s (see Lewis *et al.* (2008) for evidence of this in the British newspaper industry). Many journalists are now spread thin and are often reactive to events and sources.

Because of all this, there's a constant filtering of what gets to become 'news' and how it's reported to those consuming it (for more on the latter, see Box 7.2). To reach again for a metaphor I elaborated on at length late in Chapter 4, news journalists and editors together *construct* the news, selectively and actively re-presenting real world events (and information from sources) in line with their professional norms, and in response to many organisational pressures. To cite Brian McNair again,

> News is still what news always was: a socially constructed account of reality rather than [a carbon copy] . . . of reality itself, composed of literary, verbal and pictorial elements which combine to form a journalistic narrative disseminated through print, broadcast or online media. No matter how 'live' the news is, and regardless of how raw and visceral the account of events being brought into our living rooms appears to be, it's still a mediated version of 'reality' . . .
>
> (McNair, 2006: 6)

This (arguably) contradicts the assumption that most people would make that 'the news' *precedes* its reporting, as if it's simply waiting to be relayed by assiduous news outlets. However, in itself, the manufacture of the news is neither surprising nor, from a public perspective, especially alarming. After all, journalists and editors shouldn't be parrots: *choices must be made* about what stories to cover and how. It is therefore the *quality* of these choices that's at issue, not the fact that newsmakers have to make them in the first place.

BOX 7.2

NEWS MEDIA REPRESENTATIONS: 'FRAMES' AND 'CULTURAL PACKAGES'

For several decades now, professional media analysts have been interested in how the news media – indeed, the mass media more generally – 'frame' particular issues, topics and problems for audiences. According to Robert Entman, in a review of framing research,

> To frame is to select some aspects of a perceived reality and make them more salient in a communicating text, in such a way

> as to promote a particular problem definition, causal interpretation, moral evaluation, and/or treatment recommendation for the item described.
>
> (Entman, 1993: 56)

A decade later, he added that 'The two most important of these are ... problem definition – since [this] virtually predetermines the rest of the frame – and the remedy' (Entman, 2003: 418). In other words, frames (like those around a photograph) draw our attention to particular things in certain ways and place limits on understanding and future action. Research has suggested that newsmakers are not always entirely aware of the frames they create and use.

In part, this is because journalists and editors reach for 'cultural packages' that pervade the intellectual atmosphere of the societies to which they belong. These packages are nestled within what I termed wider 'repertoires of meaning' in Chapters 1 and 3. They comprise sets of beliefs, norms and metaphors that aid understanding and help to identify issues of contention or concern. For instance, in their influential study of how debates about nuclear power were framed in the American news media, William Gamson and Andre Modigliani (1989) identified a myriad of often rival packages. These included a *progress package* ('technical innovation and economic growth are good'), an *energy independence and security package*, a *devil's bargain package* (e.g. 'nuclear energy is risky but what choice do we have now oil supplies are insecure?'), a *public accountability package* (e.g. 'we need to ensure the nuclear industry is tightly regulated if we're to allow its further development') and a *not cost-effective package* ('it's too expensive compared with other options'). Not all news providers will utilise all packages for any given issue all, or even most, of the time. Like framing (to which it's closely related), packaging practices vary according to the stories and the particular news media outlets reporting them.

A key point about frames and packages is that they condition how 'events', 'facts' and 'evidence' are communicated in news stories. They insinuate values into the heart of all news reporting, not just because news sources have their own 'agendas' but also because newsmakers can't avoid introducing their own value judgements into reporting. This is why they matter and why news sources seek to actively shape them. Packages and frames for 'big issues' can vary over time, in part because of changing power dynamics among news sources. For instance, after the 1989 *Exxon Valdez* oil tanker spill off the Alaskan coast, a 'disaster' representation ('it was a terrible accident') gave way to a criminal narrative in the US news media ('the ship's captain was at fault') which, in turn, gave way to an environmentalist narrative ('large oil companies are needlessly endangering the non-human world') (see Daley and O'Neill, 1991). More recently, Boykoff and Mansfield's (2008) analysis of British tabloids 2000–6 shows that the reporting of climate change was framed predominantly through extreme weather events, the future of mega-fauna and arguments among personalities (like leading politicians). Frames highlighting the geographical injustice of likely climate change impacts, for example, rarely featured at all.

Reporting anthropogenic climate change and ventriloquising science

Climate change (or what was once more commonly termed 'global warming') is perhaps the best-known 'environmental issue' of our time – not least because it's been linked to a plethora of other environmental issues (like melting ice sheets and sea-level rises). The long-run emission of greenhouse gases (GHGs) into the global atmosphere over the past three centuries (but especially since 1950) appears to be increasing mean air temperatures worldwide. The volume of current and projected GHGs in the atmosphere suggests an average temperate rise of some 4°C by the end of this century.

Though this may sound like a small number to non-specialists, for environmental scientists of various stripes, it takes the earth surface well beyond the 'game changing' point in a biophysical sense. This is why 'the Anthropocene' has come into currency as a description of our epoch. As the former head of the British Tyndall Centre for Climate Change Research, Kevin Anderson, recently concluded in a co-authored evidence-based study of future scenarios:

> There is now little to no chance of maintaining the rise in global ... temperature at below 2°C ... Moreover, the [biophysical] impacts associated with [this] have been revised upwards ... so that it now more appropriately represents the threshold between dangerous and extremely dangerous climate change.
>
> (Anderson and Bows, 2011: 41)

By 'dangerous', the authors mean that anthropogenic climate change will trigger environmental transformations likely to pose a significant threat to current political–economic arrangements and settlement patterns worldwide. Much of the planet may become inhospitable for human habitation, while the biophysical character and geography of the habitable parts will change very significantly. Our entire way of life may have to be transformed in fundamental ways. Note that Anderson and Bows are scientists, they're not wearing the 'environmentalist' hat of someone like Paul Watson, and scientists aren't usually given to using inflammatory rhetoric in their peer review research papers. This doesn't mean they are to be trusted uncritically, of course, as we saw in Chapter 6 regarding West African forest loss discourses. But when thousands of earth surface scientists worldwide agree that significant environmental change lies ahead of us, in the form of the periodic Intergovernmental Panel on Climate Change (IPCC) reports, we should at least pay attention.

Clearly, if scientists like Anderson and Bows are right, then global climate change poses some momentous questions that demand profound answers and which will, in turn, require significant actions by people worldwide. As one British religious leader, David Atkinson, Anglican Bishop of Thetford, phrases it:

> Climate change is ... opening up for us ... questions about human life and destiny, about our relationship to the planet and to each other, about altruism and selfishness, about the place of a technological mindset in our attitudes to the world, about our values, hopes and goals, and about our obligations in the present and for the future.
>
> (Atkinson, 2008: 28)

In short, the ethical and practical implications of climate change on the scale projected by Anderson and Bows are extraordinarily *wide and deep* – wider and deeper than almost anything we can imagine, bar a meteor striking the Earth or a nuclear war. This is why environmental campaign groups such as Greenpeace have been echoing the claims of climate scientists in a more passionate and vocal idiom. However, alarming as these implications are, from the perspective of publics, climate change is a difficult phenomena to conceive of.

Study Task: How often do you think about 'global climate change' in your daily life? When you think about it, what images, if any, come into your mind? Do these images seem real, the stuff of fiction or about a future that's too far away to worry about?

Why is climate change hard for publics to register seriously? First, its local affects will be (are) spatially uneven and are currently barely perceptible in many places. Second, the time lags characteristic of change in large, open biophysical systems mean that current lifestyles are having major environmental effects, but ones that won't be tangible for decades. Third, the sheer difficulty of trying to predict the future of multi-scalar biogeochemical processes means there's considerable uncertainty about precisely when and how their effects will be felt. Together, these three things arguably give climate change an air of unreality in many people's daily lives.[11] Underpinning the trio is the assumption that the earth's climate (a form of 'universal nature', to use my shorthand from Chapter 1) exists separately from us, an 'external nature' existing at one remove with its own logics ('intrinsic nature') that humanity may (or may not) be altering.[12]

This is why the news media, scientists as sources of information about climate change, and environmental organisations as alarm bells about the need for action now are all so critically important. Global climate change is an *idea* rather than simply a set of 'real biophysical processes' occurring regardless of our representations of it.[13] It's a *representation of the past and future* based on current data about GHGs, historic records and deductions derived from scientifically accepted 'laws of nature'. More than most, newsmakers give shape and content to that idea in their words and images. They make (or fail to make) anthropogenic climate change 'real' and meaningful to publics by doing what the phenomenon so named cannot do on its own, namely represent itself to us here-and-now so that we can see and feel it with our own senses. While organisations like the IPCC and Greenpeace are crucial spokespersons for climate change, they must not only have their voices heard by the news media on a continuous basis, but also have their voices reported in ways that don't dilute or distort their claims. This is why I focus on newsmakers here, rather than their sources – even as I acknowledge how important it is to examine what sources say and which sources are excluded (on which subject see the next chapter).[14] Never before has the news media covered a story about nature as grand in scale or profound in its implications as anthropogenic climate change. Its role as both gatekeeper and relay of others' claims stands to be hugely consequential.

Needless to say, the fine details of climate change reporting have varied between newspapers (and other news outlets). There are now dozens of published studies of how 'claims makers' about climate change (like American scientist-activist James Hansen) have been reported in the news, which images have been selected to accompany stories about climate change, how often and how much climate change becomes 'news' and so on. Rather than attempt to summarise these studies, let me recount one to show just one way in which newspaper representations have been made sense of, drawing on the research of the Boykoffs into broadsheet reporting in America.[15]

Until relatively recently (and maybe even still today), many ordinary people have entertained serious doubts about the existence of anthropogenic climate change.[16] This is especially true in the United States, the world's largest GHG emitter per capita by some margin. Through the 1990s and much of the past decade, there's been a perceived 'climate change controversy'. This controversy arose because the 'consensus view' expressed by the IPCC was challenged by a minority of scientists. This minority – 'climate change sceptics' (or contrarians) as they became known – raised questions about (1) how reliable past earth surface temperature records are; (2) how strong the recent warming 'signal' in atmospheric temperature is; (3) whether any detected warming is part of 'natural cycles' of change; and (4) how robust future computer model-based predictions about warming and consequent environmental change are. Questions about (1)–(3) came to a head in the late 1990s in the so-called 'hockey stick controversy' in the United States, while the 'Climate-gate' and 'Glacier-gate' incidents a decade later kept the, by then flickering, flame of scepticism alive (see Box 7.3).

BOX 7.3

KEY CONTROVERSIES ABOUT THE SCIENCE OF GLOBAL CLIMATE CHANGE

The IPCC's periodic 'assessment reports' sift, sort and synthesise the findings of thousands of scientists worldwide whose research focusses on one or other aspect of atmosphere–earth surface relationships. These reports have arguably been *the* principal means of communication between climate scientists and the rest of us. The second report (IPCC, 1995) concluded that 'the observed trend in global mean temperature over the past 100 years is unlikely to be entirely natural in origin' (p. 412). The third report (IPCC, 2001) reiterated this observation and presented evidence of a 0.6°C increase in global mean temperature during the twentieth century. The fourth report (IPCC, 2007) stated that warming of the global climate system is 'unequivocal' and revised the 0.6°C figure up to 0.74°C. It stated that the observed warming was almost certainly due to GHG emissions resulting from human activity. Because so much research by so many different scientists is reviewed in the assessment reports, the findings indicate a durable consensus within the heterogeneous epistemic community that the reports have, in effect, created.

Research by Professor Michael Mann (of the Pennsylvania State University) and colleagues featured prominently in the third report. Mann *et al.* (1998, 1999) had examined large sets of existing 'proxy data' for past northern hemisphere temperature patterns going back to 1400 and 1000. Using a novel statistical technique to identify associations and patterns, and supplementing proxy data with more recent temperature observations, the authors concluded that there was a clear, and hitherto unmatched, increase in temperature from the late nineteenth century. This conclusion was represented in a two-dimensional graph showing time on the horizontal axis and temperature on the vertical axis. The graph became known as the 'hockey stick diagram' because centuries of relative temperature stability resembled a flat 'handle', while the sharp twentieth-century upward curve resembled a 'blade'. The diagram featured prominently in the publicity events organised by the IPCC to launch the third assessment report. According to science analyst Reiner Grundmann, 'after 2001 … [it] became *the* icon for the IPCC and global anthropogenic climate change' (2012: 15, emphasis added).

However, because of its prominence, Mann *et al.*'s research (1998, 1999) soon came in for criticism. Some critics took issue with certain details of the research, while a minority went further and questioned the validity of the data and analysis summarised in the hockey stick diagram. In two peer review papers, Americans Willie Soon and Sallie Baliunas, writing together and with other colleagues, surveyed already published research into past temperature records (Soon and Baliunas, 2003; Legates *et al.*, 2003). Their analysis suggested that significant northern hemisphere natural warming during the medieval period had been downplayed by Mann *et al.* (1998, 1999). Mann and many others rebutted the analysis vigorously in the science journal *Eos* (Mann *et al.*, 2003). Even so, Soon and others' criticisms were voiced in the United States Senate by James Inhofe (among others) in order to cast doubt on whether the United States had to reduce its GHG emissions. In July 2003, Senator Inhofe famously asked: 'could it be that man-made global warming is the greatest hoax ever perpetrated on the American people?' Shortly after, Canadians Stephen McIntyre and Ross McKitrick (2003) published a paper in the journal *Energy and Environment* that echoed the findings of Soon and the others.

Six years later, across the Atlantic, the Climate Research Unit (CRU) of the University of East Anglia found itself at the centre of a controversy even larger than the hockey stick one. In the lead up to the 2009 United Nations Copenhagen conference on reducing greenhouse gas reductions, thousands of private emails between senior CRU scientists were leaked. Some of the emails appeared to suggest that these scientists were suppressing (or would suppress) evidence that contradicted the scientific consensus articulated in the IPCC assessment reports. Shortly after, several cryospheric researchers challenged a statement in the fourth assessment report predicting the melting of most Himalayan glaciers by 2035 (IPCC, 2007). The IPCC chair Dr Rajendra Pachauri acknowledged the error in early 2010, but argued that it did not detract from the findings of the fourth assessment report.

Both of these recent 'scandals' have led to a set of inquiries, which have produced recommendations about the peer review process, scientific integrity and honesty, and the working procedures of the IPCC. Together, 'Climate-gate' and 'Glacier-gate' (as they've become known) have also given encouragement to the small minority of sceptics within the world of climate science, and the larger group of doubters outside it. In the United States, the Heartland Institute is a notable umbrella organisation (and facilitator) for climate change scepticism and has set up the NIPCC – the Non-Governmental International Panel on Climate Change. The Panel's compendious first report (NIPCC, 2009) has been followed by a second (interim) report (NIPCC, 2011), with both designed to challenge the IPCC 'consensus' on the 'reality' of anthropogenic climate change.

How has the 'climate change controversy' been represented by newspapers to their readerships? Let's take the American 'prestige press', that is to say broadsheet newspapers that offer relatively in-depth reporting for (on the whole) highly educated readers who mostly occupy the professions. Unlike tabloid newspapers, broadsheets aim to 'inform and edify' more than to entertain, and tend to avoid sensationalism and the overt use of rhetoric. Imagine being a regular reader of any of the *New York Times*, the *Los Angeles Times*, the *Washington Post* or the *Wall Street Journal* between 1988 and the period after the third IPCC assessment report: 1988 was arguably the first year in which anthropogenic climate change enjoyed sustained public exposure as an environmental 'issue', while by 2002, the IPCC reported that few (if any) reputable climate scientist disputed the reality of human-induced global warming. Over 3,500 separate news stories about climate change appeared in the four newspaper during this 14-year period and readers of the *New York Times* would have encountered the lion's share (some 41 per cent). Analysis of 18.4 per cent of these articles (Boykoff and Boykoff, 2004) suggests that after a two-year period (1988–90) in which climate change was frequently said to be likely and anthropogenic in origin, over half (52.6 per cent) of the reporting was 'balanced'. That is, it gave roughly equal attention to arguments and evidence for and against anthropogenic climate change. Consider, as one of the early examples of this kind of reporting, this excerpt from a 1992 *Los Angeles Times* article:

> Some scientists believe – and some ice core studies seem to indicate – that humanity's production of CO_2 is leading to a potentially dangerous overheating of the planet. But sceptics contend that there is no evidence that warming exceeds the climate's natural variations.
>
> (Abramson, 1992: A1)

Interestingly, the Boykoffs' analysis suggests that the proportion of balanced coverage didn't vary significantly between the four otherwise different newspapers. All increased their reporting of climate change, with peaks in 1992, 1997 and 2001–2. But all did so in a balanced way for around half of their reports.

This commitment to balance – one of the journalistic norms I described earlier – may seem commendable. After all, if one of science's own norms is 'organised scepticism' (Merton, 1942), then it's only right that the journalistic norm of balanced reporting should, in the four newspapers in question, have been a means for readers to learn about dissenters in the world of climate science. What's more, contemporary climate science is, unlike some areas of science, 'post-normal'. As Silvio Funtowicz and Jerome Ravetz memorably observed, it's a science in which many 'facts are uncertain, values are in dispute, the stakes are high, and decisions are urgent' (Funtowicz and Ravetz, 1993: 740). Given this, it may seem all the more necessary for news providers to have ensured balance during the IPCC's formative years – those covered in the Boykoffs' analysis.

However, one can argue that the 'balance' achieved in the four newspapers scrutinised was, in fact, a form of *bias*. One can suggest that in *re*presenting the findings of climate change scientists these broadsheets effectively *mis*represented them much of the time. Outlier views were amplified, despite 'balance' appearing to be a value-free journalistic norm. In this way, the norm in fact clashed with the other important reporting norm of 'objectivity' because, 'objectively' speaking, climate change sceptics such as Willie Soon and Sallie Baliunas (see Box 7.3) enjoyed little or no credibility among their scientific peers. The former journalist Ross Gelbspan has explained the problem of 'balanced' reporting of science thus:

> When the issue is of a political or (moral) ... nature, fairness – presenting the most compelling arguments of both sides with equal weight – is a fundamental check on biased reporting. But this ... causes problems when it's applied to issues of science. It seems to demand that journalists present competing points of view ... as though they had equal scientific weight, when actually they don't.
> (Gelbspan, 1998: 57–8)

In other words, 'balanced' reporting of debates about climate change science amounts to *decontextualising* them and thus preventing consumers of news from understanding how much (or little) importance to attach to dissenting views.

Study Task: Why, aside from respecting the norm of 'balance', do you think the quality newspapers analysed by the Boykoffs often presented both sides of the climate change debate equally over such a long period of time? Think back to the earlier discussion of what determines 'the news' in formulating your answer.

Why did 'balance as bias' (Boykoff and Boykoff, 2004) persist for so long in American broadsheets? We can only speculate. One thing we can almost certainly rule out is the kind of behind-the-scenes string pulling that Edward Herman and Noam Chomsky believe is rife in the US media. There's no evidence that climate change sceptics enjoyed privileged access to reporters working for the *New York Times*, the *Los Angeles Times*, the *Washington Post* or the *Wall Street Journal.* Alternative explanations thus suggest themselves. Arguably, balanced reporting made climate change science more newsworthy than it would otherwise have been. By focussing on challenges to the IPCC scientific consensus, reporters could write stories in which personalisation and drama were allowed to feature. At the same time, only experienced science journalists able to specialise in science reporting (like the *New York Times's* Andrew Revkin) were likely to appreciate the nuances of scientific debate.[17] The pressures that news journalists have faced over the past generation mean that many lack the time and expertise to determine whose views to give the most weight to. The default position then becomes that of a stenographer, engaged in a nominally balanced 'he said, she said' mode of reporting.[18] As Anders Hansen notes in his book *Environment, media and communication*,

> A major problem in criticism of 'media accuracy' is the notion that the perceived inaccuracy is primarily a product of sloppy reporting [or] ... downright media distortion.
>
> (Hansen, 2010: 89)

Sometimes, as the Boykoffs argue, inaccuracy can happen innocently as a result of professional norms being adhered to by busy journalists.[19]

The sustained visibility that climate change sceptics enjoyed in US broadsheets through the 1990s (and elsewhere in the news media) seems, in hindsight, especially problematic when one considers who was funding and disseminating much of the dissenting science. For instance, documents obtained by Greenpeace under the United States Freedom of Information Act reveal that the libertarian Charles G. Koch Foundation gave Willie Soon two grants totalling $175,000 in 2005/6 and again in 2010. Meanwhile, Soon received grants from the American Petroleum Institute between 2001 and 2007 totalling $274,000, and grants from oil company Exxon Mobil amounting to $335,000 between 2005 and 2010. While there's no evidence that Soon, and other scientific dissenters like him, forged, trimmed or 'cooked' data to suit the perceived agendas of these funding bodies, it illustrates graphically how closely linked politics and science have been in the contrarian camp. A number of companies that stand to suffer financially from governmental controls on current levels of atmospheric GHG emissions, and a set of conservative and libertarian think tanks/foundations, have spent considerable money, time and energy since the early 1990s doing two things, namely (1) searching assiduously for research findings that contradict the IPCC scientific consensus; and (2) lobbying the news media, notable academics (like Professor Fred Seitz (1911–2008)) and elected politicians to take notice of these findings.[20]

I'm not suggesting that the broadsheets analysed by the Boykoffs shouldn't have reported science that questioned the IPCC consensus. Respected experts such as Judith Curry (an atmospheric scientist at Georgia Institute of Technology) and John Christy (another atmospheric scientist, but at the University of Alabama) have, at times, voiced scepticism about the messages contained in the Panel's assessment reports.[21] In its desire to create clear headline messages for public and political consumption, the IPCC may have glossed disagreements among the scientists whose research it has periodically reviewed. As Daniel Sarewitz has opined in the pages of *Nature*, 'The ... idea that science best expresses its authority through consensus statements is at odds with a vibrant scientific enterprise' (Sarewitz, 2011: 7).[22] In this light, a better self-regulated press would have sought to disaggregate and properly contextualise *different forms of climate change scepticism* for readers. 'Positive scepticism' of the sort articulated by the likes of Curry and Christy is different from *cynicism*. Even the best climate science deals in probabilities not certainties. Rather than replace the 'bias of balance' with virtually no coverage of climate science contrarians, a more responsible press would have given (and should still give) scepticism its due – even while acknowledging the enormous weight of scientific evidence pointing to the reality of climate change. It would have distinguished between (1) politically motivated scepticism voiced by non-experts; (2) science-based scepticism voiced by capable non-specialists (like Stephen McIntyre); and (3) science-based scepticism voiced by specialists (like Judith Curry).

Such properly 'balanced' reporting would not, of course, pre-empt answers to the difficult questions of what to do about anthropogenic climate change (conventionally phrased as questions of 'mitigation and adaptation'), but it would better communicate messages from a large and complex epistemic community (climate scientists) whose public voice depends almost entirely on the quality of the representations offered by the news media (and the 'serious' part of the mass media more generally).

SUMMARY

This chapter has focussed on the mass media in general and the news media specifically. It's done so because the tens of thousands of epistemic workers employed by the media industry exert considerable influence on our collective understanding of nature (and everything else besides). The chapter has explored the key contribution that 'serious' parts of the media make to the existence of the public sphere and the quality of debate within it. Though 'alternative' media are more important than ever before, I've argued that the 'mainstream' media remains far more visible to most people in their daily lives. The chapter concluded with a focus on mainstream newspaper reporting of the science of global climate change. This has served to illustrate two important points. The first is that the news media is (still) *the* route along which the representations of other epistemic communities must travel if they're to capture public attention. The second is that the content of news media representations is heavily conditioned by a combination of journalistic norms and institutional pressures. In the case reported here, there was arguably a skewed representation of the findings of climate scientists. This raises questions of how better to regulate the news media, which I touched upon briefly towards the end of the chapter.

In the next chapter – the book's final one – I want to focus squarely on this question of regulation. Throughout this book, I've argued that we're epistemically dependent on others for representations that shape our values, beliefs and actions. It follows that the most influential epistemic communities operating at any one moment in history need to be regulated effectively, otherwise their influence on us can be detrimental. I want now to consider how our existential condition of epistemic dependence can be turned to our collective advantage, even as it presents the ever-present risk that we are misdirected by others.

ENDNOTES

1 These media, to reiterate Latourian language used in Chapter 3, are a combination of 'immutable mobiles' and 'immutable immobiles'.
2 This said, even the 'mainstream' mass media in a country can be highly differentiated such that people living in different areas are offered a very different mainstream media menu. This, arguably, is true of the United States, which has a highly privatised media industry in which numerous local, regional and national providers compete with public and other not-for-profit providers.
3 I place all these terms in scare-quotes because, in practice, even the 'serious' parts of the media are arguably engaged in all three things simultaneously – they're not readily separated. Two graphic illustrations of this are Kate Evans's seriously amusing comic *Funny weather* (Evans, 2007) and Randy Olson's *Global warming comedy* (2008).
4 This need is not only about information dissemination, i.e. ensuring those being governed get the necessary messages from their governments; it's also a need for social integration and cohesion, with the mass media ensuring a degree of common knowledge and experience among otherwise different and physically separated citizens.
5 For a readable survey of (largely) mainstream approaches to understanding 'media effects', see the collection edited by Jennings Bryant and Mary Oliver (2009). For a more inclusive survey, I recommend *The Sage handbook of media processes and effects* (Nabi and Oliver, 2009).
6 One of the foci of recent research into the vibrancy of civil societies is what happens when a country experiences rapid immigration and thus undergoes diversification in a religious, cultural and ethnic sense. Does civil society become weaker or does it become more fragmented, a set of islands whose inhabitants are tight knit but which exclude 'outsiders'? This is something American political scientist Robert Putnam has long been interested in, suggesting that 'bridging capital' between very different citizens can weaken, even as 'in group' 'bonding capital' remains strong.
7 Another important threat, to which I alluded near the end of Chapter 3, is a public that's cynical about, or detached from, or simply too pressed for time to focus on public affairs.
8 Doublespeak is language that deliberately disguises, distorts or reverses the conventional meanings and referents of words. It often takes the form of euphemisms (e.g. 'downsizing' for laying off workers), making the truth less unpleasant without denying its nature. It may also be deployed as intentional ambiguity or reversal of meaning (e.g. favouring 'climate change' over the more graphic terminology 'global warming'). The charge that the contemporary news media are debasing the public sphere is made with some passion by journalist Dan Hind (2010) in *The return of the public*. Non-profit organisations like the Center for Media & Democracy (based in the United States) now exist to try to expose and counter some of the more self-serving uses of spin and public relations in the corporate and political worlds.
9 Even those who disbelieve the claims made by most climate scientists that the biophysical world we inhabit will change dramatically from hereon will, I argue, have to reckon with global climate change. The discourse about it will not go away any time soon, especially if some of the predicted biophysical changes occur perceptibly in the short to medium term.
10 There's another norm I've not mentioned, but which is important to securing public trust in the news media: the norm of abiding by the law when researching and

reporting news stories. This norm was breached spectacularly in Britain during the 'noughties' when journalists and investigators employed by Rupert Murdoch's tabloid *News of the World* were found to have invaded the privacy of various celebrities and ordinary people in their desire to be news leaders.

11 In these three respects, the contrast with the 'hole' in the atmospheric ozone layer discovered in the late 1970s is instructive. This hole was perceptible in the present and could be traced with some certainty to the effects of chlorofluorocarbons in the upper atmosphere. Technically, it was a relatively 'easy' problem to solve and led to rapid and concerted intergovernmental action by 1987.

12 I agree with Julie Doyle (2011) that, as well as 'nature' being seen as something separate from humanity, we in the West cleave *time* too, regarding the future as separate – an 'elsewhere' that is somehow too big and indeterminate to try to anticipate or shape. Relatedly, I concur with philosopher Michelle Bastian (2012), who argues that Westerners have come to regard the Earth ('universal nature') as a *stable* background that supports our endeavours as essentially unchanging – meaning that the immediate *future* of the Earth will be *more of the same*, and thus not something we're causally connected to or should worry much about. For a recent summary of the research into the difficulties of communicating global climate change to publics effectively, see Susanne Moser's (2010) essay 'Communicating climate change'.

13 I mean this in *both* a descriptive and an evaluative sense. For instance, where, using the rhetoric of truth, climate scientists seek to represent the 'fact' of climate change via their discourses and images, environmentalists urge us to react to the 'fact' in particular value-laden ways. Mike Hulme's (2009) engaging book *Why we disagree about climate change* is predicated on the idea that climate change is not simply a set of physical changes but a concept that is founded upon scientific discourse and which is inciting further technical, moral and political discourse. He shows that it is a complex and contested idea, especially when one examines the various answers to the question 'what should we *do* about climate change?'

14 In their work on the news media during the 1970s and 1980s, Stuart Hall and colleagues talked about 'primary definers' – sources whose versions of an 'issue', 'problem' or 'opportunity' were reported in the media on a consistent and continuous basis. Though a great many potential sources barely have access to the news media (they are 'underaccessed' by journalists), some can win the battle for access with time, luck and effort – left-leaning environmentalists being an example since the mid-1970s. Especially vocal and innovative sources are sometimes called 'issue entrepreneurs' because they bring a new subject to journalistic attention or significantly reframe understandings of a known subject.

15 I draw here upon Boykoff and Boykoff (2004, 2007) and Boykoff (2011).

16 I say 'even still today' because surveys of public opinion in the United States and United Kingdom indicate relatively large numbers of people who believe that climate change is not anthropogenic in origin, with a small number believing that the global climate is not changing. As recently as 2007, reports of new scientific evidence that questioned the actuality of anthropogenic climate change garnered a lot of attention. For instance, in late 2007, a British writer and journalist posted a fake story on the Internet about a newly published scientific research paper that claimed to challenge the idea that humans were causing global climate change. The 4,000-word paper was entitled 'Carbon dioxide production by benthic bacteria: the death of manmade global warming theory?' and was said to have been published in the Japanese peer review *Journal of Geoclimatic Studies* which was created by Okinawa University's Institute for Geoclimatic Studies. The paper's authors were said to be Daniel Klein and Mandeep J. Gupta of the University of Arizona, and Philip Cooper and Arne F. R. Jansson of the University of Gothenburg. The paper's claim that deep water bacteria were creating large volumes of CO_2 was quickly reported on electronic mailing lists maintained by several climate change sceptics in the United Kingdom and United States, and right-wing television host Rush Limbaugh also reported the paper's findings; however, within a few hours the paper was revealed to be a hoax.

The hoax's author explained the purpose of his actions thus: 'Sometimes fiction and satire can reach places facts alone can't – in the right context,' he said. 'What the hoax showed is that there are many people willing to jump on anything that supports their argument, whether it's true or not. What we wanted to emphasize is that it's necessary to achieve scientific validity using the peer-review model. Proper climate science makes every attempt to do this, and is a constantly evolving and self-refining process, as all science is' (cited by Revkin in *New York Times*, 11 November 2007, http://dotearth.blogs.nytimes.com/2007/11/11/the-life-and-death-of-a-climate-hoax/, accessed 31 July 2012).

17 Revkin worked for the *New York Times* as a journalist 1995–2009 and now writes the Dot.Earth blog for the 'Opinion' section of the paper (see http://dotearth.blogs.nytimes.com/).

18 As Boykoff (2011) makes clear in his recent book, American broadsheets have been more accurately reporting the conclusions of the climate science community since around 2002. The 'bias of balance' is thus now less of a problem than heretofore. Indeed, on 7 December 2009, no fewer than 56 major newspapers worldwide co-published an editorial urging political leaders to act to avert 'dangerous climate change' at the Copenhagen Meeting of the Parties to the Kyoto Protocol. It's also worth noting that a major independent analysis of climate change data was recently conducted by a research team at the University of California, Berkeley. The Berkeley Earth Surface Temperature (BEST) Project participants were not associated with the IPCC and were considered by some to be overtly sceptical of the IPCC scientific consensus. Their findings, based on analysis of 1.6 billion temperature records and using a sophisticated methodology designed to correct for various possible anomalies, were issued in October 2011. The findings are highly consonant with those of the IPCC assessment reports and suggest an increase in land temperature of around 1°C since 1950. See http://berkeleyearth.org/.

19 Interestingly, independent organisations like Climate Central (http://www.climatecentral.org/), Greenwire (http://www.greenwire.org.uk/) and the Society of Environmental Journalists (http://www.sej.org/) have been set up to help reporters avoid playing the stenographic role that produces nominal (but not actual) balance. There is also now a large environmental blogosphere that journalists can learn from (see, for example, the contributions to *Grist* magazine (http://grist.org/news/) or Treehugger (http://treehugger.com)), though some skill is required to navigate between the myriad of contributors.

20 The history of climate change 'denial' in the United States is recounted by science historians Naomi Oreskes and Erik Conway in their book *Merchants of doubt* (2010), by James Powell in *The inquisition of climate science*, and by Michael Mann (2012) in *The hockey stick and the climate wars*. None of these authors are sympathetic to the contrarian cause. Washington and Cook's (2011) *Climate change denial: heads in the sand* examines the different evidence and arguments presented by climate change sceptics and systematically refutes them, while acknowledging the uncertainties involved in even the most rigorous climate science. Robert Carter (2010) tries to summarise the science from a sceptic's viewpoint in *Climate change: the counter-consensus*.

21 Christy, for example, would probably not accept the 'extreme' predictions of future temperature rises made by Anderson and Bows (2011) reported earlier in this chapter.

22 Sarewitz's point leads one to wonder if the IPCC has not become understandably defensive in the face of attacks by politically motivated climate change sceptics. By closing ranks and over-emphasising the (undoubtedly considerable) degree of scientific consensus, the IPCC may be performing *too much* of the 'boundary work' to which I referred in Chapter 3, following Tom Gieryn's thinking. This said, evidence indicates that the news media – even the broadsheet newspapers – don't find it easy to communicate the caveats entered by good scientists (otherwise known as uncertainty). This may also be a reason why the IPCC assessment reports seem not to over-emphasise extant uncertainties and focus instead on probabilities and likelihoods. I'll say more about how to address this in Chapter 8.

8 EXPERTISE, THE DEMOCRATISATION OF KNOWLEDGE AND PARTICIPATORY DECISION-MAKING: UNDERSTANDING THE NATURE OF SCIENCE

In the previous chapter, I discussed mass mediated representations of nature, taking the example of how climate change science has been reported in 'quality' newspapers. I ended with some brief observations about the need for newspaper regulation designed to serve the 'public interest'. There's a lot more I could say about press regulation – it's a manifestly important topic.[1] In this final chapter, however, I want to focus on the governance of science. If the news media, and the mass media more generally, is a crucial passage-point for non-media representations of nature, then science remains the largest producer of these non-media representations. It's at the heart of a wider phenomenon we call 'expert knowledge'. More than the words and images of poets and advertisers (say), we tend in the long run to pay far greater attention to what scientists tell us about the world (as do political decision-makers). They speak to us in a range of specialised idioms and produce cognitive knowledge in the main, though with important implications for our ethical (and aesthetic) practices. While scientists' claims typically come to us courtesy of news and current affairs outlets (as the previous chapter indicated), the way scientists manage their activities is every bit as important for us as the way news industry regulation is undertaken.

This much was obvious for the residents of Kissidougou, in the case explored in Chapter 6. But what about situations like that recounted in Chapter 4 – namely the case pertaining to human genetics and 'biosociality'? In these situations, science confronts us with very important insights (and technologies) that we're *invited*, though *not compelled*, to take seriously as citizens and/or consumers. These insights only intrude on our lives to the extent we let them. Here, the power of science is, if you like, softer than it was for inhabitants of the Guinea forest–savannah transition zone, unless and until it's acted upon by governments in the form of law and policy. But it is no less important for this fact. For instance, what if we fail to push governments to act more decisively on scientific advice that turns out to be accurate and profound in its implications? What if we endorse science-based policies that turn out to be extremely expensive and in the end, unnecessary or even harmful? What 'opportunity costs' arise when unconventional or maverick scientific research is not permitted by the scientific community to garner much public attention (e.g. by being labelled 'pseudo-science')? These questions become especially vexed in an era when public deference to the findings and recommendations of scientists is arguably at a historic low in many countries. In part, this is because science has appeared unable to anticipate or deal effectively with several major incidents, including an outbreak of bovine spongiform encephalopathy (BSE) in British cow herds in the 1990s. Today, science is not simply, to use a minimal definition, an especially prominent form of 'systematic inquiry'. It's also, in effect, 'a [key] representative body in which a few speak for the many' (Fuller, 2000: 8). This is because credentialised scientists – a tiny minority of the world's population – are licensed to make cognitive claims about 'nature' in every conceivable aspect. This is an extraordinary capacity.

I begin this chapter by presenting an extended case of scientists' own regulatory practices being exposed to the glare of public scrutiny. It relates (once more) to climate change and I touched upon it very briefly in Box 7.3. The so-called 'Climate-gate' and 'Glacier-gate' affairs, which were revealed to publics worldwide in late 2009 to early 2010, arguably constituted a crisis of legitimacy for international climate science. They cut to the heart of perennial questions about the role of science in democratic societies: should we trust scientists' representations of nature, especially when they have far-reaching implications for us and future generations; and how should we act on scientists' pronouncements supposing we place our trust in them? I will use the two affairs to provide possible answers to these important questions before moving to consider further cases. The answers speak to the issue of how far, and in what capacities, scientists should be allowed to be self-governing. The chapter thus draws this book to a close by exploring what we, in our daily lives, should do about our existential condition of epistemic dependence. While we cannot eradicate our dependence, we need robust mechanisms in place to ensure we don't lose whatever independence of thought and action we possess. A study of our complex relations with professional science shows us why.

CLIMATE CHANGE SCIENCE UNDER ATTACK

Since the late 1980s, the international scientific consensus that global climate change is (1) occurring, (2) caused principally by anthropogenic GHG emissions, and (3) likely to have significant effects on non-humans and people has grown steadily. As I noted late in Chapter 7, the periodic IPCC reports codify this consensus in words and images. The IPCC was established by the World Meteorological Organization and the UNEP in 1988. To quote directly from its website, it is

> a scientific body [that] ... reviews and assesses the most recent scientific, technical and socio-economic information produced worldwide relevant to the understanding of climate change. It does not conduct any research nor does it monitor climate-related data or parameters. Thousands of scientists from all over the world contribute to the work of the IPCC on a voluntary basis. Review is an essential part of the IPCC process, to ensure an objective and complete assessment of current information. The IPCC aims to reflect a range of views and expertise.
>
> (http://ipcc.ch/organization/organization.shtml, accessed 30 September 2012)

In the IPCC, scientific research into climate change that has, for the most part, passed through 'blind' peer review is analysed *en masse* by a team of leading Earth/environmental scientists and synthesised into reports into the 'state of the art' knowledge about the subject. These scientists' own research publications are among the many surveyed. The assessment reports are intended to be taken notice of by political leaders, the global business community, publics worldwide and those (e.g. environmentalists) in the 'third', or voluntary/charitable, sector. The first scientific assessment report was published in 1990 and the most recent (the fifth report) released in 2013. Such was its standing 19 years after its creation that the IPCC was co-awarded the 2007 Nobel Peace Prize (along with former US Senator and environmental campaigner Al Gore). Despite concerted efforts by 'climate change sceptics' in the United States to challenge the IPCC's messages, by 2009 virtually no climate scientist was prepared to dispute points (1)–(3). Consequently, even in the Western countries where the sceptics have commanded greatest public attention, i.e. the United States, Canada and Australia, the quality press had begun to refrain from the 'bias of balance' discussed towards the end of the previous chapter. Related to this, numerous surveys showed that publics worldwide were increasingly convinced of points (1) and (2), if not necessarily (3).

Given this context, 2009 should have been an important year for climate change science and climate change politics alike. Over 2,000 climate change researchers met in Copenhagen in March 2009 at a 'Climate Congress' designed to distil key scientific messages that could be fed into the December 2009 meeting of international policymakers to agree further action to reduce GHG emissions. The peer reviewed synthesis report published by the Congress organisers reiterated the conclusions of the IPCC fourth assessment report.[2] The most recent (post-2006) research surveyed by the organisers did nothing to contradict the clear 2007 Panel conclusion that:

> Warming of the climate system is *unequivocal*, as is now evident from observations of increases in global average air and ocean temperatures, widespread melting of snow and ice, and rising global average sea level.
>
> (http://ipcc.ch/publications_and_data/ar4/syr/en/spms1.html, emphasis added, accessed 30 September 2012)

On this basis, many hoped that the December meeting of the Kyoto Protocol signatories would at last lead to decisive political action globally to reduce the risk of global mean temperatures exceeding 2°C in the next two to three generations.

However, this meeting produced no such action, and yet 2009 proved to be an important year for climate change science nonetheless – though for reasons both unanticipated and damaging. As we'll see, the potential power of climate science to trigger meaningful worldwide political action failed to actualise.

Two surprising 'scandals'

The CRU at Britain's University of East Anglia is among several important contributors to global climate change science. The conduct of this science is, like all science, normally hidden from public view. However, a few weeks before the December 2009 Copenhagen meeting, over 1,000 private emails sent and received by leading CRU researchers were made public without authorisation (along with many other documents and files). Importantly, these scientists all had frontline involvement in the IPCC, rather than simply having their research surveyed by others charged with writing the periodic assessment reports. The emails were almost certainly hacked by persons sympathetic to the arguments of climate change sceptics. They stretched back to the period leading up to the publication of the IPCC third assessment report in 2001. A headline, evidence-based message of that report was that:

> The Earth's climate system has *demonstrably changed* on both global and regional scales since the pre-industrial era, with some of these changes attributable to human activities.
>
> (http://www.grida.no/publications/other/ipcc_tar/?src=/climate/ipcc_tar/, emphasis added, accessed 30 September 2012)

The leaked emails appeared to reveal malpractice occurring at the heart of the climate science community. Three issues loomed large. First, some emails suggested that CRU scientists and their associates (such as Michael Mann at Pennsylvania State University) had abused the peer review process in order to reject research submitted for consideration by scientists questioning the idea of earth surface warming induced by human activities. Second, some emails apparently evidenced a reluctance to share research data (and analytical techniques) with certain others seeking to verify interpretations of temperature change based on this data. Third, some emails appeared to suggest that authors of the IPCC scientific reports 'smoothed out' contradictions in the available evidence in order to reduce doubt among politicians and citizens that anthropogenic warming is, in fact, occurring. Importantly, the veracity of the emails was not questioned by the CRU scientists involved, or by their correspondents elsewhere in the United Kingdom and overseas. This scandal was quickly followed by a second one: an error was detected in the IPCC fourth assessment report of 2007. A claim that Himalayan glaciers would likely all melt away by 2035 was found not to be supported by any credible scientific research. This suggested potential sloppiness or a tendency towards exaggeration among the report writing team. A timeline of both scandals can be found in Box 8.1.

BOX 8.1

'CLIMATE-GATE' AND 'GLACIER-GATE': A DURING AND AFTER TIMELINE

2009

19 November. Unauthorised release of over 1,000 emails and other files involving senior CRU scientists.

23 November. University of East Anglia (UEA) announces an inquiry to be led by Sir Muir Russell.

1 December. Professor Phil Jones, whose emails were hacked, steps down as Director of the CRU.

9–15 December. International meeting of the political representatives of countries committed to a global climate change mitigation agreement in Copenhagen. It fails to ramp up transnational efforts to reduce GHG emissions.

20 December. Pennsylvania State University announces a review of the research of Professor Michael Mann, whose email correspondence with Phil Jones was hacked.

2010

18 January. It's revealed that a statement about future Himalayan glacier melting included in the IPCC fourth assessment report (2007) was baseless.

22 January. UK House of Commons Select Committee on Science and Technology (SCST) launches an inquiry into 'Climate-gate'.

10 February. The Pennsylvania State University inquiry into Mann's research concludes he's not guilty of any professional wrongdoing.

10 March. The UN Secretary General and the IPCC Head request the Inter-Academy Council (IAC) to reveal the working methods of the IPCC.
22 March. UEA asks Lord Oxburgh, former chairman of the UK House of Lords SCST, to examine the scientific claims made by CRU researchers.
24 March. A leading British forest scientist files a complaint with the UK Press Complaints Commission alleging misreporting of IPCC fourth assessment report findings by *The Sunday Times* newspaper. A month later, a Canadian counterpart files a libel suit against *The National Post* for misrepresenting climate science findings.
31 March. The SCST inquiry criticises UEA for fostering a 'culture of withholding research data', but finds no wrongdoing among CRU scientists.
14 April. The Oxburgh Inquiry concludes that the published scientific findings of CRU researchers were robust.
6 July. An independent inquiry into the content of the IPCC fourth assessment report, instigated by the Dutch Government, finds few factual errors; however, it observes that the report tends to highlight 'worst-case scenarios'.
7 July. The UEA Muir Russell Inquiry upholds the integrity of CRU scientists but highlights problems with sharing research data and information about analytical techniques with outside parties. Jones subsequently reinstated as CRU Director. The US Environmental Protection Agency also issues a report that finds no evidence of scientific malpractice in the 1,000-plus hacked emails.
15 August. A US National Science Foundation inquiry into Mann's research reaches similar conclusions to those of the Pennsylvania State University inquiry. The same week, the former climate change sceptic Professor Bjorn Lomborg (2010) publishes *Smart solutions to climate change* in which he acknowledges the seriousness of future anthropogenic warming of the Earth's surface.
30 August. The IAC report recommends a reform of the IPCC working methods to introduce new checks and balances. These are designed to ensure robust review of existing evidence about climate change and increased transparency to outsiders.
14 October. The IPCC agrees to implement many of the IAC recommendations immediately.

2011
10 July. The BBC releases an independent report, authored by geneticist Professor Steve Jones, into how the corporation reports the latest science.

22 November. A week before the UN Climate Change Conference in Durban, several thousand hacked emails withheld in 2009 are released without authorisation, almost certainly by the climate change sceptics who instigated 'Climate-gate'.

The fallout of 'Climate-gate' and 'Glacier-gate' was immediate and significant. They made headline news worldwide and cast a dark shadow over the Copenhagen Kyoto Protocol parties meeting. Through newsletters, email lists, websites, blogs and so on, climate change sceptics used the scandals to rekindle public doubt that global warming is either occurring, anthropogenic or significant. In response, the American Association for the Advancement of Science (AAAS), the American Meteorological Society and the Union of Concerned Scientists released statements supporting the scientific consensus that the Earth's mean surface temperature had been rising for many decades. No fewer than six official inquiries were launched into the two scandals on both sides of the Atlantic in order to ascertain if wrongdoing had occurred and how to prevent it happening again. A series of public opinion surveys nonetheless revealed, as late as mid-2011, that significant sections of the British and American publics (1) distrusted climate scientists, and (2) believed anthropogenic climate change was not happening.

In short, the scientific consensus about climate change suffered a significant crisis of public credibility – even though only an extremely small number of climate scientists were implicated in the two scandals. At the time of writing, the many official inquiries into both affairs have largely absolved the scientists of any wrongdoing (see Box 8.1). In the wider climate science community, the claims of sceptics like Sally Baliunas and Stephen McIntyre are still not considered sufficiently convincing to challenge the consensus view that global warming is real, anthropogenic and likely to be significant looking ahead. It will probably take some time before public trust in climate science reaches, let alone exceeds, the levels evident just before the two scandals became headline news. When scientific knowledge travels into the public realm, it typically fails to 'perform' if serious questions arise as to its trustworthiness. Unlike other discourses, that of scientists exerts power only to the extent that people are assured it accurately represents 'reality'. In this way, its 'power over' is conditional.

Does this matter? *Yes*, it most certainly does. If Kevin Anderson and Alice Bows's predictions are right (see p. 234 of this book) – and *many* climate scientists believe they *are* – then continued public scepticism about the causes and consequences of global warming will prove very costly in the long term (in all conceivable senses of the word). Given the profound economic, demographic and ecological implications of global warming predictions, however, it may be equally costly to accept – and act on – the word of scientists uncritically. What, then, are we to do?

LEARNING FROM THE CONTROVERSY OVER CLIMATE CHANGE SCIENCE

As the previous paragraph intimated, 'Climate-gate' and 'Glacier-gate' reveal graphically a plethora of important issues pertaining to the content and effects of scientific representations of nature in the twenty-first-century world. Before I try to itemise these issues why don't you have a go by completing the study task below?

Study Task: Much scientific research has no *direct* bearing on our lives and we are scarcely aware of it because it seems irrelevant (see the residents of Kissidougou). However, climate science is one of several examples of research that indirectly stands to affect virtually everyone on the planet (or at least very large numbers of people). Other examples are biomedical science (discussed in Chapter 4), biotechnological research, epidemiology, the science of risk calculation, conservation biology and research into energy technologies. Think a little about the major characteristics shared by these otherwise different kinds of scientific endeavours from the perspective of an ordinary person. Compare them with the high-profile but fundamental (non-applied) science involved in the Hadron Large Particle Collider. What, in your view, do they have in common once you look past their specialised discourses, techniques and findings? Equally, are there any major divisions that cross-cut them?

What do the two affairs tell us about the most publicly prominent areas of contemporary science? First, climate science is addressed to 'the public', civil society and their political representatives in the *widest possible sense*. This is for the obvious reason that its subject matter is global in scale. What's more, it's not a 'pure science' that's only designed to satisfy our curiosity. This is for the equally obvious reason that, depending on the degree of change ahead, global warming is likely to be consequential for most people (though we don't know the fine details of quite how). The implication is that climate scientists are 'representatives' in a very grand way: they speak of changes said to be germane to most, even *all*, of humanity. Though few other scientists do this, there are, nonetheless, other contemporary sciences that seek to make claims on very large numbers of people – the new human genetics discussed in Chapter 4 being a prime example. Second, climate science, like all science, is normally inscrutable to outsiders. The training, techniques and judgements required to represent past, present and future climatic trends are free from external inspection or regulation. This is no different to other esoteric areas of science, of course. But it means that, from a public perspective, there's a need for assurance that the science is of a very high quality. In practice, this usually resolves to the question: are scientists adhering to their own professional norms?

Third, climate science must reckon with extremely complicated open systems in which, to use the jargon, positive and negative feedbacks, thresholds, blockages and spatio-temporal lags may all be in play. These systems are assumed by climate scientists to exist 'out there' waiting to be studied as objectively as possible and they are very challenging to understand in ways that ensure any reasonable degree of certainty. Again, this fact is not unique to climate science, but it contrasts with sciences that study, and successfully control and construct, 'closed systems'. These sciences typically achieve high degrees of explanatory, predictive and practical certainty. Yet climate science is one of several forced to caveat its claims about nature, and where the qualifications – if too cautious or else insufficiently so – can have life or death consequences in the medium to long term. As I said near the end of Chapter 7, what we call 'climate change' is very obviously an epistemic construction. It's an amalgam of specialised scientific attempts to depict past, present and future alterations to the Earth's surface processes. Without these attempts there would, quite literally, be no such thing as climate change so far as we're concerned – just as there'd be no 'genes' without the claims of microbiologists (see Chapter 4) or 'deforestation' in West Africa without the discourse of silviculturalists (see Chapter 6).

If climate science arguably exemplifies the power and high profile of scientists in the contemporary world, the 2009–10 scandals reveal continued anxiety about whether this power is warranted and how it's used. Soft though this power is, when the stakes are as high as they are in climate science, it's no wonder that some want to peer inside the citadel to see what the epistemic workers are really up to. They worry that we may be unduly subject to the power of science rather than empowered by its findings. Two big, and far from new, questions arise from the two 'gates'. First, can we trust scientific representations of nature when their implications stand to be highly significant? Second, if we can, then how should we as citizens respond practically to science's insights? As I said, science (like the mass media) does not usually *oblige* us to do things: the force of its insights is rarely that strong. Rather, as a 'governmental' institution in Foucault's expanded sense, it employs the power of expertise to direct our attention to various problems and possibilities. It enjoins us to take notice of certain things and invites us to consider how best to respond to them. This doesn't mean the power of science is weak. On the contrary, the case of climate change demonstrates why it's so important that the governance of science and our relationship to science are in the best possible health.

The governance of science

The professions, such as law and accountancy, have a long history of self-regulation. Science is no different. Scientists have created norms and institutions designed to govern their professional behaviour. Of course, this doesn't guarantee them absolute freedom of thought and action. For instance, during the presidency of George W. Bush, stem cell research in the United States was stymied by federal government because of its highly charged ethical implications. This is one of many examples where science impinges visibly on society and invites external regulation of its practitioners' activities. There have been more than a few times when science must be held to account, especially when funded by the public purse. In the case of climate science, however, no such outside intervention in the black box of scientific practice has applied, even in the period after the two scandals of 2009–10. This is because the research surveyed by the IPCC is not, in itself, considered ethically problematic by the majority of national governments or citizens. On the contrary, this research is largely considered to be of vital importance. In the 'post-gate' period, the discussion has therefore focussed on improving the self-governance of climate science. It's the working practices of scientists that have been at issue, rather than the wider implications of their research. Before I recount some of this discussion, complete the study task below.

Study Task: Why, as citizens, must we rely on the claims made by the epistemic community of climate scientists if we're to anticipate future environmental change? In your view, are there any feasible alternatives to this reliance? If so, would these alternatives be designed to regulate climate scientists or to produce new knowledge independently that would be compared with the representations produced by climate scientists?

'Climate-gate', as we've seen, pointed to a potential abuse of scientific peer review, data manipulation and an unwillingness to share empirical evidence and methodological information with non-members of the credentialised climate science community. 'Glacier-gate' suggested a lack of rigour or even potential dishonesty in the communication of climate science insights to politicians and publics. Virtually none of the official inquiries in Box 8.1 denied that errors of scientific judgement were made, even as they cleared the scientists implicated of malpractice. Why did such failures in professional self-governance occur? We can discount the charge levelled by some climate change sceptics that the 'climate change consensus' represented in the recent IPCC reports is a 'scientific conspiracy' or, as American Senator James Inhofe once (in)famously opined, a 'hoax' perpetrated against the public. Though scientists operate in extended professional networks, it's simply infeasible that thousands of climate specialists worldwide would – or could – form the largest cabal in history (see Plate 8.1). So, can we instead conclude that a few senior climate scientists rebuffed sceptics in order to, perhaps, protect their own influential research and hard-won reputations from challenge? This conclusion isn't unreasonable, though no evidence exists to support it. We could just as well point to the possible effects

Plate 8.1 Disreputable scientists playing politics?

This clever and amusing 2009 cartoon by Arthur 'Chip' Bok represents the view of many climate change sceptics. The classic image of scientists as trustworthy men in white coats is here referred to subversively. It is depicted as a front for corrupt practices of 'cooking' data so as to create the impression of serious anthropogenic global warming ahead (depicted by the lighter held under a correspondingly hot thermometer). The tag line 'An inconvenient truth' is a witty riposte to Al Gore, whose docufilm of this name was predicated on trust that the 'climate change consensus' among qualified scientists is honestly arrived at. The cartoon can equally be interpreted as satire, mocking the absurd idea of a worldwide scientific conspiracy to use science to pursue a political agenda to regulate 'petro-capitalism'. It is worth comparing climate change science with the forestry science discussed in Chapter 6. The 'environmental myth' of West African deforestation could arguably survive in ways a climate change 'hoax' surely could not because of the small numbers of experts involved who were routinely shielded from criticism.

on CRU members of the IPCC of the febrile political environment surrounding climate change science early in the new millennium. Arguably, a combination of naïvety and defensiveness made the likes of Professors Phil Jones and Keith Briffa deal unprofessionally with sceptics who, in the United States, were successfully questioning scientific wisdom in the public domain.

What lessons can we draw for scientific self-governance and science communication? As you may have concluded from the previous study task, one lesson we should surely *not* draw is this: science would be more trustworthy if we were to more frequently open its inner workings to widespread public scrutiny. This is not the most feasible or credible way to keep scientists honest. Additionally, there are more appropriate ways to allow citizens to have a say about science. To be specific: public involvement in science 'upstream' is arguably important in many cases (e.g. we could collectively debate whether to permit scientific research into cloning humans) – more on upstream interventions later in the chapter. It's also vital to foster 'downstream' discussion about the implications of research for society and environment. But this is not the same as advocating for public assessment of how, to take examples from climate change science, researchers combine temperature data from far-flung weather stations or code computer models that simulate future air–land–water dynamics. The best way to ensure trust in the conduct of esoteric science is to *utilise the existing resources of the scientific enterprise*. To do otherwise is simply to displace responsibility on to others who may not be best equipped to judge the science. As one commentator observed of 'post-gate' arguments, climate scientists should submit voluntarily to 'extended peer review' of their research methods and findings by diverse stakeholders:

> if several auditors reached conflicting conclusions, then somehow a judgement would have to be made about *their* respective competence. And who should make that judgement? Presumably a group of suitably qualified, honest individuals with a proven track record in a relevant discipline – in other words [the climate scientists being audited!].
>
> (Corner, 2010: 37, emphasis added)

Quite aside from this, the lack of evidence of *widespread* malpractice in the climate change – or any other scientific – community means that it's an over-reaction to insist on brand new regulatory arrangements.[3]

One lesson of the two 'gates' is thus that climate scientists need to do a better job when it comes to 'quality control'. If science, as Robert Merton (1942/1979) famously said, is a form of 'organised scepticism' in which investigator experience, logic, trial-and-error and evidence combine, then it must be seen to apply in a fair and comprehensive way. One hopes that a positive outcome of the 2009–10 scandals is to remind climate scientists and the wider scientific community that professional norms cannot be adhered to selectively. The credibility of science as a powerful genre of representation is at risk if systemic malpractice is perceived – all the more so for **'post-normal' science** where the stakes are high and decisions urgent. Accordingly, blind peer review needs to be utilised virtually without exception, and data and methodological procedures made publicly available where possible. There can be no suggestion of arbitrariness, manipulation or evasiveness. Peer review 'works' because it applies the norm of dispassionate assessment to the research of others without fear or favour. Relatedly, transparency about evidence and how it's analysed helps to ensure maximal rigour and honesty among scientists.

A second lesson relates to science communication directed to politicians, the press and publics. Climate change science is one of several where epistemic uncertainty looms large. Yet, the successive IPCC reports have created the impression that scientists are now relatively sure that anthropogenic warming is occurring and will intensify in future. For years, climate change sceptics have actively exploited the uncertainty in order to challenge the consensus view in the public arena. To the extent they've succeeded, it is arguably because of a recent failure of climate change scientists to adequately represent uncertainty and to acknowledge publicly, however critically, any evidence that appears to challenge the IPCC consensus. The use of the word 'denialist' to characterise many climate change sceptics is perhaps symptomatic of an over-confidence in recent IPCC communications about climate change. As Daniel Sarewitz (2011: 7), not a climate change sceptic, has sagely noted, 'a claim of scientific consensus creates a public expectation of infallibility that, if undermined, can erode public confidence'. In this case, Sarewitz argues that a better way to represent Earth's surface processes

> might borrow a lesson from the legal system. When the US Supreme Court issues a split-decision, it presents dissenting opinions with as much force and rigour as the majority position. Judges vote openly and sign their views, so it is clear who believes what and why – a transparency [usually] absent from expert consensus documents.
>
> (ibid.)

In Sarewitz's view, the climate science community would be well advised to foreground epistemic uncertainty and to give contrary evidence its due. This would, he believes, strengthen rather than undermine the IPCC message that global warming is real and potentially dangerous for many people. It's worth remembering here that climate change sceptics have frequently voiced their doubts in the name of 'sound science'. 'Climate-gate' and 'Glacier-gate' all too easily appear to outsiders as instances of 'unsound science' where censorship and group think are at work.[4]

Some climate scientists might worry that foregrounding the error bars and counter-evidence will further delay meaningful action to mitigate the degree and effects of global warming. But one can argue the opposite: an honest, suitably caveated presentation of the future possibilities and probabilities of environmental change can put the onus on societies worldwide to take meaningful precautionary action *now*. Uncertainty can as easily *encourage* action when the stakes are high as inhibit it. As part of this presentation, the climate science community would be entitled to ignore the claims made by unqualified sceptics like Christopher Monkton.[5] Equally, however, it would consider seriously any research authored by certified non-climate science specialists and sceptics like Ross McKitrick and Douglas Keenan when it passed through blind peer review.[6] It would also take seriously the original research by certified specialists who question the climate change consensus (like American Willie Soon and Australian Garth Paltridge). When or if this research is found wanting, the reasons would have to be robust and recorded clearly. Finally, the climate science community would – or should – welcome systematic individual and team-based research by certified specialists seeking to double-check existing evidence or computer models. Notable here is the massive Berkeley Earth Surface Temperature Project led by a now 'converted' former sceptic, Professor Richard Muller.[7] It re-examines patterns detectable in the mass of temperature data available for the previous 1,000 years or so.

These suggestions mean that climate scientists could usefully enlarge the epistemic field in which they perform 'boundary work' and secure their own epistemic identities. But they don't amount to a proposal for new governance arrangements designed to let people like you and me monitor or assess the nitty-gritty of research: most of us simply lack the wherewithal to perform this role. Box 8.2 itemises some of the proposals about how to alter the self-governance of climate science made in some of the official 'post-gate' inquiries. They're all consistent with arguments made in this sub-section that current self-governance and communication practices could be improved rather than replaced.

BOX 8.2

REBUILDING TRUST IN CLIMATE SCIENCE: FINDINGS AND SUGGESTIONS OF THE 'POST-GATE' INQUIRIES

- *Recommendation of the UK House of Commons Science and Technology Select Committee*, March 2010. In a press release Committee Chair, MP Phil Willis summarised the findings as follows:

  > Climate science is a matter of global importance. On the basis of the science, governments across the world will be spending trillions of pounds on climate change mitigation. The quality of the science therefore has to be irreproachable. What this inquiry has revealed is that climate scientists need to take steps to make available all the data that support their work and full methodological workings, including their computer codes.
  >
  > (http://www.parliament.uk/business/committees/committees-a-z/commons-select/science-and-technology-committee/inquiries/uea/, accessed 30 October 2012).

- *Observation of the Oxburgh Inquiry*, April 2010. The report noted that 'there were important and unresolved questions that related to the availability of environmental data-sets . . . this is unfortunate and seems inconsistent with policies of open access . . .' (http://www.uea.ac.uk/mac/comm/media/press/CRUstatements/SAP; 5, accessed 30 October 2012).
- *Recommendation of the Muir Russell Inquiry*, July 2010. The report suggested that in future 'the CRU should make available sufficient

information, concurrent with any publications, to enable others to replicate their results' (http://www.cce-review.org/pdf/FINAL%20REPORT.pdf; 14, accessed 30 October 2012).

- *Some recommendations of the Inter-Academy Council Inquiry into the IPCC*, August 2010. In a press release, the Council summarised its report suggestions as follows:

> To enhance its credibility and independence, the executive committee should include individuals from outside the IPCC or even outside the climate science community.
>
> The IPCC chair and a proposed executive director ... should be limited to the term of one assessment in order to maintain a variety of perspectives and fresh approach to each assessment. Formal qualifications for the chair ... needs to be developed, as should a rigorous conflict-of-interest policy to be applied to senior IPCC leadership and all authors, review editors, and staff responsible for report content.
>
> Review editors [contributing to the IPCC assessment reports] should also ensure that genuine controversies are reflected in the report and be satisfied that due consideration was given to properly documented alternative views. Lead authors should explicitly document that the full range of thoughtful scientific views has been considered.
>
> The committee also called for more consistency in how the Working Groups characterize uncertainty. In the last assessment, each Working Group used a different variation of IPCC's uncertainty guidelines, and the committee found that the guidance is not always followed. The Working Group II report, for example, contains some statements that were assigned high confidence but for which there is little evidence.
>
> [A new communications] ... strategy should emphasize transparency and include a plan for rapid but thoughtful response to crises. The relevance of the assessments to stakeholders also needs to be considered, which may require more derivative products that are carefully crafted to ensure consistency with the underlying assessments.
>
> (http://reviewipcc.interacademycouncil.net/ReportNewsRelease.html, accessed 30 October 2012)

Responding to science: avoiding the misuses of epistemic dependence

We should welcome scepticism in and about science. 'Climate-gate' and 'Glacier-gate' point us to better ways in which scepticism can be institutionalised in the scientific study of what we call nature. However, the self-governance of science is one thing. How we choose to respond to what scientists tell us is another entirely. Institutionalised scepticism, and public involvement 'up-' and 'downstream' of science can all help to protect a society from being rendered too vulnerable to epistemic dependence on scientists. But ultimately, societies must decide whether and how to act when well-governed science represents things like climate change to us. Here, it seems to me, the two 'gates' are highly instructive. Both incidents, as we've seen, focussed enormous attention on what happens inside the black box of science. But, equally, they *should have instigated a searching examination of how societies currently use science in post-normal situations*. That they did *not* tells us much about our current willingness to defer decision-making for fear of choosing badly. Before I explain what I mean, complete the following study task.

Study Task: Think of all the ways in which you pay conscious attention to what scientists say in your own life. By 'scientists' I mean experts who study the natural world (broadly conceived) and/or figure out how we can protect, adapt to or modify it. After reflecting on a range of examples where you voluntarily, or by force of circumstance, listen to what scientists say, ask yourself how you 'consume' science. For instance, are you impressed by science in the main? Do you largely trust what scientists say? Does it impinge directly on your own health or well-being? Are you happy to take instruction from science? Can you think of situations where science would be relevant to something that matters to you, but insufficient to help you decide how to proceed?

As I noted earlier in this book, scientists use the rhetoric of 'truth', 'reality' and 'objectivity' when presenting claims about the world that they want us to notice. We take that rhetoric seriously because it 'performs' successfully: our illnesses get cured or a predicted hurricane duly arrives. But our social contract with science is such that we normally take responsibility, often through our elected politicians, for the value decisions that are underpinned by the 'facts' of science. For instance, if I'm told I have a 25 per cent chance of surviving cancer, it's for me to decide if I want radiotherapy or not. I'm not suggesting that science is value-free. For instance, creating artificial life forms is shot through with profound value judgements. But once scientists commit to a line of inquiry, like understanding past global temperature trends, we want and expect them to assemble the best evidence and to analyse it scrupulously. The fact that scientists are not automatons exercising stern logic and pure rationality doesn't make them dishonest. We can accept the STS insight that science 'constructs' knowledge (refer to Box 3.3) without concluding that scientists simply fabricate findings to win respect, research money or a pay rise,[8] hence the furore when the two 'gates' initially implied other things were going on. In short, scientists, and the rest of us, work together to assure ourselves that 'politics' occurs *outside* science as much as is possible. This is part of our collective commitment to policing the dualisms listed in Figure 1.5.

Throughout this book, I've insisted that *all* forms of representation (including science) are political and performative. Representatives of various kinds both 'stand for' (depict) and 'stand in for' (speak on behalf of) others. Using different genres, they seek to affect us with a range of ends in mind. I stand by these claims, but now want to argue that 'politics' is undertaken in different registers that ought not to be confused. In the case of climate change science, we've allowed a number of sceptics to *wilfully use science as a means of hiding contestable value judgements about whether and how to respond to global environmental change*. These sceptics aren't committed to open deliberation about possible alternative courses of action in response to future ecological transformations; instead, they prefer to conceal their political convictions by highlighting continued gaps and contradictions in climate science. They thus seek cynically to use (1) our epistemic dependence on science and (2) the rhetoric of scientific 'rigour' against those in society who disagree with their unwillingness to 'decarbonise' capitalism. Relatedly, they make use of 'is-ought' reasoning that the philosopher David Hume showed to be faulty some 250 years ago. For them, because 'the facts' are not clear-cut, it follows that we shouldn't aggressively reduce GHG emissions. But this is to accord the facts too much normative power. The strategy has worked (especially in the United States) because of a collective refusal by politicians and citizens to make important societal decisions in the face of continued, indeed chronic, scientific doubt. As long as we continue to expect science to always trade in certainty, we'll offload our responsibility to decide how to act on important, if necessarily fallible, scientific findings.

In his excellent book *The honest broker*, science commentator Roger Pielke (2007) argues that the findings of scientists only 'compel' non-scientists to act under very special (arguably rare) circumstances. These are situations where epistemic certainty is high and a political consensus over what the implications of science are is very strong. Pielke uses the metaphor of 'tornado politics' here, where

> participants in the decision-making process share a common objective – in this case the goal of preserving one's life – and the scope of choice is highly restricted – stay or go.
>
> (Pielke, 2007: 41)

He contrasts such 'is-ought' reasoning with that used in what he calls 'abortion politics' situations. Here there are fundamental disagreements about how to act because there are fundamental underlying value disagreements. For instance, if I value women's freedom, I might support their right to terminate an unwanted foetus; however, if I value life I might seek to make abortions illegal. In these situations, science cannot tell us what to do, even though it can help us understand what's doable in a technical sense. The 'is-ought' link is very weak indeed – 'under-determined' to use philosophical jargon – and yet, Pielke argues, this is often forgotten, deliberately or otherwise. Attempts are often made to pretend that 'abortion politics' is really 'tornado politics'. Reference to science is made to justify very particular courses of (in)action, as if these are mandated by the evidence (or by uncertainties in the data). This is a misuse of science as a genre of representation.[9]

Interpreted in Pielke's (2007: 7) terms, many climate change sceptics can be seen as 'stealth issue advocates'. They pretend to separate 'science' and 'politics', only to use uncertainties in the former to surreptitiously advance their own preferred agenda (e.g. burn more oil, gas and coal). In essence, it's an anti-democratic manoeuvre in which reference to knowledge about nature (the climate system) is used to delay important decisions about our collective future.[10] The sceptics have politicised science by suggesting that climate research is not scientific enough, even as they try to conceal their own politically motivated use of epistemic uncertainty. This is one example of a wider practice, summarised by Steve Fuller, wherein

> the invocation of scientific findings – almost any findings will do – has turned out to be the most ideologically palatable means of [effectively] coercing the populace . . .
>
> (Fuller, 2000: 104)

I can find no better way to illustrate the folly and perniciousness of this than as follows. Imagine it's the year 2050 and recent temperature records, combined with extraordinarily powerful predictive climate models, reduce uncertainty about future environmental change to virtually zero (so far as climate scientists are concerned). Would this enhanced scientific certainty in itself resolve political debates over the 'what is to be done?' question? I suggest not (see Oreskes, 2004). While doing nothing might be ruled out-of-court by virtually everyone (a 'tornado politics' is-ought connection), beyond this we'd still need to debate some profound ethical issues about the value of humans and non-humans (i.e. engage constructively in 'abortion politics'). At some point, agreement on these issues, however challenging, would compel more or less drastic actions to adapt to climate change. Such agreement would not be reducible to questions of what the scientific evidence tells us about the future state of the world. At best, as Pielke argues, post-normal scientists can together act as 'honest brokers' who point out the range of practical implications of their findings, without telling us how to act on them (see Figure 8.1).

The current climate change sceptics' argument that we need more or better evidence for, analysis of and models for climate change is thus ultimately a diversion, given the consensus represented in the most recent IPCC assessment reports. It involves the exercise of soft power using the rhetoric of

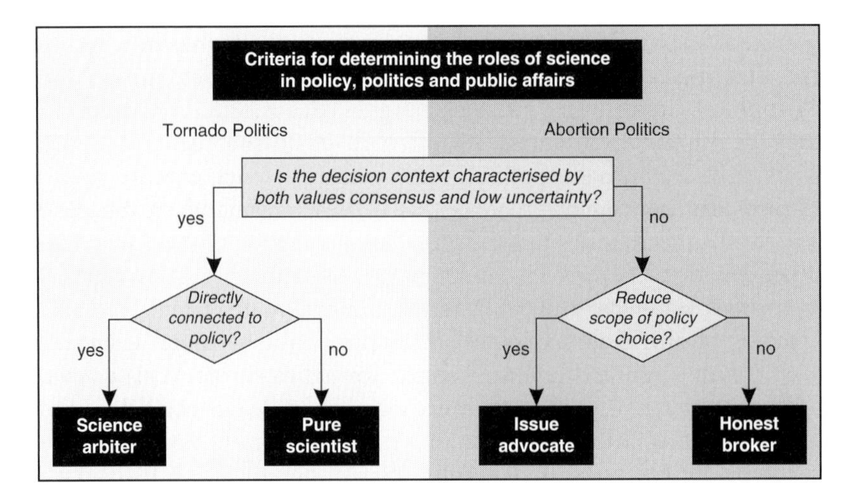

Figure 8.1 Possible roles for scientists and their representations in modern societies

Roger Pielke (2007) identifies four available roles for scientists. As individuals they would play these roles *variably* over time, depending on the situation. The 'pure scientist' provides information with no concern for its utility in politics or public life. The 'science arbiter', by contrast, presents the full suite of evidence potentially relevant to the available policy options. They also help decision-makers identify the best option in light of agreed values and possible resource constraints. In the context of 'abortion politics' situations, the 'issue advocate' makes arguments for just one course of action that follows from available scientific findings. Here science and politics are connected overtly and organically: the scientists must justify a particular 'is-ought' connection. Finally, the 'honest broker' presents information about the pros and cons of all available policy choices, without advocating for any one. These options may embody a wide range of value positions. Note that Pielke is not making any assumptions about whether or not citizens get a say in what scientists do research on (an 'upstream' consideration) or how they do it (a 'mid-stream' consideration). Instead, his focus is on how scientists relate to 'downstream' issues of what practical actions follow for citizens and their political representatives from science's insights.

'good science' as a weapon. A key lesson for politicians and publics resulting from the two 'gates' is that they need to better understand what 'post-normal science' can – and cannot – do. The real issue is not that a few leading scientists 'played politics', but that others have used climate science to avoid the difficult work of persuading the rest of us that *their* interests are *our* interests.

Arguably, the most profound questions about global climate change today are *non-scientific* ones. For instance, do we really need a binding global agreement like the Kyoto Protocol or should we aim for a plethora of local/regional/national solutions? To answer a question like this we need the widest possible range of ideas, plans and arguments across the full spectrum of extant communicative genres – art, literature, newspapers, current affairs magazines, cinema and so on. Ideally, we should approach them with an attitude of positive scepticism, rather than the negative scepticism (cynicism) of many climate change contrarians. Ultimately, we need to recognise that difficult decisions about how to respond to climate change, and also other biophysical 'realities', oblige us to first debate options. More than most issues, climate change throws the debate about means and ends wide open. It speaks to what many regard as the ultimate human question, namely 'how should we live?' While we necessarily rely on various epistemic workers and their modes of representation to help us understand the options and corresponding value commitments, we have to assume responsibility for instructing politicians, businesses, NGOs, etc. on how to proceed on our behalf. As stakeholders we have rights and responsibilities. To pretend otherwise is to regard epistemic dependence as simply a curse rather than a resource we can use actively to fashion our thoughts, actions and, ultimately, our person-hood. However imperfect our democracies are, the collective power of citizens to deliberate and decide on societal futures is real enough (if often weakly actualised). It's one way we utilise our epistemic dependence so as to make it work in our favour – or not as the case may be (see Box 8.3).

BOX 8.3

BEYOND SCIENCE: THE NON-DEBATE ABOUT HOW TO RESPOND TO CLIMATE CHANGE

In the West, and probably beyond, there's currently no political consensus about how to respond to the prospect of significant global climate change. Rhetorically, many government leaders worldwide express serious concern about the prospect, as do many citizens and NGOs. In practice, however, piecemeal reforms, like the introduction of carbon trading schemes, suggest a default to the socio-technical and economic status quo. This is to be expected because the prospect of serious climate change invites 'abortion politics' in Pielke's (2007) sense of the term. The worst-case biophysical scenario for the twenty-first century (that presented by Kevin Anderson and Alice Bows, 2011) incites profound reflection about the widest conceivable range of possible responses predicated on the widest conceivable range of values. While a plethora of thoughtful responses have been suggested, and a whole variety of underpinning values proposed and justified, most of these arguably remain ghettoised. They've either been invisible to most people or not taken seriously by them. There's a whiff of 'post-politics' here (see Box 3.4), especially in the tendency of some climate change sceptics to pretend that the 'real issues' are about the quality of climate science rather than a societal failure to properly consider how to respond to what the science is telling us.

Why has there been an unwillingness to engage in meaningful 'Abortion politics' *viz.* climate change in government arenas and the public sphere? We can only speculate with the help of research. Recent studies (e.g. Hobson and Neimeyer, 2012; Leiserowitz *et al.*, 2012) suggest that public scepticism about global warming in the 'post-gate' period has grown largely among those *already* disposed to be doubtful based on their preferred values (e.g. individual freedom is good, limited government intervention in society is desirable). Those holding different values have had their trust in climate scientists less affected by the 2009–10 scandals. This indicates that, in many modern democracies, citizens inhabit face-to-face and virtual networks that shield them from deliberating about how their own values articulate with – and may stand to be changed by – those of diverse others. So much for the political benefits of living in the 'information age'!

As Yale University's Dan Kahan notes,

> if the cost of having a view of climate change that doesn't conform with the scientific consensus is zero, and the cost of having a view that is at odds with members of one's cultural community can be high, what is a rational person to do? In that situation, it's perfectly sensible for individuals to be guided by modes of reasoning that connect their beliefs to ones that predominate in their group.
>
> (Kahan, 2012: 255)

Kahan's point is that the majority of citizens who are either climate change sceptics or non-sceptics *all* tend to inhabit cultural silos. This prevents them from discussing with each other the whole range of 'mainstream' options for climate change mitigation and adaptation (see Hoffman, 2011). It also shields them from taking seriously 'extreme' options hailing from either the far right or the far left of the political spectrum.

As one of the CRU scientists whose emails were hacked, Mike Hulme (2009: 322), put in his book *Why we disagree about climate change*, 'Just as the transformation of the world's physical climates is inescapable, so is the idea of climate change unavoidable.' He goes on:

> We should use climate change as both a magnifying glass and a mirror. As a *magnifier*, [it] . . . allows us to conduct examinations – both more forensic and . . . honest than we have been used to – of each of our human projects: whether they be projects of personal well-being, self-determination, . . . localised trade, poverty reduction . . . [etc.]. Climate change demands that we focus on the long-term implications of short-term choices, that we recognise the global reach of our actions, and that we are alert both to material realities and cultural values. And as a *mirror*, [it] . . . teaches us to attend more closely to what we really want to achieve for ourselves and humanity.
>
> (Hulme, 2009: 363, emphasis added)

Hulme's insights are acute. Sadly, research by Kahan and others suggests that many modern democracies lack the means for citizens and even politicians to 'magnify' and 'mirror' through dialogue rather than mere introspection. The news and current affairs media are, in theory, a means to foster dialogue; however, as newspapers demonstrate, the news media is itself typically segregated such that audiences have their preferences satisfied by the news channels they select. What's more, special interests like the oil industry use the media as a vehicle for expressing preferences rather than debating options in a civic (or public-minded) register.

NO ROOM FOR CITIZEN SCIENCE?

Taking the case of climate change, I've posed – and sought in brief to answer – key questions about our relationship to science. I've focussed on 'post-normal science' of the kind that has potentially large and wide implications for most of us in our daily lives. Using 'Climate-gate' and 'Glacier-gate' as instructive cases, I discussed the self-governance of science as part of a concern that it is accountable to the publics it ultimately affects. I made passing mention of the need for civic debates about science 'upstream' before major research programmes are embarked upon. But I focussed mostly on how to ensure trust in science once a programme is underway. Governance issues aside, I also explored the relationship of citizens to scientific representations of nature 'downstream' of the latter's creation.

In all this, it may seem to some readers that I'm far too willing to grant scientists the right to produce knowledge without external intervention once the citizen-informed upstream debates have occurred. This may still appear to grant science too much power to represent the world, limiting the rest of us to deciding how best to respond after the fact. It may seem to license a sort of 'expertocracy' (at worst) or 'benign professionalism' (at best) that ultimately keeps the black box of science sealed shut to outsiders. So, am I saying that there are never situations where citizens can be 'co-producers' of scientific knowledge? As it turns out I'm not. However, as I'll now explain, it depends very much on context. A generalised democratisation of scientific knowledge production is not, as many otherwise sympathetic commentators have argued, possible or desirable. In our daily lives, most of us rarely find ourselves in the situation of the Kissidougou residents, where the science–policy link affected everyday life in direct and very tangible ways. But if and when we are so affected, there *is* a case to be made for **citizen science**. Here, the epistemic content of scientific representations is not dictated by trained specialists alone. We involve ourselves in the research process, rather than wait (possibly in vain) for sympathetic spokespersons to represent our interests to scientists (like James Fairhead and Melissa Leach did in the Kissidougou case).

Study Task: Before I make, and illustrate, a qualified case for citizen science, it's useful for you to think about whether there are circumstances in which *you* could conceive of becoming a citizen scientist. Building on the previous study task, can you identify a situation where you, or someone you know well, would be highly motivated to involve yourself in collecting, synthesising or even creating information or knowledge that might shape the conduct or conclusions of a specialist science? It may help to think of this in terms of the Kissidougou case explored in Chapter 6. Rather than be subject to the power of scientific claims, as many Kissidougouans were, in what circumstances might you want to shape science *inside* the 'black box'?

Flood management apprentices

To exemplify my point, let's consider the interesting case of Pickering, a historic market town near the city of York in northern England.[11] Like many localities in the United Kingdom, Pickering has suffered from recurrent and acute flooding after heavy periods of precipitation. A large stream, whose catchment extends to the rainy hills of Ryedale to the north, runs through the town. Flood events can exact high economic and emotional costs: homes and businesses are ruined, people displaced and extensive repair work necessary. In Pickering, as elsewhere in the United Kingdom, immediate responsibility for flood risk assessment and flood prevention falls to the Environment Agency (EA), a national government body. After serious flooding in 1998 and 2000, the EA subsequently paid for flood defence works in several affected Yorkshire towns – except for Pickering.

This was not for want of trying. The EA submitted a planning application to Ryedale District Council in mid-2001. It requested permission to build a flood wall through the town designed to contain high water. This request was based on 1-D computer models of the water retention effects of walls of different heights in different rainfall and snowmelt scenarios. The application was subsequently withdrawn after a series of local objections. These included worries that the wall would actually make some properties more (not less) vulnerable to inundation, that it would be a potential eyesore, that other flood mitigation options hadn't been properly considered, and that too few local stakeholders had been consulted by the EA. The Agency then commissioned a private engineering consultant (Babtie, Brown and Root) to reassess the options. Its 2003 report concluded that a flood wall was still the best means of preventing serious flooding in Pickering. Again, computer models calibrated against data from observed flood events were the cognitive basis of the preferred solution. By this time, however, a special pot of central government money made available to Pickering and other flood-prone Yorkshire towns had been withdrawn. Accordingly, very heavy rains in 2007 caused serious flooding in the settlement. Several locals then complained loudly that those who'd challenged the EA's recommendations were ultimately responsible. If only the Agency had been listened to, they argued, Pickering would almost certainly have been flood-free in 2007 and beyond.

This, it seems, is a case of stalled 'tornado politics'. Pickering residents clearly wanted to eliminate flooding, as did the EA (though not at any monetary cost because it must work within an annual budget). A number of possible flood prevention options were identified by the Agency, but one option was recommended twice at different points in time. The recommendation didn't command widespread public assent in Pickering and the chance to pay for a flood wall vanished; consequently the town suffered another serious inundation. No one, it seems, benefited from the failure to act on expert advice. How, then, was the impasse dealt with?

The Pickering situation attracted the attention of a set of academic geographers funded by the Economic and Social Research Council. Their interest was in how residents who had experienced floods first-hand could engage with epistemic workers licensed to analyse and predict floods, and whose representations of future hydrology underpinned EA policy recommendations. Two of the five academics were flood modellers and, with eight volunteer Pickering residents, they formed the Ryedale Flood Research Group (RFRG) in 2007. The Group met several times over a nine-month period, and its working methods were inclusive (non-hierarchical). It aimed to interrogate the EA hydrological models by adding local knowledge of flood events and exploring alternative policy options in light of stakeholder preferences. The result was a computerised 'bund model' that represented the flood prevention effects of mini-dams if they were built in select parts of the upper Ryedale catchment. As with the one-dimensional models used by the EA, the 'bund model' was calibrated using existing hydrological data. The model was presented by way of a local public exhibition and a report, which received local news coverage. It combined the authority of science, courtesy of the two hydrological modellers, with the legitimacy conferred by being the result of local stakeholder participation. Subsequently, it was acted upon: central government money was expended on this 'demonstration project' and a multitude of state agencies co-operated in its delivery. A representation of one possible hydrological future thus served to make that future present by mobilising funds, materials and labour. The previous impasse disappeared and a large degree of local consent formed around the 'bund solution'. The eight Pickering locals had deliberated together and also with 'experts' in the RFRG. They'd found a way to rearticulate the 'public interest' in and through a new specialised representation of flood risk and prevention.

The power of lay epistemology

In their germinal book *Rethinking expertise*, the British science analysts Harry Collins and Robert Evans have used cases similar to the Pickering one in order to pinpoint when and how citizen science is relevant (Collins and Evans, 2007). Their objective has been (1) to move past the science–citizen dichotomy that so often holds people at a distance from scientific practice, while (2) nonetheless insisting that it's not always appropriate to involve citizens 'mid-stream' (i.e. in the actual 'doing' of research). They usefully disaggregate the concept of 'expert' and identify three kinds of expert who can, in context-dependent ways, interact to shape the conduct and outcome of scientific research:

1. **Contributory expertise.** This refers to people whose knowledge and experience allow them to make claims about some aspect of the world that others might consider to be credible. Importantly, Collins and Evans consider contributory expertise to extend beyond any 'core set' of credentialised epistemic workers. One need not, for instance, have done doctoral research to contribute substantively to the understanding of a particular problem, event or phenomenon of interest to credentialised scientists. For Collins and Evans, we should consider contributory expertise to be possessed by some lay actors, depending on the situation.
2. **Interactional expertise.** This refers to people who, while not contributory experts, have enough understanding of what they do to interact meaningfully with them. Collins and Evans make an instructive comparison here with art critics and artists: experienced critics possess interactional expertise, even if they may lack the skills to create great art themselves.
3. **Referred expertise.** This refers to people who may possess contributory or interactional expertise in a specialist area, but who lack it in other areas in which they may have a professional or personal interest. Nonetheless, their understanding of what it is to possess expertise may help them appreciate its nature and use in these other areas. Equally, though, referred expertise does not qualify one to pronounce with confidence on research by communities of contributory or interactional experts to which one does not belong.

Quite who determines which experts fall into which categories in any given situation is not discussed by Collins and Evans. It would, presumably, be a process of negotiation. Regardless, in light of their triad, try to categorise the various knowledge makers involved in the Pickering case by completing the study task below.

Study Task: In Pickering, the original, twice-presented flood wall proposal was made on the basis of contributory expertise from EA-solicited scientists with higher degrees. Local stakeholders were consulted, but about the practical policy options rather than the analysis of flooding as such. The RFRG, however, worked differently. It was a heterodox and temporary epistemic community composed of flood modellers, credentialised experts with little knowledge of flooding, and local (sometimes highly educated) people with experience of flooding but no experience in flood analysis or mitigation. See if you can apply Collins and Evans's categories to group the 13 members of the RFGR. Speculate about how the expertises may have interacted.

In the Pickering case, it seems to me that a complex dance of expertises took place to produce a technically effective solution to flood threats that also received widespread local support. First, two of the five RFRG academics possessed contributory expertise and recognised credentials; however, importantly, they were not the same contributory experts earlier retained by the EA. They came into the Group with open minds and a willingness to use their technical skills after sustained dialogue with others. Second, the other three academics possessed some combination of interactional and referred expertise in flood analysis and prevention. Third, while the lay members of the Group largely lacked contributory expertise, their experiences of flooding and knowledge of Pickering's geography were nonetheless formative for the experts (and vice versa). In particular, the interactional and referred expertise of three Group members allowed them to translate between lay members and the modellers. In short, the production of hydrological knowledge and policy recommendation was, in this case, collective. By playing different roles, Group members created something new (the bund model) that had not previously been on the table in the expert-led approach of the EA. To use Silvio Funtowicz and Jerry Ravetz's term, the bund model was not so much a more 'truthful' representation of water flows and their management as a more socially 'robust' one (Funtowicz and Ravetz, 1993). Alternatively, if it *was* more 'truthful', then it's in the pragmatist sense discussed at the end of Chapter 4.

We can turn this around and ask what happens when 'lay epistemology' is *excluded* in situations where, with hindsight, we can see that it was required. In a now classic study, Brian Wynne (1992) – like Collins and Evans, an analyst of science-led policy – examined a case where non-credentialised 'contributory experts' were *not* recognised by credentialised ones employed by the UK Government. As in the Pickering case, it was a 'tornado politics' situation. After the 1986 meltdown of the Chernobyl nuclear power station in the former USSR, wind-deposited radioactive material was detected in the hills of Cumbria in north England. Acting on the advice of scientists employed by the Ministry of Agriculture (MoA), the government announced a three-week ban on the sale and movement of sheep and lambs. This was presumed to allow radiocaesium time to decay to safe levels according to its measured half-life in livestock; however, in a stunning about-turn, the MoA then imposed a blanket ban in July 1986. This ban seriously threatened the livelihoods of Cumbrian hill farmers. It was subsequently softened to allow the sale and export of sheep and lambs as long as their meat did not enter the human food chain. Even so, the market for mutton and lamb was restricted, and already precarious farming lives made more financially difficult. The blanket ban was imposed because livestock testing revealed high levels of radiocaesium present well *after* the 'unsafe' three-week period. It turned out that MoA scientists had made general inferences from tests conducted in areas of clay soils, whereas acid peaty soils predominate in much of Cumbria. In the latter soils, radiocaesium remains chemically mobile for weeks and available for uptake in plant roots and thence into grazing sheep and lambs. Hence the long period of livestock contamination.

By late 1986, the area to which the ban applied covered a crescent-shaped area much smaller than the original one. This area was adjacent to the Sellafield nuclear power plant. Formerly called Windscale, the plant had experienced a serious fire back in 1957, which was known to have discharged radiocaesium into the atmosphere. By early 1987, farmers began to ask MoA officials difficult questions about the possible combination of radiocaesium traces from 1957 and 1986. The officials were evasive and didn't make available requested data about pre-Chernobyl levels of radiocaesium in the crescent-shaped area. Had there been a long-term cover up? Unsurprisingly, trust in these officials among Cumbrian hill farmers was fragile nearly a year after the first ban was announced.

In the meantime, MoA officials continued to monitor radiocaesium levels in the hills and valleys in both livestock and soil-vegetation. Wynne discovered that farmers looked on bemused at scientists' measurement practices and inferences. For instance, with little or no consultation with farmers, the scientists set about taking point readings of radiocaesium in random locations and trying to aggregate the results. However, large variations in the readings suggested problems with averaging or merging data from different sites, and yet the scientists proceeded to average and merge, and to pronounce with confidence that the ban should still apply. As Wynne (1996: 66) noted, farmers' intimate understanding of micro-variations in soils, relief, run-off, vegetation and land management practices was 'wiped-out in the scientific knowledge and the ignorant or insensitive ways it was deployed'. In short, the science was not robust because a fairly epistemic boundary was enforced. The findings and policy recommendations of MoA officials would have benefited from the insights of hill farmers possessed of contributory expertise – perhaps in the inclusive mode of the RFRG. As Collins and Evans put it,

> the body of expertise that should have emerged in respect of Cumbrian sheep [ecology] was a combination of the separate contributory expertises possessed by . . . scientists and farmers . . . to produce the optimum outcome, the scientists needed to have the interactional expertise to absorb the expertise of the farmers. Unfortunately, they seemed reluctant either to develop or to use such expertise.
> (Collins and Evans, 2002: 256)

Unlike the Pickering case, the 'local' problem of meat contamination had much wider public implications beyond Cumbria. The fact that 'citizen science' was stymied in this case reveals what can be lost if only accredited scientists get to represent nature. The MoA experts, as central government employees, were tasked with using science in order to devise measures intended to uphold 'the public interest'. The farmers were stakeholders: their livelihoods were at stake and they had a personal interest in reducing the length of any ban on the sale and export of mutton and lamb; however, the farmers *also* had the potential to be spokespeople for British consumers more widely. It wasn't in their long-term or reputational interest to mislead Ministry scientists in order to declare mutton and lamb safe to eat prematurely. The farmers' detailed ecological knowledge of hills and valleys would've produced a more robust outcome in both an epistemic (speaking of) and political (speaking for) sense. This was true at both the local scale and the wider national scale.

HOW BEST TO DEMOCRATISE SCIENCE?

In the previous section, I discussed situations in which people who don't inhabit one of science's many epistemic communities nonetheless have something important to offer in the making of scientific knowledge. It's unlikely that any of us would find ourselves in such situations during a typical lifetime, but when they arise, it is important that we have opportunities to participate or that others can represent our interests. It's sub-optimal for these opportunities to arise serendipitously, as in the Pickering case (see also Box 8.4). Arguably, it is far better if many credentialised scientists begin, as part of their modus operandi, to appreciate that (1) lay actors do sometimes possess expertise and (2) it can make scientific knowledge and related policy more robust. Clearly, this didn't happen in the Cumbrian case analysed by Wynne. In many parts of the world, however, such problems of lay exclusion have led to a number of well-publicised attempts to democratise science mid-stream, as well as upstream, since the mid-1980s. These attempts show how one can institutionalise public participation beyond the 'deficit model' of science–society interactions. Dominant through the twentieth century, this model presumed that 'ignorant' people needed to better understand the results of science before being left to make 'downstream' value-based decisions about how best to act.[12]

BOX 8.4

AUTO-DIDACTS AND CONTRIBUTORY EXPERTISE: WHO GETS TO BE NATURE'S SCIENTIFIC MOUTHPIECE?

There are now many examples of lay actors having to fight to both resist and utilise the power of established scientific knowledge. In the Pickering case, local people were invited to join the Ryedale Flood Research Group, while in the Cumbria case knowledgeable farmers were bemused by their treatment by Ministry of Agriculture experts. By contrast, it's sometimes the situation that lay actors have self-educated about a topic to a high level, and subsequently demand their voices be heard. Here are some now famous examples:

1. In 1994, a college chaplain, Sharon Terry, learnt that her two children had a rare inherited disorder called pseudoxanthoma elasticum (PXE). PXE causes tissue in the body to 'mineralise', damaging normal function because calcium and other minerals are not where they should be in the correct quantities. By the time sufferers are in their 30s, PXE has been shown to lead to blindness. With her husband Patrick, Terry collected DNA samples from over 1,000 PXE sufferers. They also set about fundraising so that research on the samples could be conducted. This research has now isolated the gene whose mutation leads to PXE, and has led to experiments with treatments to manage the condition.
2. In the mid-1980s, an Italian-American couple refused to accept the grim medical diagnosis they received when Lorenzo, their son, was found to have adrenoleukodystrophy (ALD). People with the disease are usually young boys who would gradually become mute, deaf, blind and paralysed before dying. Augusto and Michaela Odone began to undertake research and to create, and experiment with, various treatments designed to slow down the progress of Lorenzo's ailment. They did so with help from a credentialised neurochemist, Professor Hugo Moser, but most medical professionals were initially sceptical that two ordinary people could understand, let alone treat, a genetic illness. The Odones set up the international Myelin Project, which supports research into mutations that waste away myelin, a sheath that protects the human nerves.
3. In the 1980s, many gay men began to catch a then new disease called acquired immunodeficiency syndrome (AIDS). It turned out to be caused by a human immunodeficiency virus (HIV). This is

transmitted primarily via unprotected sexual intercourse (including anal and oral sex), contaminated blood transfusions and hypodermic needles, and from mother to child during pregnancy, delivery or breastfeeding. There is no current cure or vaccine; however, antiretroviral treatment can slow the course of the disease and may lead to a near-normal life expectancy. The illness was first discovered by the Centers for Disease Control and Prevention in the United States. The purposeful development of 'expertise' among gay men in San Francisco and New York played a significant role in changing the methodology of clinical drug trials designed to ascertain how to treat HIV/AIDS.

Among the best-known procedures for facilitating citizen engagement with science upstream and mid-stream are the following:

- **Citizen juries.** Between 12 and 20 randomly selected citizens are assembled to hear expert witnesses testify about a pre-selected issue of wide citizen relevance. For instance, the issue could be whether cloning humans is desirable, even if it's technically possible. The jury deliberates over a period of days and uses testimony to form a considered view. This may be expressed as a report.
- **Citizen consensus conferences.** As with juries, a citizen panel is chosen to be demographically representative of the wider public. Members of the panel are given written or audio-visual material prepared by a facilitator in order to gain a basic understanding of the issue at hand. The panel then participates in a conference lasting 2–4 days. Over the duration of the conference, the citizens' panel participates in a Q&A session with experts where they hear opposing views. Citizens then prepare a final document containing their stances, arguments and recommendations for the issue. On the last day of the conference, the panel then discusses their final document with policymakers and stakeholders.
- **Science shops**. Often attached to a specific department of a university or an NGO, these 'shops' provide independent participatory research support in response to concerns expressed by civil society actors. It's a demand-driven and bottom-up approach to research. Science shops were first established in the Netherlands in the 1970s and their main function is to increase both public awareness and provide access to science and technology to laymen or non-profit organisations.

These and other cognate procedures are inspired by the ideals of 'deliberative democracy' (without being reducible to them). Unlike the procedures comprising a 'representative democracy' (such as periodic elections and parliaments or, in the scientific realm, a body like the IPCC), they foster *direct* interaction between citizens and those whose actions will affect their lives or the lives of their many peers. They aim to foster dialogue, not token 'participation'. The presumption is that lay-scientist discourse (in the interactive verb sense) may alter the eventual focus and content of scientific discourse (in the noun sense). This is why a lot of thought goes into the procedural details, e.g. by what criteria will citizens be selected for a consensus conference? This is all to the good. It takes citizens' right of speech seriously, rather than, in Amaryta Sen's (1999: 288) sharp words, treating them 'as well-fed, well-clothed and well-entertained vassals' on whose behalf others make the important decisions. The value of scientific expertise is not negated here, but citizen involvement in science upstream, and sometimes mid-stream, serves to better anchor it in the considered views and (when appropriate) expertise of ordinary people.[13]

However, for all its utility, this kind of involvement tends to be rather piecemeal. It usually takes an issue-by-issue form – an example being the UK Government-instigated GM Nation debate in 2002.[14] This means that the *overarching* patterns of scientific inquiry prevalent at any one time are not subject to public discussion. The absence of such macro-level discussion reflects a failure to appreciate the full implications of the fact that 'science both governs and is governed without being formally constituted as a government' (Fuller, 2000: 8).[15] This is regrettable for a number of fairly obvious reasons. First, in aggregate, professional scientists produce representations of nature (and other things) whose sheer size vastly eclipses that contributed by citizen scientists. These representations circulate far and wide, permeating society via not just the news media but a whole range of other mobile and fixed media too. Second, in those countries where scientists comprise a sizeable workforce, the annual costs of research (wages, buildings, equipment, etc.) extend into billions of dollars/euro/yen. Importantly, taxpayers cover a significant percentage of these costs. Third, at any one time, several particular areas of science command the lion's share of science funding, leaving other areas relatively under-funded. This 'Big Science' imposes opportunity costs on society as a whole: what alternative topics are not researched or questions not asked because so much resource is going into a few high-profile areas (like cancer research)? Fourth, and relatedly, these areas tend be dominated (and championed) by only a minority of the wider epistemic community of scientists (for instance, those based in 'elite' research universities or government research laboratories). Fifth, a lot of modern science has applications designed to satisfy the commercial motives of private companies. We saw this in Chapter 5 in the discussion of patents on biotechnological 'inventions'. This raises questions of how far the commercial and public interests can coincide.

For these five reasons, modern societies would arguably benefit from institutionalised opportunities to take stock of, and possibly alter significantly, the diet of scientific knowledge that makes it on to their metaphorical plates. The flip side of this entails public reflection on what (1) scientific knowledges are currently marginalised in the mainstream (such as 'intelligent design' theory in biology), and (2) what questions about nature aren't currently being investigated by scientists. In turn, this entails reflection, if only implicitly, on the current mix of science funders and their various agendas. These funders include central government and private firms primarily, but also various research-focussed think tanks, foundations, charities and NGOs. Their motivations for paying for scientific research vary.[16] The first of these, obviously, has the legal and regulatory power to alter the current parameters for science funding (including its own funding decisions). Relatedly, it can also act on behalf of national or local publics to alter the current mix of scientific knowledge that lays claim to our collective attention in daily life. So, we're talking here about potentially far-reaching procedures to intervene in science upstream that would go beyond piecemeal forms of citizen engagement like juries or the Pickering RFRG.

I'll offer some hypothetical examples of what these procedures could look like momentarily. They're hypothetical because few countries currently have coordinated mechanisms in place to permit public scrutiny of the whole body of national scientific inquiry. But first let's consider the ends to which they'd be directed. According to what principles should a country's approach to the macro-governance of science be referenced? Analysts of science have provided a range of answers. I want to follow sociologist Steve Fuller's lead because his 'republican' political philosophy is designed to ensure a rich 'semiotic democracy' of the sort I discussed briefly in Chapter 3.[17]

'Republicanism's underlying idea', Fuller writes (2000: 13), 'is that true freedom requires the expression, not merely toleration, of different ... [perspectives].' It's thus the enemy of two ideal-type societies, namely 'communitarian' and 'liberal' ones. The former are committed to social cohesion and stability. Here, freedom of speech and action are restricted in order to prevent any section of the population being offended against. By contrast, liberal societies value change insofar as it's a result of individual creativity or hard work'. However, they may have a weak sense of any 'public interest' beyond the aggregated self-interests of myriad individuals. Additionally, because some individuals command considerable wealth, liberal societies can, paradoxically, permit de facto censorship because the less well-off are rarely heard. In both communitarian and liberal societies, 'freedom' is unduly circumscribed when viewed through republican lenses.

By contrast, a 'pure' republican polity would foster what a famous rebel once called 'permanent revolution'. Unlike liberalism, it values deliberation as a mechanism for mutual learning and the mutual transformation of citizens. It regards society as more than the aggregation of individual interests or preferences. But it recognises that mechanisms may need to be designed that allow a myriad of perspectives to flourish and to garner wider attention.[18] Though often mischaracterised as a conservative, the famous philosopher of science Karl Popper, advocate of the 'open society', was republican in this sense. For him, science would be at the heart of such a society, not just because it embodied republican ideals but because it thereby produced a plethora of knowledges that helped instigate social dynamism (Popper, 1945). Its power would thus be productive, variously useful for most or all social constituencies because it was tethered to the needs of no one elite or majority group. In his rather too optimistic view, by applying the 'critical attitude', scientists would together resist the 'excesses' of communitarianism and liberalism in their own epistemic domain. I say 'too optimistic' because Popper assumed scientists would be 'spontaneous' republicans if left to their own devices. He devoted too little attention to whether external regulation might be required if the republican spirit was not to be honoured in the breach.[19]

What form might such regulation take in order to make good on republican ideals? Before I offer an example or two, there's one last study task you might usefully undertake.

Study Task: Representative democracies have a well-known procedure at the national and sub-national levels designed to permit change (or continuity) in the agendas pursued by elected rulers. The procedure is the election. Think now about the various sciences of human and non-human nature. Can you conceive of any new procedures that will permit public engagement with these sciences in a *systematic* way, as opposed to *ad hoc* arrangements like an occasional consensus conference?

Among Fuller's most radical answers to my question is that the state–science relationship be reformatted in virtually every country that calls itself a democracy. He proposes that central governments relinquish their power to influence the focus and content of scientific research, except for a few agreed areas of strategic importance. Fuller suggests that through tax breaks and other government-sanctioned measures, various institutions (independent universities, NGOs, charities, firms, etc.) should be free to conduct more scientific research than currently. The state would aim to level the playing field so that the wealthiest institutions don't produce science that, by virtue of its sheer scale, eclipses that conducted by others. The state's role would then be less to fund research from the taxpayer purse and more to 'expose alternative scientific perspectives [and findings] once they have been developed [independently]'. The state thus takes on the role of helping to *distribute*, rather than indirectly *create*, scientific representations in order to enrich public life (Fuller, 2000: 103–7). This could be done through a whole *network* of science shops. It would also involve making more visible *already published findings* that, for whatever reason, have not been noticed in the public domain. Mining the existing body of publicly available knowledge would be as important as showcasing new scientific knowledge.

Another of Fuller's suggestions relates to peer review, which I discussed earlier in relation to the two 'gates'. Let's suppose the previous suggestion is considered far too radical in all democratic countries. One option would be to widen peer review beyond its currently 'technical' orientation. At present, when scientific research proposals seeking outside funding are reviewed, the peers in question are usually other scientists in the relevant field. The same is true for research papers. In each case, the focus is on the potential or actual 'quality' of the research defined in esoteric terms more-or-less set by one set of epistemic workers. Fuller suggests that, when scientists request very large sums of public, charitable or private money, a 'non-technical' layer of peer review upstream accompanies the current mid-stream location. They could be compelled to justify the potential value of intended research to a panel of peers from entirely different subjects. For instance, a team of microbiologists requesting $10 million from America's National Institutes of Health might be cross-examined by a panel of historians, archaeologists, musicologists and geologists. Though none of these would possess 'contributory expertise', their 'referred expertise' would enable them to act as proxies for wider public inspection of the intended science. Indeed, the point is that these other academics '*themselves* constitute part of the lay public for every branch of knowledge that goes beyond their [own] speciality . . .' (ibid.: 143, emphasis added). They could be tasked with asking questions about the values embedded within a programme of research, as well as the various social (or environmental) ends the results of the programme might conceivably serve. Such questioning would be especially valuable in cases where a scientific research programme eventually becomes relevant to 'abortion politics' situations.[20]

SUMMARY

In this final chapter, I've focussed on the governance of contemporary biophysical science. I've scrutinised science because it's an extraordinarily productive and visible maker of knowledge about nature and its collateral referents. I've focussed on governance not just because of science's social importance, but also to illustrate the wider stakes of regulating institutions with significant power to represent the world to all of us. These institutions also include the news media (my focus in Chapter 7), the wider mass media and educational organisations (from primary through to tertiary levels).

The chapter has considered whether and where it's important for scientists to pay close attention to the wider societal dimensions of their enterprise. Beginning with 'Climate-gate' and 'Glacier-gate', I focussed on the questions of how we can trust science when most of us aren't scientists, and how we might use scientific representations of the world once they journey into the public domain. These questions are especially acute when 'post-normal science' is involved and 'abortion politics' is the context in which its representations are used by actors outside science. I argued that, in many situations, existing procedures of scientific self-governance have sufficient potential to ensure the integrity of these representations. However, in other situations it is, I argued, appropriate that actors lacking scientific credentials play a role in the creation of scientific knowledge. For the most part, though, lay involvement in science is best focussed 'upstream' and 'downstream', rather than 'mid-stream'. In the former case, external measures to ensure public involvement have been necessary since the 1970s. However, because these are often *ad hoc*, I concluded the chapter with some reflections on how to ensure systematic public engagement with the wider body of science at any given moment in a society's history. These suggestions for change were predicated on the normative arguments contained in a republican political philosophy. These arguments are, I readily admit, rather idealistic. They don't suggest an end to the epistemic dependence that's an existential fact of twenty-first-century life, but they do suggest new possibilities for redressing the potentially acute imbalances between scientific experts and lay consumers of scientific knowledge. What would it mean to reform current regulations so that *all* the major institutions producing information, knowledge and signs were governed according to republican principles? Now *there's* a question.

ENDNOTES

1 Indeed, at the time of writing, the huge independent inquiry into malpractice in the UK press issued its report. The Leveson Inquiry in 2012 constituted a wide-ranging review of the infringements of personal privacy that became regular practice for tabloid journalists in the early 2000s. Its final report made a series of recommendations for new governance arrangements underpinned by law.

2 For details, see http://climatecongress.ku.dk/pdf/synthesisreport.

3 An instructive contrast here is with the tabloid press in the United Kingdom leading up to the Leveson Inquiry into journalism and newspaper regulation in 2012. Because of widespread malpractice among tabloid journalists and their sources, the official Inquiry recommended important reforms to the manner of press regulation in the United Kingdom.

4 An interesting example of what I mean relates to the Royal Society, Britain's most prestigious representative body for scientists. In 2007, it published 'Climate change controversies: a simple guide'. This guide took strong issue with claims made by various climate change sceptics, drawing on the research of the many non-sceptical scientists cited in the IPCC assessment reports. However, 43 Royal Society Fellows subsequently complained about the over-certain – even strident – tone of the guide. They considered the style insufficiently guarded and asked that uncertainty and counter-evidence be foregrounded. The Society recanted and published a revised guide in late 2010.

5 Monkton, a British hereditary peer and former politician, is one of several climate change sceptics who lack expertise in climate science but who nonetheless write and speak volubly about the 'myth' of anthropogenic global warming.

6 As I said late in Chapter 7, there are a group of highly educated climate change sceptics who are not trained in any of the fields germane to understanding past, present or future climate change (e.g. dendrochronology). Despite this lack of training, they have been able to self-educate and use their prior education to offer credible analysis

of findings published by climate change specialists. They might be said to possess 'interactional expertise', at a minimum, and possibly 'contributory expertise'. I will define these terms later in the chapter.

7 For details, see http://berkeleyearthorg/.

8 Despite most STS scholars being social democrats politically speaking, some have accused them of being no better than the most extreme libertarian or conservative climate change sceptics. The reason is that they, like the latter, have been seen to call the authority of mainstream science into question by highlighting its all-too-human dimensions. Thus the 1990s 'science wars' in the United States (to which I referred in Chapter 3) are seen by some as a left-wing attack on science coincident with a right-wing attack led by commentators funded by big oil and tobacco firms.

9 Philosophers have long debated the proper relationship between 'facts' and 'values', 'evidence' and 'belief'. The debate is very complex indeed. I don't propose to summarise it; suffice to say that I think it's rare that 'an issue can *only* be decided by evidence' (Taverne, 2005: 280, emphasis added). This may be the case when addressing 'how to?' questions, like the fuselage and engine design necessary to reduce an aircraft's fuel consumption by 15 per cent. For questions where people perceive values or feelings to be at stake, however, evidence is rarely enough to compel a change of belief or associated action. For instance, 'proving' that GM foods pose fewer health risks than non-GM foods might not sway the convictions of a Greenpeace activist.

10 Some argue that mainstream climate scientists have foolishly played politics too. For example, leading American scientist James Hansen has actively campaigned for the end of coal, oil and gas as main energy sources in the United States; however, unlike many climate change sceptics, Hansen's political advocacy is overt. It's a personal decision, based on his view of how best to act in the face of current scientific evidence. To suggest that he is 'politicising science' is to misunderstand his personal choice to be a political advocate based *on* science not by *manipulating* science.

11 I draw here on research conducted by geographer Sarah Whatmore and others (Whatmore and Landström, 2011; Lane *et al.*, 2011).

12 Lay participation in science is not, of course, necessarily a panacea for epistemic dependence. As Jason Chilvers (2009) reminds us, many ostensibly inclusive discussions of science can *reinscribe* existing power inequalities, merely rendering them more legitimate because they are glossed with a patina of public participation.

13 It's worth noting here that what constitutes 'citizen science' is a matter of some debate. For instance, people interested in species conservation now routinely provide information about local sightings of various insects, birds, etc., but providing such information to conservation professionals is not exactly participating meaningfully in 'science'.

14 This country-wide 'debate' involved a cross section of the public being given a range of 'factual' and interpretive–evaluative information about GM foods, and then debating the environmental, health and wider ethical implications.

15 The same can, of course, be said of other prominent representational institutions, like the news media and wider mass media.

16 For instance, state agencies can fund research designed to meet specific national 'strategies' and 'interests', while private companies will have product development, new consumer markets and ultimately profit in mind.

17 Fuller's arguments are intended to go beyond science and extend to other important makers or disseminators of information, like the news media and wider mass media.

18 Put differently, republicans are doubtful that the public sphere and civil society will be 'strong' simply by virtue of the spontaneous actions of citizens. Instead, the 'framework conditions' for such strength must be designed in from the outset.

19 In one of the most famous books about science ever published, the American historian Thomas Kuhn (1962) euphemised the tendency towards communitarianism in science with his famous term 'paradigm'.

20 A sample of these and other ideas can be found in part three of Fuller's brilliant but very difficult book *The governance of science* (2000).

9 CONCLUSION: MAKING BETTER SENSE OF SENSE MAKING

This book's been predicated on a seeming paradox: there's no such thing as nature, I've argued, but 'nature' matters all the same. As one of the major concepts used in Western discourse, what we call nature continues to organise thought and action in a plethora of arenas. This is because we've naturalised the concept to the point that it appears a *necessary* part of our collective vocabulary. I've sought to help you make sense of what we call nature by denaturalising it, along with its collateral terms. In so doing, I've followed in the footsteps of many social scientists and humanities scholars writing over the past 40 years. Despite these scholars' considerable efforts, nature's apparent naturalness remains 'common sense' in early twenty-first century-Western societies – at least outside the academy. This being so, my principal claims have been as follows and were elaborated in summary form in Part I.

First, nature is routinely made sense of for us by a myriad epistemic workers operating across every conceivable communicative genre using a range of discourses and a variety of communicative media. Second, I've suggested that their representations of human and non-human nature – often couched as claims about 'race', sex, genes, biodiversity, climate change, heredity and more – are politics by other means. They contain contestable value judgements and 'perform' in a myriad of intersecting registers – linguistic, numerical, pictorial, cartographic and so on. Third, I've argued that these representations are a major part of the complex and incessant process whereby our opinions, convictions, sentiments, values, prejudices and actions are governed. They render us epistemically dependent: we're at once reliant on them as resources for enlightenment, discovery and self-improvement, and yet also potentially passive recipients of others' creations. Either way, the immense stream of representations we consume under the large semantic umbrella of 'nature' and its collateral terms has a subject-forming role in our lives – often in relation to momentous issues like global climate change or the character of our genetic code. Power, accommodation and resistance are at work here, though analysts don't agree on what precise forms they take. Finally, I've suggested that how others make sense of nature for the rest of us is central to the dynamics of contemporary democracy. The formal political apparatus of the modern *demos* must reckon with important scale issues: the few can only legitimately govern the many if mechanisms exist that ensure the will of 'the people' affects incumbent and aspirant governments. The wider political health of any democracy, however, depends on the existence of a vibrant 'semiosphere'. Representations of nature and its collateral terms, I've argued, are very important here. We can ask questions about the provenance, variety and intended effects of these representations: why do we notice some more than others, what value commitments do they contain, and are we aware of how even seemingly 'innocent' ones seek to interpellate us as subjects of a certain kind?

These four argument have allowed me to bring otherwise disparate themes, issues and examples into a single topical space. If you've made it this far you will, I hope, be able to see why it makes sense to discuss things like Fraser Island dingoes, the Human Genome Project, tree loss in Guinea, bestiality in Washington State, *Stone butch blues*, intellectual property in seeds, newspaper reporting of climate change, IPCC assessment reports and citizen science in Pickering at one and the same time. I've provided a set of tools that will hopefully enable you to interrogate the myriad of ways in which 'governmentality' (broadly defined) requires diverse and repeated representations of everything from DNA to melting ice-caps. You (and I) are together subject to these representations. But we can also use them as resources to change ourselves and the world for the better. The question is: how?

There's no magical bullet here that will turn unavoidable epistemic dependence to our personal advantage or that of our fellow citizens. Writing at a time when detailed divisions of mental and symbolic labour were in their infancy, philosopher Immanuel Kant famously opined that

> Enlightenment is man's [*sic*] release from his self-incurred tutelage. Tutelage is man's inability to make use of his understanding without direction from another. Self-incurred is this tutelage when its cause lies not in lack of reason but in lack of resolution and courage to use it without direction from another. *Sapere aude* – 'Have the courage to use your own reason' – that is the motto of Enlightenment.
>
> (Kant, 1784/1992: 90)

In the centuries prior to Kant, a combination of social inequality and social stability had been achieved by powerful social actors (e.g. religious leaders) recycling a limited number of cognitive, moral and aesthetic messages. Kant hoped that the new knowledges proliferating in eighteenth-century Europe might empower more than a small elite, if people could only learn to use it of their own accord.[1] Writing 138 years later, the American Walter Lippman was far from sanguine. 'The real environment', he wrote in his book *Public opinion*

> is altogether too big, too complex, and too fleeting for direct acquaintance. We are not equipped to deal with so much subtlety, so much variety, so many

> permutations and combinations. And although we have to act in that environment, we have to reconstruct it on a simpler model before we can manage it. To traverse the world men must have maps of the world.
>
> (Lippman, 1922: 16)

For Lippman, early twentieth-century life was so labyrinthine that, contrary to Kant's hope, people positively *required* 'direction from another' in order to comprehend and shape the wider world. He saw it as inevitable that the 'maps' citizens navigate by would be created by a minority of thought-shapers. He wasn't optimistic that ordinary citizens would, by virtue of these maps, possess sufficient 'reason' to use it independently.

Nearly a century on and we might have cause to challenge Lippman's pessimism about the fact of epistemic dependence. After all, in the West far more people stay in formal education for longer than ever before, and the general semiosphere we inhabit appears to be composed of more voices than previously, courtesy of alternative media (like blogs). Surely we're in a better position than our forebears? To use Kant's terms, we might say there are tools aplenty for us to use our reason, if only we have the courage to do so. However, what I've argued in this book is that it's easy to think we're more literate about the world than we actually are because we're inundated with so many claims about it. Focussing on nature and its collateral terms, I've sought to illuminate some of the ways we're invited to make sense of the world that demand closer scrutiny.

Part II was intended to exemplify the sort of 'positive scepticism' that's a bulwark against undue forms of epistemic dependency; however, in itself such scepticism will never be enough. We cannot get 'the courage to use our reason' from this alone. As Part III made plain, we also need the various epistemic communities discussed in this book to self-regulate – and be externally regulated – with that protean thing called 'the public interest' in mind (if not all the time, then at least *more* of it). Otherwise, the way nature is made sense of for us (and much else besides) becomes a vehicle of weakly, even anti-, democratic rule. This is especially true in circumstances where most people, spending large periods at work, engaged in domestic matters and so on, have neither the time nor inclination to engage meaningfully with their semiosphere. Who'd have thought that claims made about beluga whales, natterjack toads, Siberian tigers, hermit crabs, orangutans and other 'natural' phenomena could be so important? The authors upon whose work I've drawn here, clearly, but the message needs to be broadcast more widely. Despite its considerable length, this book constitutes only a limited engagement with the many issues and questions it has sought to conjoin. After all, I have cast my net very wide. But I hope it has given readers new to the idea of 'denaturalising nature' the tools to make sense of the world in order to better facilitate the courageous use of their reason. It may even have reminded those well familiar with these tools that they are still of use, even though academic fashions move on.

ENDNOTE

1 In Chapter 1 of his book *The march of unreason*, Dick Taverne (2005) offers a readable potted history of this flowering of Enlightenment thought in Europe.

GLOSSARY

Aesthetics One of the so-called 'three faculties', it pertains to human emotions and the study thereof. An aesthetic experience occurs when we see, hear, smell or touch something and it elicits some sort of emotional response from us – pleasure, pain, euphoria, disgust, fear and so on. This 'something' can be a manufactured object (e.g. a painting), a special site (e.g. a visit to Lourdes) or some aspect of the non-human world (e.g. an ascent of Mount Everest). *See also* cognition and morality.

Ambivalent categories Words whose meanings contain diverse, often contradictory elements. For instance, whereas 'black' and 'white' are antonyms (mutually exclusive), a word such as 'human' bridges the 'nature and society' dualism. Ambivalent categories challenge the otherwise dichotomous vocabulary Westerners use to make sense of the world.

Audiencing The process whereby a particular act of one-to-many communication (e.g. a television programme or university seminar) is intended for, and helps to reproduce, a specific kind of audience. Audiences do not exist prior to being addressed. On the contrary, the various ways they are addressed calls forth audiences.

Biosociality A mode of thinking about oneself and one's relations with others that accents biological characteristics, such as one's genes, skin colour or susceptibility to certain illnesses.

Citizen science Any form of systematic inquiry and resulting information or knowledge that's conducted by ordinary people who don't belong to the epistemic communities normally tasked to undertake such inquiry. Often, citizen scientists are 'auto-didacts' who learn about science in a given domain by reading the works of credentialised practitioners. Citizen scientists enter the scientific process 'mid-stream', rather than simply 'upstream' or 'downstream'. They can play a wide variety of roles in shaping the conduct or results of scientific inquiry.

Civil society The various relations between citizens designed to promote (or address) certain causes, goals and issues. Civil society is part of the public sphere but not synonymous with it. It comprises things like community organisations, non-governmental organisations and new social movements.

Civil society groups can be very informal (e.g. a parent-run children's sports team), but also very institutionalised (e.g. the National Audubon Society).

Cognition The ordering of sense-data in the mind so as to describe and explain the world in which we live. Cognition involves the mental representation of what we take to be 'reality' (social and natural). It also involves the use of logical reasoning. Cognitive knowledge and information is communicated daily to us, especially through 'realistic' genres of communication, such as broadsheet journalism or science publications. Cognition is considered one of the three human 'faculties', the others being aesthetics and morality (or ethics). Although this makes cognition sound like a universal similar process common to all humans, there is good evidence to suggest that cognition is varied according to a person's milieu.

Collateral concept Any widely used word whose meaning is shared with another keyword. Collateral concepts need not always be synonyms of other keywords, but much of the time they are. For instance, at least one of the meanings of the concept 'nature' may be intended when the word 'race' is used by some people. Likewise, the word 'wilderness' is often used as a synonym for one or more of the meanings of nature.

Connotative reference Any written, verbal, visual or other act of referring to the world that connotes something not referred to explicitly. Most references are *denotative* (or literal). For instance, if I say 'Contemporary polar bears have a smaller habitat than their forebears', I'm obviously denoting polar bears and their Arctic environment; however, intentionally or not, I'm *also* making a connotative reference. I am connoting the melting of Arctic ice because of global warming caused by greenhouse gas emissions. For some listeners I'm also connoting a criticism of the destructive aspects of economic growth. Product advertising frequently utilises connotative reference, though it's hardly alone.

Cultures of nature Sets of ideas about what nature is and how it should be valued that are shared among members of a particular society. Anthropologists have shown not only that such cultures vary, but also that the very idea of 'nature' is itself not a cultural universal.

Denaturalisation The process whereby an analyst shows that something normally said to be natural can plausibly be shown to be the opposite, i.e. a product of society, economy or culture. *See also* naturalisation and re-naturalisation.

Discourse The process of communicating via words or other media that signify the meanings and referents of words. 'To discourse' is an act, while the results of that act may be 'a discourse' (e.g. a set of terms used by plant biologists). The act is itself structured by the terms made available by prior discourse, meaning that there is a recursive link between discourse in the verb and noun senses through time. Humans are thoroughly discursive: we communicate through language day-in, day-out. Some claim that the various languages humans have created have a common structure, while others point to their differences. Regardless, any particular discourse (in the count noun sense) must be tethered to the conventions of the language of which it is a part. For example, in Anglophone societies, the specialist discourses of meteorologists or environmental economists cannot circumvent the 'semantic rule book' of the English language. This rule book says, among many other things, that nature and culture are antonyms, so too reality and fiction, truth and opinion. Discourses in the plural thus exist within any single major language (discourse in the singular or mass noun sense).

Environmental myth A set of beliefs about environmental change in a particular place or region that lack a strong evidential base and yet which are widely held. Proper scrutiny of large-scale environmental change requires large amounts of evidence that is regularly updated. Environmental myths typically arise when small amounts of evidence are used to make generalisations, as if it's reasonable to infer from a few local cases to the wider regional or national scale.

Epistemic boundaries The barriers to entry created and enforced by epistemic communities in order to both identify and exclude 'outsiders'. These barriers take many forms. For instance, in science, credentialisation is very important. People without a PhD in the 'right' subject are typically accorded far less legitimacy as knowledge creators or challengers than people with one.

Epistemic community A group of epistemic workers who share a particular specialism. Different epistemic communities gain their distinctiveness, and sense of self-identity, through a mixture of their value-set, ontological beliefs, questions of interest, objects/domains of concern, methods of inquiry, the criteria favoured for determining worthy ideas, knowledge or information, and their chosen genre of communication. This mixture determines how specialised a given epistemic community is, how tight-knit it considers itself to be and how distinct it is from both other communities and the 'lay public'. Some epistemic communities are highly credentialised, others much less so. Typically, each of them speaks to the rest of us using specific media and rhetorical tools. *See also* genres of communication and mode of representation.

Epistemic dependence A situation in which the majority of people are reliant on a minority to provide representations of the world, including ones of their own minds and bodies. The doctor–patient relationship is one of many examples of epistemic dependence. It arises because of specialisation. There's now a large and diverse cadre of epistemic workers whose job is to produce and disseminate knowledge, information, arguments, pictures, symbols and so on. The majority of people are consumers of the latter rather than their creators.

Epistemic identities The particular sense of self created by membership of any given epistemic community. It entails socialisation over a period of time into the norms and practices of the community in question. This socialisation invites epistemic workers to play one or more roles recognised by their peers.

Epistemic workers Those who are employed by others or of their own volition to create and disseminate various kinds of knowledge, information or imagery. Epistemic workers range from natural scientists to environmental campaigners.

Epistemology In philosophy, the study of how we know the world and what counts as 'knowledge'. For instance, if I insist that the existence of something can only be demonstrated by repeated observations of that something that others could replicate, I'm making an epistemological claim. For decades, philosophers searched for a foolproof way to guarantee the accuracy of knowledge. In so doing, they distinguished knowledge from opinion, speculation, metaphysics, fiction and so on. However, in this book I use the term in a wider sense to encompass *all* forms in which we make reference to what we call nature. So my interest is in nature poetry as much as biology, wildlife documentaries as much as environmental news reporting. All these media are social in the sense that the things they communicate are the result of work performed by various epistemic communities. The term 'social epistemology' was first used by the library scientists Margaret Egan and Jesse Shera in the 1950s but, in more recent times, the sociologist and philosopher of science Steve Fuller (1988) has fleshed out a research programme based upon it in great detail. Working with a republican, democratic political theory, Fuller's principal interest is in how the production and dissemination of knowledge should be organised in a world where most people are not authors of their own beliefs, values or goals. Like Fuller, in this book I take very seriously the fact that knowing, believing, valuing, feeling and acting all occur in a world structured by *highly unequal* access to knowledge, resources and practical opportunities for action. *See also* ontology.

Essentialism The practice of characterising a person or thing in terms of a single, supposedly 'defining' characteristic. When essentialism involves reference to nature or its collateral terms, it is often designed to suggest *fixed* or *intractable* characteristics. Naturalisation or re-naturalisation often involves making essentialist claims.

External nature All those phenomena taken to be natural, in degree or kind, that are seen to exist 'out there' in the world beyond our doorsteps. To talk of nature as external is to presuppose an ontological dualism that also underpins the belief that 'human nature' is a substrate upon which the metaphorical houses of 'culture' and 'society' are built.

External regulation A form of governance in which an organisation external to a specific institution or epistemic community monitors the latter's conduct. It contrasts with self-governance.

Genres of communication The different styles conventionally utilised to convey meaning in any given society. For instance, poetry uses language in a way that's distinctly different from that of a scientific research paper. These diverse uses are linked to perceived differences in aims: poetic discourse typically seeks to move our hearts and minds, while scientific discourse typically seeks to impart 'facts' or 'truths', however provisional.

Governance The norms, habits or rule employed by organisations or communities to manage their own behaviour. Governance need not have any formal connections to government. It is often quite informal, but can also be very codified. The professions, such as law and medicine, have a long history of self-governance.

Government The process whereby elected (or self-appointed) political leaders and institutions discuss and implement laws, policies and regulations in the territories over which they have authority. Otherwise independent national governments can, of course, create and enforce cross-border laws, policies and regulations.

Governmentality In Michel Foucault's famous definition, governmentality is 'the conduct of people's conduct'. Governmentality is a rather inelegant term that, in geographer Stephanie Rutherford's apt words, 'makes some scholars tremble with anticipation and leaves others cold at the thought of inscrutable text and . . . high level abstraction' (2007: 291). Foucault was interested in the way subjects get enlisted in the project of their *own* rule as part of their (apparent) 'freedom' from the dictates of others. In this book, governmentality is used in a broad and loose way. *Making sense of nature* looks beyond the actions of governments and other institutions committed to the 'strong' shaping of thought and action. For me, governmentality usefully describes the way in which sometimes interacting, sometimes non-interacting, epistemic communities (and their institutions) aim to 'impose' their different 'rationalities' (i.e. ways of knowing, arguing, representing and communicating) and their knowledge-claims upon the diverse audiences they conjure into existence. While I accept that certain communities get a lot of exposure and may exert a strong influence on the social imaginary, I make no empirical claims as to the degree of influence in any given case. Foucault, at many points in his career, was interested in epistemic communities and their institutions (though he didn't use the former term) that are *explicitly* focussed on the 'the conduct of conduct' (like prisons and the legal profession). In effect, my tack is to utilise the idea in a *methodological* rather than *substantive* sense. Mitchell Dean's (1999) book remains the original and best introduction to Foucault on governmentality.

Interpellation The process by which a person is encouraged to occupy certain subject-positions over extended periods of time. Interpellation is social, arising from discourse among individuals and associated practices. Individuals are spoken to – face-to-face or otherwise – in ways that make normative assumptions about the sorts of people they should be. Interpellation is a continuous process and it begins to shape the self-identity of people from the moment they are born.

Intrinsic nature The idea that entities said to be natural have properties or exhibit behaviours that are definitional of them and/or difficult to change. *See also* essentialism.

Keyword A term that appears with high frequency in a range of otherwise different discussions and arenas. Keywords should be distinguished from other very promiscuous terms like 'and', 'people' or 'rain'. Whereas the latter have relatively simple meanings and are 'ordinary' words, keywords typically have plural or contested meanings. They can perform diverse semantic work for this reason, often becoming a focal point for debates about important issues. While some keywords go out of fashion (e.g. 'globalisation'), others are far more enduring (e.g. genes or life).

Metaphor A reference to something that points to a similarity to something else that it's otherwise unrelated to. For instance, I may use the metaphor of 'mother' when referring to the natural environment (as in 'Mother Earth').

Metonymy Any written, verbal, visual or other act of referring to the world that uses particulars to denote something bigger or larger. Metonymy can often be a form of connotative reference.

Mimesis The attempt to represent something accurately or faithfully in some medium. Mimesis rests on a presumed dualism between reality and representation. It is central to the meta-genre of 'realist' forms of representation that links together science, serious news journalism, documentary photography and much else besides.

Mode of representation The ensemble of media in and through which representations within a particular genre are communicated. For instance, by disseminating my academic arguments in this book, I'm observing the conventions of my profession. You would be very surprised indeed if *Making sense of nature* was published as, say, a cartoon strip.

Moral regulation Statements intended to produce 'proper' behaviour among individuals in respect of moral issues. These statements are codified in law, but are also day-to-day ones uttered as part of quotidian discourse in a range of arenas. For instance, we might think bankers who pay themselves huge annual bonuses are morally flawed and we might say so, even if existing law regards their behaviour as morally acceptable. In everyday life, statements about im/moral behaviour often 'read off' from what are said to be 'facts'. For instance, if one thinks heterosexual acts are natural and homosexual ones aberrant, it's easy to classify the latter as 'immoral' because they 'go against nature'.

Morality Values, their justification and the actions they serve to engender. Sometimes known as ethics, morality pertains to what is considered 'right' and 'wrong', 'just' and 'unjust', 'fair' and 'unfair' and so on. In everyday life, we rarely bother to justify our moral decisions and actions. This is because 'lay morality' usually rests on a real but unarticulated consensus about 'proper behaviour'; however, professional ethicists in fields like philosophy and law devote their lives to debating alternative moral principles and practices. Often, this is because new technologies challenge existing moral norms (e.g. cloning humans). Nature and its collateral referents have long been both subjects of moral concern and a basis for the justification of moral actions.

Naturalisation The process whereby a person, a group or a whole society come to regard some phenomenon as a part of nature or possessed of natural attributes. Naturalisation thus involves the assignment of meaning: one or more of the four principal meanings of nature identified in Chapter 1 of this book are attached to the phenomenon in question.

Nature effect A class of metonymic reference in which a particular natural phenomena (like a glacier) is used to signify a much wider or larger set of natural phenomena – or indeed 'universal nature'. *See also* metonymy.

Normative Any argument or assertion that is critical of current arrangements or practices and which, implicitly or explicitly, posits a future alternative that the speaker prefers. Some forms of normative reasoning take an 'is-ought' form, that is, they state (or imply) that current reality's perceived flaws dictate that a change is necessary, not optional.

Ontological gerrymandering An attempt made to relocate the boundaries between real world processes or phenomena so as to advantage those making these attempts. These attempts may then be disavowed, as if the boundaries in question are 'real' ones that inhere in the world. Re-naturalisation is arguably a form of ontological gerrymandering. Things once thought to be social or cultural are said, 'in fact', to be natural.

Ontology A set of statements or unstated beliefs about the fundamentals of existence. For instance, if I believe that there is no God and that humans are an accident of natural history, then my ontology departs in at least one significant respect from that of those who think that a deity created all life.

Ontology is necessarily linked to epistemology. This is because statements or beliefs about what exists require us to make statements or hold beliefs about *how we can know what exists*.

Politics The process of debating, justifying and acting on different values, means and ends. In a formal sense, politics is conducted inside institutions, and in arenas, expressly designed for the job, e.g. the public sphere or a political party conference. However, politics happens in most other institutions and arenas too. Regardless of its location, there is a difference between politics carried out in the open and politics that proceeds by stealth, i.e. in situations that are said be 'above', 'outside' or otherwise separate from politics. Ideally, politics conducted openly always has two aspects. The first is *bargaining* with others to get what one wants. The second is *deliberating* with others so that one might, perhaps, change one's belief in what one wants.

Polyvalent discourse A discourse produced by an epistemic community that has effects on distinctly different groups of people outside the community simultaneously. For instance, a piece of ground-breaking research into rates of permafrost melting may influence the environmental ministries of Canada and Russia, while affecting public perceptions about the rate of global climate change.

Positive scepticism A disposition among individuals to question knowledge, information, argument, etc. hailing from other people. Unlike cynicism, positive scepticism empowers people to hold to account those whose claims about the world may have serious implications for their day-to-day life. It is a facet of 'active citizenship' as opposed to passive citizenship.

Post-normal science Any kind of scientific inquiry whose findings are hedged around with uncertainty, whose topical focus is manifestly profound in its implications for whole societies, and where the need to act practically on the science is urgent. Climate change science is quintessentially post-normal because it analyses highly complex and physically large 'open systems' and helps humanity divine the distribution and degree of environmental change on the medium-term horizon.

Post-politics A condition in which political debate and policy decisions are narrowly circumscribed. Some critics argue that contemporary 'democracies' are scarcely worthy of the name because far too much is excluded from political discussion.

Pragmatism An approach to understanding all forms of reference (linguistic, graphic, aural, etc.) that inquires into their practical goals. For instance, pragmatists challenge the idea that 'truthful' statements are linguistic 'mirrors' of the physical properties of the social or natural worlds (see mimesis). They argue that 'truth' is a word we attach to statements to allow us to achieve things that, for now at least, enough of us deem to be desirable or unproblematic. Pragmatists argue that genres of communication differ not because they attend to a pre-existing 'reality' in different ways. Instead, they differ, so pragmatists claim, because they seek to realise different human goals (e.g. the appreciation of beauty, the cultivation of 'moral' conduct and so on) by talking about the world in ways that articulate those goals.

Productive power Any form of social power that works by making things possible for social actors rather than expressly prohibiting or punishing. Typically, but not always, soft power is productive. It works by presenting opportunities for thinking and acting that may not be perceived by those taking advantage of them as tools of power. Critics point out that if power is seen as productive, it bleeds into *resistance to power*, rather emptying the concept of social power of substantive meaning. Additionally, we must acknowledge that any form of social power that's productive must *also* have a prohibitive dimension in order to work. This is because no form of social power can make all things possible for all social actors: there must be implicit restrictions on the field of available thought and action.

Public pedagogy The process whereby members of any public are daily 'educated' about the world in which they live. Like formal schooling, public pedagogy has a 'curriculum', albeit an unplanned one controlled by no single authority. This curriculum ensures social reproduction and stability, and economic growth and stability. It's largely 'taught' by prominent epistemic communities inhabiting the worlds of television, print journalism, universities, advertising and film.

Public sphere A virtual domain in which citizens are able to express and discuss matters of common concern. In democratic societies, the public sphere is relatively autonomous from the domains of the family, market and state. For decades, the news media (print and broadcast) have been a vital ingredient in the public sphere's existence. Without a public sphere it would be hard for citizens to form views that could then inform their voting decisions or other actions towards an elected government. A vibrant public sphere is thought to be a precondition and result of a strong civil society.

Referent A part, set of parts or the totality of the material world to which signifiers and signifieds are attached by convention. For instance, the word 'snow' conjures up a mental image of white, soft, frozen water falling from the sky. That image is then used to make sense of certain meteorological phenomena, as if snow is snow in all times and places. Far from being the 'objective' anchors of our signifiers and signifieds, we arguably only 'see' these anchors *via* the latter.

Re-naturalisation The process of treating as natural some aspect of the world previously thought to be social in character or else caused by social processes. It involves applying one or more of the four meanings of nature, possibly through the term's collateral concepts. These meanings are then thought to inhere in the things signified.

Representation (1) The act of making reference to what is presented as some definite aspect of the world in images, writing, speech, etc. (i.e. to *re*-present); (2) The particular content and form resulting from such an act (e.g. a photograph, a book or a daily newspaper report). Representation in sense (1) involves people acting as representatives (or spokespersons). They need not be 'political actors' in a formal sense (e.g. animal rights campaigners) to so act: *all* representation in sense (1) involves people acting as 'proxies' for that which they claim to re-present. This means that representation in sense (2) covers every conceivable genre of communication. Representation does not need to be only 'realistic' to count as representation (see mimesis). In the West, the idea that we represent the world rests on a taken-for-granted distinction between representatives and representations on the one hand, and things existing outside representation available to be represented on the other.

Semiotic democracy If a 'semiosphere' is the mixture of discourses, signs and symbols that affect the quality and quantity of political debate in democratic societies, then a semiotic democracy is where that mixture is highly diverse. Democracies cannot thrive if only the formal apparatuses of a democratic political system are in place (e.g. periodic elections). Citizens need to be able to engage in a wide range of debates and be exposed to a variety of arguments, propositions, ways of living and so on. While censorship is necessary in all societies, a balance must be struck between semiotic closure and openness. Too much formal and informal censorship breeds conservatism and squashes the voices of those with alternative voices, goals or lifestyles.

Signified The meaning denoted or connoted by a particular signifier or combination of signifiers. Signifieds are real but immaterial. They exist in the mind and are conjured by words, sounds and images, but affect our emotions and actions.

Signifier A word, sound or image that denotes, or possibly serves to challenge, an established meaning (a signified). For example, a picture of a foetus floating in amniotic fluid might signify 'life', while the word 'green' denotes a particular colour that has shades of dark or light.

Social constructionism An approach to understanding a wide range of phenomena that highlights their social origins and content. Social constructionists question the belief that some phenomena can be understood 'objectively' or are registered by people the same the world over. They argue that 'objectivity' is simply one of many terms we use to convey the impression that the world 'out there' can speak for itself – if only we let it. Social constructionists look for the social causes and content of a wide range of discourses, including the supposedly value-free discourses of science. Critics argue that some social constructionists are contradictory in the sense that they give ontological weight to 'society' (which 'constructs') but deny the autonomy or ultimate knowability of non-social phenomena.

Social power The capacity to influence others in a range of ways arising from the reputation, technologies and/or resources that a person, group or institution possesses. There's no agreed definition of social power among its analysts, though there's a broad consensus that social power comes in several distinct forms in modern societies. For instance, legal power differs from the power of product advertising in fairly obvious ways. A good deal of the social power we are subject to today is 'soft'. It works through inducement and persuasion rather than by force or sanction. Indeed, for many analysts, 'hard power' is better described using terms like violence, force or intimidation. Power in all its forms is both potentiality and action, i.e. power *to* (which may be unactualised) and power *over* (which is). Chapter 6 of this book defines and analyses social power. *See also* soft power.

Socialisation The process of inculcating a person to a set of conventional ideas, values and habits. Socialisation occurs in all walks of life, from families and schools to workplaces. Socialisation forms the background against which some organisations and people seek to challenge. Certain ideas, values and habits can only be 'new' if they emerge out of a different set of established ones. Social scientists have long debated how 'free' people are to escape the power of their socialisation.

Soft power Any form of social power that involves using persuasion, argument or inducement to alter the thinking or practice of various people. Arguably, soft power is writ large in contemporary democratic societies. The word 'soft' is not intended to mean that power in such societies is necessarily *weak*.

Spatialisation The practice of associating particular phenomena or events with particular locations or territories. Spatialisation is mental, an imaginative geography, but it's anchored in what we perceive to be 'realities'. For instance, when I say the word 'wilderness', I might think of Siberia or Alaska. This is because both regions appear to be barely inhabited or domesticated when compared with, say, the farmlands of the Canadian Prairies.

Subject-position A metaphorical place from which one can speak, leading to commensurate actions. Subject-positions are socially created. For instance, what it means to be 'butch', 'feminine' or 'androgynous' will differ in the detail historically and geographically. Because a large number of subject-positions are 'available' for us to occupy in most modern societies, the 'self' is these days a complex *amalgam* of such positions. There can be acute contradictions between them, which we 'manage' in a variety of ways.

Subjectification The creation of people's sense of self and world (their subjectivity) over time. Through numerous routine acts of interpellation we are all fashioned into certain sorts of people; however, we are also free to work with the repertoire of available subject-positions to take a measure of control over the people we become.

Super-ordinate nature The idea that there are forces intrinsic to natural phenomena that cross-cut a myriad of otherwise discrete individuals or systems. These forces include the 'programming power' of genes and the conservation of energy across air and water masses.

Temporalisation The practice of associating particular phenomena or events with particular periods of time – past, present or future. For instance, it has become conventional for modern Westerners to consider 'wild nature' to be a thing of the past. It's increasingly conventional, courtesy of climate change science, to think of the future as a period of dramatic biophysical change.

Universal nature The idea that nature is everywhere, rather than simply external to humans. This is because humans are considered to be 'natural' animals, as much as creatures of culture and society.

Visuality The culturally specific and learnt ways in which people see the world through their eyes. It's been argued that vision is not the same the world over. Instead, there are 'ways of seeing' that reflect the values and norms of each society. Additionally, visual media, like film and photography, is said to present the world in specific ways that 'organise the view' for image consumers.

KEY SOURCES AND FURTHER READING

This section is for those (probably professional researchers rather than students) who are keen to trace the main sources of my various claims and arguments. I've deliberately refrained from suggesting chapter-by-chapter further readings for Part 1 of this book. Readers new to the ideas and arguments presented will, in my view, have more than enough to chew on without me suggesting that they take on a prodigious load of additional reading spanning multiple disciplines – and often rendered in very abstruse or technical language. If I were pressed to recommend one thing that resonates strongly with what I say in Part 1, it would be chapters 1 and 2 of David Delaney's brilliant book *Law and nature* (2003) – these chapters contain ideas and messages with implications well beyond the law. An accessible book with a looser connection to this one, but which contains some of its breadth and its strong interest in representation, is Julia Corbett's (2006) *Communicating nature* (read chapters 1 and 5–10). Despite its age, I should say that Kate Soper's (1995) *What is nature?* remains an indispensable source for those wishing to explore the relations between 'constructivist' and 'realist' approaches to what we call nature. These aside, the recommended further reading per chapter begins from Part 2 of this book.

Those wishing to situate this book at the heart of a degree module should consult the 'How to use this book' section.

CHAPTER 4 UNNATURAL CONSTRUCTIONS Key readings relating to three cases and the discussion of 'social constructionism'

1. Clayoquot Sound and visions of the forest

The key primary sources here are the essays by Bruce Willems-Braun (1997) and by Braun and Wainwright (2001). The latter explains the post-structural theory used in the former study to analyse the Clayoquot dispute, but note that one does not have to be a post-structuralist to show that 'nature' is an effect of discourse. These two studies aside, an essential reference is Warren Magnusson and Karena Shaw's edited book *A political space: reading the global through Clayoquot* (2003). The chapters in this collection are noteworthy: Shaw's recounts the story of political conflict over Clayoquot through the 1990s; Catriona Mortimer-Sandiland's chapter reveals the 'culture of nature' made manifest in the environmentalist campaign; and the chapters by Tom Kuehls and by Gary Shaw both detail the (non-)involvement of indigenous peoples in the Clayoquot case. The website linked to the Magnusson and Shaw book contains a lot of interesting detail (http://web.uvic.ca/clayoquot/aPoliticalSpace.html).

Rossiter's (2004) essay provides insight into the Great Bear Rainforest case and its public representation. Shepard Krech III's (1999) readable book *The ecological Indian: myth and history* takes issue with the assumption that first nations North Americans are intrinsically 'in harmony' with the non-human world. Though not about British Columbia, Mark Spence's *Dispossessing the wilderness* (1999) details the corralling of first nations North Americans that lay behind the creation of national parks in both the United States and Canada. Though focussed on the Great Bear Rainforest, Richard Clapp's (2004) attempt to chart a way past the conflicts between foresters, environmentalists and indigenous groups has much wider relevance. So does Andrew Baldwin's (2009) research into how Canadian forest scapes are imagined though, unlike Clapp, he sees few prospects for moving beyond a white, implicitly racist discourse, even when it appears to be 'pro-aboriginal'. Lest the material referents of the term 'nature' seem to be wholly ignored here, Jessica Dempsey (2010) offers a corrective, focussed on the living, breathing grizzlies in the Great Bear Rainforest.

2. Human genetics: biology and identity

Nikolas Rose's (2001) suggestive essay explores the concept that ideas about human genes are now a key part of a new mode of self-governance in twenty-first-century neoliberal societies like the United Kingdom and the United States. For critical assessments of the metaphor of a gene map, see the essay by Ed Hall (2003) and the earlier essay by Lippman (1992). For an excellent collection looking at gene mapping as both metaphor and a socio-technical practice, see *From molecular genetics to genomics: the mapping cultures of 20th century genetics* (Gaudillière and Rheinberger, 2004). Though he focusses on mapping as a literal, not merely metaphorical, thing, Denis Wood's *Rethinking the power of maps* (2010) is perhaps the best contemporary attempt to defamiliarise cartography. It's both a summary of research and an original intervention in the debates on mapping. Wood and Fels's *The natures of maps* (2008), though it doesn't focus especially on human genetics, does explore the links between maps and 'nature' (in its many and varied phenomenal forms) as re-presentational couplings infused with politics, interests and values. On how numbers have been used to describe the human genome, see Tora Holmberg's (2005) essay: note how the journalists and scientists Holmberg refers to consider the significance of the numbers, rather than what lies behind them being attached to the genome in the first place. More broadly, Jonathan Mark's (2002) witty and wise book *What it means to be 98% chimpanzee* reminds us how easily numbers get misused in discussions of contemporary human biology. Adam Bostanci (2006) explains some of the behind-the-scenes complexities that are belied by concepts like 'genes': he shows how the idea of the human genome is, in actuality, rather unstable (by focussing on the donors who were used for, and differences between, the two draft human genome maps of the HGP and Celera Genomics).

On the recent return of 'race' to discussions of group-level biological differences among humans, the interdisciplinary edited collection by Barbara Koenig *et al.* (2008) is excellent, and broadly critical of the use of 'race' in this context. The same can be said of Ian Whitmarsh and David Jones's (2010) rich collection *What's the use of race? Modern governance and the biology of difference*. More specifically, Anne Fausto-Sterling (2004) summarises the debates about the research of Risch *et al.* (2002) and Rosenberg *et al.* (2002), adopting a calmly critical perspective. So too does John Hartigan (2008), adopting a Latourian approach to both sides. David Skinner (2006) sets the research reported in Hartigan and Fausto-Sterling's essays in a broader contemporary scientific and cultural context. A number of critics have been prominent in the rejection of the idea that 'race' is biological – notably Kenan Malik (in England) and Troy Duster (in the United States). In her brilliant anthropological research, Amade M'charek (2005) shows the range of ways in which scientific knowledge of 'genes' maps on to particular notions of 'race' in the twenty-first century, within a wider discourse of 'human diversity'. Steven Pinker (2002: chapters 9 and 11), in his wide-ranging best-seller on 'human nature' *The blank slate*, tries to inject some (as he sees it) balance and sense into the debate. However, for all his brilliance, Pinker pays too little attention – in my view – to the way that scientific research is infused with the content of its putative 'outsides' (politics, values, prejudices). Though she goes beyond the topic of 'race', and though her book is a little dated, I recommend Barbara Katz Rothman's (1998) critique of molecular genetics because of its attentiveness to these inevitable 'inside–outside' transgressions.

The most wide-ranging exploration of how our understanding of 'nature' is being at once reproduced/confirmed and modified/challenged by the 'new genetics', i.e. both scientific knowledge of genes and new technologies and products consequent upon this knowledge, is the edited book by Sarah Parry and John Dupré entitled *Nature after the genome* (2010). In her monograph *The poetics of DNA* (2007), Judith Roof shows that genetic science is not immune from its sociocultural context and examines the metaphors and metonymies that pepper geneticists' discourse.

3. Fraser Island and encountering 'wild nature' first hand

My analysis is inspired by Kevin Markwell (2001), and by the work of anthropologists Adrian Peace (2001, 2002) and Stephen Healy (2007). This said, I think that Peace's analysis is a little forced in places, while readers not *au fait* with the ideas of Bruno Latour and Donna Haraway may find parts of Healy's essay quite hard to digest. Markwell focuses on Borneo not Australia, but analyses 'first-hand' tourist experiences of non-human nature in the same way that I've done here. Though also not about Australia, Joe Hermer's (2002) book *Regulating Eden* usefully treats nature reserves and parks as 'technologies' for actively organizing human experience and conduct. Far from being simply an 'escape' from the rules, restraints and conventions of social life, Hermer shows that experiences of nature are tightly regulated. Jennifer Carter (2010), focussing on Fraser Island, both describes and challenges the spatial compartmentalisation of its terrain into 'natural' and 'social' spaces (though she doesn't home in on dingoes like Peace and Healy do). Background on Fraser Island can easily be found through a Google search, though it's important to note that these web sources comprise governmental and tourist sites in the main. This means that they present rather basic factual descriptions that lack depth or much in the way of critical content; they are also, of course, key participants in the ecotourist discourse of 'wild nature' and 'first nature' that I've questioned in my discussion of the 'dingo problem'. Note that, in my analysis of ecotourism, I've paid no attention to the way that tourists, and the tourist industry interpellating them, project on to the non-human world very specific values and desires as if they were 'externally given' aspects of the biophysical world. I've simply focussed on the creation of the effect of 'nature out there' (i.e 'the environment'). Bruce Braun's (2003) excellent analysis of North American adventure tourism and outdoor recreation reveals how specific notions of white masculinity are almost synonymous with 'the great outdoors'. Nature tourism, when analysed in detail, also shows graphically the relations between how the non-human world is represented and how human subjectivities get reproduced (or altered). For good examples of what I mean see Susan Frohlick and Lynda Johnson's (2011) research in Costa Rica and New Zealand, and Kathryn Besio *et al.*'s (2008) inquiries into dolphin tourism in New Zealand.

4. The social construction of 'nature'

The idea of 'social construction' is closely associated with 'critical' (that is, politically left-wing) social science, and is usually traced back to Peter Berger and Tom Luckmann's classic work *The social construction of reality* (1966). Work on the social construction of nature is now as large and diverse as it is old (going back 20 years), and has – as I intimated – been rather eclipsed by the 'post-constructionist' work of Latour and others. Rather than provide a long list of 'representative readings' from across the disciplines, I'll simply focus on those published in my own subject, human geography. This will at least permit some focussed reading and thinking – fortunately, the debates of nature's social construction in geography also resonate strongly with those in other disciplines, so the content of the following pieces is not especially geography-specific. I'll also focus on agenda-setting theoretical/debate pieces rather than more empirically focussed ones. These have inquired into whether, and to what degree, it's appropriate to think of 'nature' (the apparent antithesis of a social construction) as something that's 'constructed'. Most of these do not discuss 'human nature' but, instead, non-human nature – academic geographers have not participated in the 'nature versus nurture' debates in the ways that many biologists and psychologists have (e.g. Rutter, 2005). This is no surprise, given that the 'nature' most physical geographers examine relates to Earth surface processes and landforms. (This said, a few human geographers, like Kay Anderson and Robyn Longhurst, have looked at issues pertaining to 'human nature', such as 'race' and the embodied character of our living.) To the readings then. Jim Proctor (1998) gives a solid insight into the constructionist aims and claims. David Demeritt of King's College, London, has published a string of very illuminating essays in which he seeks to clarify what, by the turn of the millennium, had become a rather confused and confusing debate in geography about nature's social construction. He aimed to disentangle different meanings of 'construction' and 'nature' so as to clear the muddied intellectual waters (Demeritt, 1998, 2001, 2002). In the meantime, other geographers were determined to move 'beyond' the constructionist paradigm, notably Sarah Whatmore (1999). Jonathan Murdoch's review essays do an excellent job of showcasing the 'post-constructionist' approach' to 'society' and 'nature' (1997a, 1997b). Bruce Braun (2004, 2009) updates the discussion very lucidly in two equally programmatic pieces. My recent review (Castree, 2011) tries to be even-handed where Braun favours relational, post-natural approaches (note that I also discuss Marxist ideas about the material *production* of nature, ideas that do not feature in the present book).

CHAPTER 5 ENCLOSING NATURE: BORDERS, BOUNDARIES AND TRANSGRESSIONS Key readings relating to the three cases

1. Intellectual property in seeds and the politics of semantic boundary-drawing

Two essential readings on how the boundary between nature and non-nature is applied sharply by biotechnology firms to their commercial advantage are by Thom van Dooren (2007, 2008). He uses the ideas of 'socio-natural networks' and 'biosocial emergence' to criticise the semantic and practical 'cuts' made by the likes of Monsanto. Michael Carolan (2008) provides an overview of the legal cases I mentioned in Chapter 3, plus some other relevant ones besides. Both Carolan and van Dooren take theoretical inspiration from Bruno Latour's (1993) critique of the 'modern constitution', and they both explore alternatives to patents that might recognise and reward the biological 'inventions' of indigenous and peasant farmers. For more on the Schmeiser versus Monsanto patent infringement case, see van Dooren (2010). In this essay, van Dooren uses the trope of 'leaks' to explore how Monsanto and legal professionals tried to enforce proprietary claims to 'inventions' even as they seek to maintain the idea of 'products of nature' into which those claims intrude.

2. Bestiality, zoophilia and the Enumclaw case: regulating and re/defining 'appropriate' sex and behaviour

The key source here is an essay in the journal *Environment and Planning D* by Michael Brown and Clare Rasmussen (2010). Equally important is chapter 10 of Delaney's monograph *Law and nature* (2003), entitled 'Fear of falling'. Kay Anderson (2000) presents a historical survey of the way references to animals and 'animalistic behaviour' have been used in the West to regulate people's thoughts and actions and to punish offenders. Relatedly, the introduction to Ham and Senior's book *Animal acts: configuring the human in Western history* (1997), plus many of the chapters therein, offers rich insight into how Westerners have referred to animals when defining and policing their own 'humanity'. Though he doesn't talk about bestiality, Steve Baker's *Picturing the beast* (2001) remains perhaps the best introduction to how references to animals as 'others' function to define various forms of social identity today.

3. Transsexualism and transgender

Susan Stryker's (2006) introduction to her co-edited book on transgender studies is an excellent starting point. Readers should then dip into, and preferably read all of, the re-issue of *Stone butch blues* (Feinberg, 2003). My analysis of the book is inspired largely by Jay Prosser (1998: chapter 5). A recent, wonderfully rich collection of essays on 'queering' conventional subject-positions and social practices is *Queer ecologies* (2011), edited by Mortimer-Sandiland and Bruce Erikson. This book explores how conceptions of nature can be deconstructed, with meaningful consequences, in tandem with rethinking what counts as 'permissable' sex and sexual identity.

4. Other studies of nature's borders and boundaries: semantic, administrative, legal and spatial

In section two of Chapter 5, I made the point that attempts to create and enforce borders and boundaries in respect of 'nature' and its collateral referents are very widespread. It is instructive, I suggested, to see how 'border work' and 'boundary work' are performed in ostensibly different arenas of social thought and practice. With this in mind, the following studies are worth reading comparatively: Jim Proctor and Stephanie Pincetl (1996) about spaces for owl preservation on the Pacific Northwest; Bjorn Sletto (2002) on Nariva Swamp in Trinidad; Tom Mels (2002) on Swedish national parks; van Hilvoorde *et al.* (2007) on attempts to keep elite sport 'natural'; David Sibley (1995) on the 'purification' of English rural space; and Paul Robbins (alone and with a co-researcher) on land cover categories in India (Robbins and Maddock, 2000; Robbins, 2001a, 2001b). These studies aside, two notable issues where nature/non-nature boundaries have been the focus of intense surveillance are (1) the management of so-called 'invasive species', and (2) the management of natural resources within a 'fortress' approach to environmental conservation.

CHAPTER 6 THE USES OF NATURE: SOCIAL POWER AND REPRESENTATION

Key readings relating to the discussion of social power and the West African case study

1. Definitions of, and debates about, social power

Because I begin the chapter trying to help readers make sense of the various definitions of, and debates about, social power, I choose not to list additional readings on this large and complicated subject. For those especially keen to know more *The handbook of power* by Stewart Clegg and Mark Haugaard (2009) covers the various different theories and perspectives on the subject, with an emphasis on critical social science contributions. I'd urge readers consulting this book not to presume that social power is one of those things awaiting a single, 'correct' definition and analysis. As I argue in Chapter 6, the debates about the nature and effects of social power showcased in Clegg and Haugaard are *participants* in the drama of social power's recognition, operation, successes and failures. The lack of analytical consensus on the 'realities' of social power is symptomatic not of academic failure, I'd argue, but of contests over whose analysis will prove to be the most persuasive or efficacious within (and without) the world of the social sciences and humanities.

2. Deforestation discourse in West Africa's forest–savannah transition zone

As I make plain in Chapter 6, the case study of deforestation discourse in Guinea and, more specifically, in Kissidougou, is inspired by the research of British anthropologists James Fairhead and Melissa Leach. Though quite old now, the lessons of this research are relevant today nonetheless for they inquire into what can happen when influential epistemic workers go unchallenged. Their essay in the edited book *The lie of the land* (Leach and Mearns, 1996) is a good place to begin: it is pithy but comprehensive. It should be followed by a reading of the final two chapters of *Reframing deforestation* (Fairhead and Leach, 1998a), which discuss the creation, persistence of and possible ways of moving beyond West Africa's deforestation discourse. For insights into the 'realities' of forest dynamics and use in the Guinea transition zone see Fairhead and Leach's chapter in *The social life of trees* (1998b), edited by Laura Rival. To learn more about the way deforestation discourse was powerful, yet without winning over ordinary land users in Kissidougou and elsewhere, Leach and Fairhead (2000) is the key reading.

CHAPTER 7 NATURE'S PRINCIPAL PUBLIC REPRESENTATIVE: THE MASS MEDIA

Key readings relating to the discussion of the role and effects of the mass media, publics and environmental reporting by the news media

1. The mass media

The survey literature on how the mass media has been studied, and what its audience effects are, is voluminous. Because I focus on the news media, I recommend Max McCombs *et al.* (2011). *The news and public opinion: media effects on civic life*.

2. Publics and democracy

Alastair Hannay's book *On the public* (2006) is essential reading, as is the essay by Clive Barnett (2008) on communication, virtuality and the creation of publics.

3. The news media and environmental issues

All of the following provide readable, up-to-date introductions to how and why 'environmental news' gets made in the ways it does: Max Boykoff *Who speaks for the climate?* (2011), Anders Hansen *Environment, media and communication* (2010), and Libby Lester *Media and environment* (2011). Boykoff's (2009) review essay is perhaps the place to begin before dipping into one or more of these books. Thereafter, the following are key sources on the 'bias of balance' discussed towards the end of Chapter 7: Antilla (2005, 2010), Boykoff and Boykoff (2004, 2007), Carvalho (2007), and Freudenburg and Muselli (2012). Jennings and Hulme (2010) criticise news media preference for drama in their analysis of reporting of 'abrupt climate change'.

CHAPTER 8 EXPERTISE, THE DEMOCRATISATION OF KNOWLEDGE AND PARTICIPATORY DECISION-MAKING: UNDERSTANDING THE NATURE OF SCIENCE Key readings relating to 'Climate-gate' and 'Glacier-gate', scientific self-governance, public use of science and direct public involvement in science

1. The two 'gates' and their implications for future scientific self-governance

Though there's a lot of credible web material on the two 'gates', investigative journalist Fred Pearce's (2010) *The climate files* is a reliable and well-researched primer on the 'scandals'. More widely, Sarewitz's (2004) short essay shows how 'post-normal' science is always liable to be shrouded in controversy. There's a wealth of informed reflection on the implications of the two 'gates' for future scientific self-governance of knowledge production and communication. Food for thought can be found in Hulme *et al.* (2010c), Anderson and Bows (2012), Houllier (2012), Hulme (2010b), Rapley (2012) and Sarewitz (2011) (short pieces all) and, from the perspective of science 'outsiders', Beck (2011), Grundmann (2012) and Grundmann and Stehr (2012: ch 1, 4, 5). Those interested in the implications of the two 'gates' for environmental news reporting, linking back to Chapter 7, should read Holliman (2011).

2. The relative roles of 'certified' and non-certified specialists in the shaping of post-normal science

The 'gates' raise big questions about whether and how non-certified specialists can represent the interests of stakeholders, or the wider public, by being able to shape or check post-normal science as conducted by certified scientists. The essential resource here is the discussion paper by Harry Collins and Robert Evans (2002) followed by a set of commentaries. The set appeared in the journal *Social Studies of Science* 32 (2).

3. The uses of post-normal science by politicians and other stakeholders

Roger Pielke's (2007) book *The honest broker* is a very good place to begin in order to identify the varied ways science can be used in wider political and public debates. Sarewitz (2004) succinctly details the misuses of science in ways that echo Pielke. It is instructive to think of climate change sceptics' use of science in light of Pielke's and Sarewitz's arguments. A useful aid here is Naomi Oreskes and Eric Conway's (2010) book *Merchants of doubt* or, alternatively, Washington and Cook's (2011) *Climate change denial.* For a shorter discussion of scepticism, see Douglas (2009).

4. Lay involvement in science up-, mid- and downstream: how to democratise science?

For more on the Pickering case, see Landstrom *et al.* (2011) and Lane *et al.* (2011). It's worth reflecting on how far the situation arising in Pickering might be replicated in other places for other issues (not just flooding), and how possible it might be to institute the 'experiment' expert and lay actors participated in. A good deal of citizen participation in expert science has been piecemeal and local, arising around contentious issues. The previously mentioned work by Collins and Evans proposes a model for the public monitoring and checking of science; however, it tends to focus on who is qualified to intervene 'mid-stream' and so fails to give wider attention to citizen participation up- and downstream. Such participation has been discussed in terms of (1) who should be involved from the public and (2) how best to involve them. The diverse conclusions reached by analysts on these two issues reflect different views on 'real democracy'. The following authors have written in interesting ways on lay involvement in science up- or downstream: Elam and Bertilsson (2003), Hagendijk and Irwin (2006), Munton (2003) and Yearley (2012). Part I of *Science and citizens* by Leach *et al.* (2005) is also worth reading; however, the indispensable source for the wider vision of democracy, participation and knowledge I introduce at the end of Chapter 8 is Steve Fuller's (2000) *The governance of science.*

HOW TO USE THIS BOOK

Those (few) who might read this book of their own accord for no other reason than because it interests them can complete the chapters' various study tasks as they wish. However, those teaching courses on the major themes and arguments covered in *Making sense of nature* will no doubt ask how this book can be *used* as opposed to simply *read* by them and their students. These students will almost certainly be either upper-tier undergraduates or master's degree candidates. What formative and summative activities can allow such students to make this book central to a wider course of study that extends beyond its pages? I don't aim to be prescriptive. Below is a suite of suggestions that can inspire (1) reflective learning exercises within class or in a student's own time, and (2) summative assessments.

In all cases, my presumption is that no tutor would give week-by-week lectures or seminars based *directly* (or mechanically) on each of the book's chapters. That practice would be more fitting for introductory level degree modules, but this book is not pitched at new undergraduates. Most would find it confusing or simply too big to digest. Instead, tutors might ask their students to read part or all of a chapter before (or after) a lecture or seminar in which a tutor presents *independent* material that explores the key issues in the chapter. Some of this material might be case- or problem-led, other material more discursive, theoretical or debate-centred. Some of it might be drawn from academic literature, but some of it from everyday life – newspapers, adverts, magazine articles, literature, movies, art exhibitions, documentaries, zoos, radio programmes or poetry, say. In my view, a course that covers the many related topics explored in this book works best when it's *personal*. Rather than talking about the issues in general terms, or by way of examples/cases that seem far removed from students' daily lives, it's preferable to anchor any course in students' mundane experiences of 'nature' – that is to say, in and through everyday media like tourist literature, Greenpeace leaflets, a film like *The Cove* (2009), organic-produce adverts and so on. When this can't readily be done, I often use a simple technique: I ask class participants to imagine themselves in a place or situation I want to make the focus of discussion.

In summary, this book does not provide a tight-knit 'framework' that must somehow be used lock, stock and barrel; instead, it provides a toolkit of ideas, concepts and arguments that can be employed as part of a wider course of study designed to cultivate the sort of critical literacy I call 'positive scepticism'. *Making sense of nature* is thus intended to be a resource rather than a manual. I like to think the themes are broad enough that they can, in future, be explored by course teachers through new cases and examples as they appear in academic and non-academic fora.

CHAPTER 1 ***Key themes:*** when we talk about 'nature' we're talking about ourselves in significant measure; discourse is irreducible to the material world it describes, explains or assesses

1. The early part of the chapter itemises the various meanings of 'nature' and then links them to nature's 'collateral concepts'. It makes the point that nature and its filial terms are not natural but, instead, semantic concoctions. The following is tried-and-tested material for a good first seminar focussed on the question 'is nature natural?' Chapters 1–3 of John Takacs's highly readable book *The idea of biodiversity* (1996) detail the emergence of one of nature's newest collateral terms, 'biodiversity', and describe the epistemic community principally responsible. By taking a historical approach, the chapters show readers that language has a certain autonomy from the world it purports to 'cut at the joints'. In discussing Takacs's work, one can pose and answer the three questions I address early in Chapter 1, namely 'what is nature?', 'where is nature?' and (most curiously for students) 'when is nature?' One can then loop back to 'biodiversity' and ask: does it pre-exist being named as such?
2. The middle part of the chapter introduces the idea that Western thinking is structured by a family of parallel dualisms, of which nature–society is just one. I follow other writers in suggesting that it's within these that most signifier–signified–referent chains are organised. An old but rich and thought-provoking essay inspired by cross-cultural research is that authored by British anthropologist Tim Ingold: 'Hunting and gathering as ways of perceiving the environment' (1986). The essay shows well how the world looks when one dispenses with dualisms – making it a difficult read for those of us schooled in relying on them! Ingold's essay allows students to appreciate the contingency of the Western 'mental map' while acknowledging that it can't be readily replaced with a new one.
3. The middle part of the chapter also makes a distinction between discourse and discourses, and introduces the idea of epistemic workers upon whom we all routinely depend. These workers talk about nature in different idioms, using different media. They help us make sense of nature day in, day out. Provide students with a newspaper article on a topic (say biodiversity loss) and a scientific paper on some aspect of the same topic (both are word-dominated forms of representation) and ask them to identify any shared meanings of nature or one of its collateral terms, but to also itemise the ways in which the discourses differ – not so much

in content as in style and purpose. Ask them to reflect on the reasons why the discourses differ, and the situations (context, frequency and significance) where they are likely to consume such discourses in their daily lives. This anticipates material dealt with later in the course/book.

4. Towards the end of the chapter, I make the argument that many references to what we call 'nature' are metonymic: references to particulars evoke 'wider realities'. Again, I suggest that this isn't natural. Although it doesn't focus on nature, students can read chapter 1 of Tim Mitchell's (1988) *Colonising Egypt*, which usefully introduces the relationship between 'seeing' as a contingent practice and the equally contingent discourses discussed earlier in the chapter. This relationship is discussed further in Chapter 2. Students can be asked to read Mitchell's chapter and take notes in response to a set of tutor-devised questions about its content. As with the Ingold piece previously, it's a good test for student skills of close reading because its contents are counter-intuitive.

CHAPTER 2 ***Key themes:*** 'nature' is represented for us by others whose representations work in two registers at once, the descriptive and the political; what we call nature must be represented

1. The early part of the chapter introduces the idea of 'epistemic dependence' and an expansive conception of 'epistemic communities'. In class, ask students to itemise all the epistemic communities they come into contact with. Which are the most prominent and why? Which are the most important for them and why? What different relations of dependency are involved in each case between them and the epistemic workers in question (e.g. trust, scepticism, necessity, chance, etc.)? What are the goals of the relationship (e.g. selling commodities, as in advertising organic foods; warning us of danger, as with flood risk experts, etc.)? Finally, how many of these workers make references to nature a core part of their occupational discourse?
2. The middle part of the chapter focuses on the character of 'representation'. A readable introduction to this subject is Stuart Hall's (1997) introductory essay to his well-known book *Representation*. Hall touches upon the difference between literal (denotative) and connotative reference, flagged late in Chapter 1. It also links discourse with putatively non-linguistic forms of representation, such as images. In class, Plate 2.1 is an excellent vehicle to explore how representation 'works', including its conjoined epistemic and political moments. It is also a good vehicle for answering the question 'why must "nature" be represented?' (i.e. why can't it represent itself?). This question is answered systematically by philosopher John O'Neill (2001) in an accessible essay that students can read and take notes from in their own time. Again tutors may want to pose study questions to focus student reading of this piece.

CHAPTER 3 *Key themes:* references to 'nature' and its collateral terms are, in various ways, mechanisms of 'rule'; the identity and influence of epistemic workers is actively secured through 'boundary work'; although not all references to nature are important, others can make a big difference to our thinking and our worldly activities

1. The bulk of the chapter sketches the relationship between different genres of communication, associated representations of 'nature' and their intended effects on our opinions, beliefs, sense of self and actions. Rather than have students read Foucault on 'governmentality', it's perhaps better to direct students to the introduction and chapter 1 of Stephanie Rutherford's (2011) *Governing the wild*. The material therein can be used to stage an in-class discussion of whether and how students might entertain the idea that they were being 'governed' when visiting natural history museums at various moments in their lives. At base, what constitutes 'government' is the question here, and many students will (justifiably) argue that a museum visit is a 'free' act and the visitor experience scarcely governed at all. If this does not appeal, another option is to have students watch the 1997 Hollywood movie *Gattaca*. This should be done before class, with some question prompts about representation, nature and identity. In class, a managed discussion can be had of the story, characters and the 'take-home' messages about 'biopolitics'. Focussing on a film also anticipates Chapter 3's discussion of the 'work' done by different genres of communication.
2. The middle section describes the sort of 'boundary work' performed by epistemic workers prior to their representations 'travelling' elsewhere. In the United States, attempts by neo-Darwinian biologists to reject the claims of 'intelligent design' (ID) advocates show well how epistemic communities seek to police genre distinctions. In his essay 'Public education and intelligent design', philosopher Tom Nagel (2008) explores how boundaries between 'science' and 'religion' get drawn in order to discredit ID proponents. Students can be asked to read this with a question in mind, namely 'who has the right to represent life on Earth to the public: neo-Darwinian evolutionists or ID advocates?' This question obliges them to consider how the right to represent is struggled over and ultimately a social determination.
3. The last part of the chapter links references to nature by various epistemic workers to wider issues about democracy and its quality. This is returned to in some detail in Chapter 8. For now, and in anticipation of that chapter, have students read the British columnist for *The Telegraph* newspaper Christopher Brooker. One of his pieces criticises the way the BBC, a public broadcaster, has represented climate change to British audiences. Whatever one thinks of his arguments, he makes it clear how much is at stake for modern societies when a minority get to represent 'big issues' to the rest of us (see Brooker, 2011). Of course, Brooker himself is part

of this minority, having weekly access to thousands of middle-class readers. So, too, is my colleague Erik Swyngedouw who comes at this from a more radical angle than Brooker. He claims that pretty much all mainstream representations of climate change in the media, by ENGOs, etc. are 'post-political', that is to say contributory to 'thin democracy'. One of his many articles and chapters on this subject usually makes for lively class discussion about the relationship between discourse and democracy.

CHAPTER 4 *Key themes:* what we call nature is 'constructed'; to what extent and in what ways is the claim of nature's constructedness defensible?

1. Most of this chapter focusses on different groups representing nature and its collateral terms using varied genres and media (words, images, numbers). It suggests not only that representations are constructed but also that the world they purport to represent is not itself 'independent' of representation. This latter point is explored well in chapter 5 of Susan Davis's (1997) excellent book *Spectacular nature: corporate culture and the Sea world experience*. Have students read chapter 1 first and then chapter 5, taking notes or having an in-class discussion about how visitors experience an 'artificial nature' that is paradoxically consumed as somehow 'natural'. Invite students to pay close attention to what is constructed, detailing the various physical and symbolic aspects of Sea World. Davis's book overlaps with Stephanie Rutherford's and raises questions about who is doing the 'constructing' and to what ends. Chapter 4, however, is focussed less on who and what, and more on how 'construction' occurs as a routine practice. Keep the students' focus here. Alternatives to Davis's book are the fine essays by Jeffrey Sasha Davis (2005, 2007) on the re-naturalisation of Bikini Atoll. Invite students to imagine they are going on holiday there, working back in time to how the 'pristine' environment was made to appear so after the nuclear testing period.
2. The Davis work can then be situated in wider discussions about the construction of 'nature'. One or both of the essays by David Demeritt (1998, 2001) offer rich food for thought, though there are many other synoptic pieces available on the character and limits of 'constructionism'. Students can be asked to reach a considered view (orally, on paper or both), based on the Demeritt essays, on what it sensibly means to insist that nature is socially constructed.

CHAPTER 5 *Key themes:* how do boundaries between the natural and the non-natural get utilised; what do transgressive representations tell us about these boundaries and how might we judge them?

1. This chapter can be read before or after a practical class in which the photography of Ingrid Pollard is a focus. Take some images from her

1987 'Pastoral Interlude' series, fronted with some contextual information about Pollard's biography and the cultural norms prevalent in late twentieth-century Britain. The discussion can focus on what Pollard's photographs communicate to us, with the two key theme questions in mind. It is also an opportunity for students to see how words and images work together because there is no non-discursive way of interpreting Pollard's wordless images. Furthermore, students can explore the ideas of collateral concepts, metonymy and connotative reference effectively by way of Pollard's evocative stills. Tutors looking for help interpreting Pollard images can start with Kinsman (1995). It's well worth asking students if they ever pay much attention to 'alternative' representations of the world, like political photography, as part of their daily lives. This can inspire some discussion on how some things become visible to the masses and some not. It can also inspire discussion of the social role of different genres of communication, such as poetry, reports by NGOs, government inquiries, Disney films and so on.

2. If this doesn't appeal, an alternative is to design a session where students look at the 'parks versus people' approach to land-based nature conservation. This was especially prominent in eastern and southern Africa throughout the twentieth century. It has given way to a less exclusionary approach that to some extent dissolves the nature–people dualism. Students can be given two or three readings that, between them, examine the origins, effects and criticisms of the earlier approach.

CHAPTER 6 *Key theme:* the 'nature' of social power

This chapter asks 'what is social power?' and explores a case where reference to forests becomes a means for a community of scientist-bureaucrats to intervene in the affairs of ordinary people. The work of Fairhead and Leach used in the chapter can be given to students independently prior to class, leading to a discussion about how social power operated in this case. Alternatively, this is a good moment to make *Gattaca* (1997) a focus of classroom discussion, if not used earlier. This film, a work of stylised fiction, raises questions about the kind of 'biopower' many social scientists regard as an important aspect of contemporary life. It's also a chance to revisit the themes of freedom and constraint that may have arisen in discussions of Stephanie Rutherford's book.

CHAPTER 7 *Key themes:* how and why does the news media report environmental issues in the ways it does; why is news reporting a vital aspect of public life; how do we normally 'consume' news stories?

This chapter's subject matter is very 'close to home' because all students will, at some point each week or month, cast eyes over a newspaper or a newspaper website. This is an opportunity to discuss and develop students' media literacy. One possibility is to have students read Chapter 7 first. In class, they could be given two newspaper excerpts on the same issue (e.g. GM foods, artificial life, or geoengineering). One would be a regular article, the other a column or editorial. They could be from the same newspaper or from two different ones (e.g. a 'left' and a conservative paper). Students could be invited to analyse the articles focussing on style, use of metaphor or analogy and overall message. They can reflect on whether the article is 'obviously' more objective than the column or editorial. Thereafter, a facilitated discussion of how students typically consume or use the media can ensue. This will enable a link back to the questions of representation, nature and social power explored in Chapter 6.

CHAPTER 8 ***Key themes:*** how should prominent epistemic communities self-govern, or be governed, to ensure that they serve the public interest and enrich the 'semiosphere'?

1. Where the previous chapter focussed on the news media, this one takes the case of another highly significant representative of 'nature', namely science (with a focus on climate science, which is quintessentially 'post-normal'). It works well to get students to research 'Climate-gate' and 'Glacier-gate' on their own with this question in mind: what changes to the governance of post-normal science, if any, are suggested by the two incidents? There is ample commentary on the two 'gates' in *Nature*, *Science*, the *Scientific American*, *New Scientist*, broadsheet newspapers and websites of climate sceptic think tanks like the Heartland Foundation. The distinction between up-, mid- and downstream governance is a useful heuristic here. Tutors looking for a different case than the two 'gates' might consider the battle between neo-Darwinian biologists and intelligent design proponents. A set of mid- and downstream issues arises here. Should ID be taken seriously by non-proponents in the science community? Should ID be discussed in the public mainstream, even included in school curricula (as Darwinian evolution is)? How should lay people regard ID in light of Darwinism's dominance as an account of natural history – with open minds or disbelief? The first question is tackled by Tom Nagel in his earlier mentioned essay, but has implications for the other two because Darwinian scientists are currently empowered to determine ID's *prima facie* legitimacy in the public domain. All three questions were addressed in the fascinating court case *Kitzmiller versus the Dover Area School District* in 2005. There's a good Wikipedia page describing the trial, whose most notable feature, arguably, was the appearance of sociologist Steve Fuller as an expert witness for ID proponents. Fuller, as I show near the end of Chapter 8, is an advocate of free speech and a strong critic of 'group think'. His readable book *Dissent over descent* (Fuller, 2008) explores his support of ID's right to be taken seriously, even though he's not an ID advocate as such. To use my terms, he was adopting 'positive scepticism' towards the Darwinian orthodoxy, unlike

world-famous science critics of ID such as Richard Dawkins. Reviews of this book, plus other commentary on Fuller's controversial 'support' of ID, are not hard to find and they debate answers to the three questions posed earlier. A particular issue was Fuller's status as an 'expert' whose testimony was underpinned by a republican political philosophy. As an STS scholar, was his 'expertise' relevant in a court case about ID teaching in public schools? Arguably it was because his considerable esoteric knowledge was being shared in what he felt was the wider public interest. Some of his STS colleagues demurred, as is plain in the special issue of *Social Studies of Science* (36, 6: 2006). See also the discussions of Fuller and his response to them in the journal *Philosophy of Social Science* volume 40, issue 3, September 2010 (Forsman, 2010; Shearmur, 2010).

2. The two 'gates', the *Kitzmiller* case and other examples considered in Chapter 8 can usefully be related to Harry Collins and Robert Evans's (2002) model for expert–lay relations, which attracted several responses in the journal *Social Studies of Science*. Carolan's (2006) short essay in *Society and Natural Resources* is also useful to think about here. The edited book by Selinger and Crease (2006) *The philosophy of expertise* contains some rich essays on how society can use and regulate experts.

FREE-STANDING SUMMATIVE ASSESSMENT TASKS

1. *Traditional term papers:* There are a great many topical options linked to one or more of the book's major themes. For instance, it's not hard to assemble social science literature on how climate models (representations of global nature in the future) get designed and operationalised. Publications by David Demeritt, Helene Guillemot, Spencer Weart, Mohd Hazim Shah bin Abdul Murad, Mikeala Sundberg, Myanna Lahsen, Gabriele Gramelsberger and Joel Katzav are useful. This relates to the topic of epistemic 'construction' and allows students to (1) see science as a 'craft' (which is what it usually is), and (2) consider what reasons we have for (dis)trusting scientists. In short, what follows if we acknowledge that science is never 'objective' in the naïve sense that it is usually taken to be? On this same theme (constructionism), the well-known debate inspired by Bill Cronon's essay on wilderness (published in his edited book *Uncommon ground*, 1996a) is a good way for students to think about the value of the 'denaturalising' approach adopted in this book. Responses to Cronon can be found in the journal *Environmental History* (Cronon *et al.*, 1996b). On this same theme once more, students can be invited to read the work of 'cultural climatologists' like Mike Hulme (2010a, 2011, 2012). The question here might be: what are the benefits of accepting that 'climate' is more than a physical reality existing independently of our thinking about it?

Another term-paper candidate is 'invasive species': how and to what ends do they get represented in scientific, policy and public discourses? This topic relates to boundaries and bounding, as discussed in Chapter 5. It's a good vehicle for exploring how social decisions about what belongs and does not get naturalised. In the twenty-first century, a number of human (and non-human) representatives have come to the fore as spokespeople for 'nature' – Al Gore, David Attenborough, Jane Goodall, Bjorn Lomborg, Diane Fossey and Steve Irwin are good examples. Of late, this is part of what's called 'celebrity environmentalism'. Students can be invited to consider what's gained and lost when a small number of prominent individuals command widespread public attention as nature's representatives, in ways that blend glamour, entertainment and seriousness. This links to the themes of representation (explored in Chapter 2) and politics (explored in Chapters 3 and 8). In my experience, students really enjoy researching this topic. Excellent general pieces on celebrity and politics are those by Mark Wheeler (2012) and Graeme Turner (2004: chapters 1, 5 and 7). For a focus on celebrity and nature, Dan Brockington (2008a, 2008b, 2009) is the indispensable source. Reviews of the last of these (easy enough to find) provide some critical engagement with Brockington's theses. Finally, a topic that reveals graphically the politics of representation and the high stakes frequently involved is that about 'greenwashing'. This was the subject of much academic analysis in the 1990s and focussed on the relations between message manipulation and honesty in representations of environmental degradation and protection. Were large companies appropriating the rhetoric but not the substance of 'sustainability'? Were ENGOs exaggerating the magnitude of the 'environmental crisis'?

2. *'Practical' term papers:* Rather than opt for traditional literature-based essays, some tutors may wish to give students something with a more 'practical' feel. Students can be invited to select one representation of 'nature' and to subject it to close, critical analysis. They can choose something mainstream (like a Disney film or children's book) or something alternative (like political cartoons that don't appear in mass circulation newspapers or magazines). Alternatively, they could choose to compare and contrast both in relation to a common topic or phenomenon (e.g. climate change, animals, genes). In either case, students can be reasonably expected to show some wider understanding of the history and purposes of the genre of representation that their selected case(s) belong(s) to. For instance, if a student was studying how a new or older wildlife documentary represents birds or mammals, they might well have read works by Derek Bouse or Gregg Mitmann for context. Likewise, a student focussing on how animals are represented on antivivisection websites and in pro-hunting literature might profitably read books authored by Steve Baker. Beyond this, they can also be reasonably expected to have some 'method'

for analysing the content of their chosen representation(s). This does not necessarily need to be too formal, but it could be should you or they wish it. For instance, there are now plenty of books about how to analyse movies, novels, newspaper articles, etc. A search online also quickly reveals several primers to image or textual analysis that can be adapted for student use by tutors.

A related option is to ask students to research and write a book review essay. The tutor could decide on the book list from which student selections are made. It might include notable texts that use 'nature' as a key point of reference, such as Francis Fukuyama's *Our post-human future* (2002) and Kenan Malik's *Strange fruit* (2008) – and those that do not (such as Emma Marris's *Rambunctious garden: saving nature in a post-wild world* (2011)). Students can be directed to find reviews of these books, but also wider academic reading about nature and culture.

3. *Exam questions:* Finally, some tutors may wish to set a traditional 1–3 hour exam in which students get 60 minutes to answer a question from a set of five or more. The questions would obviously reflect the topical mix of the taught course. Below you'll find some more general questions that can be set, supposing tutors created relevant reading lists beforehand. These are taken from my own 'Making Sense of Nature' degree module.

a. 'The idea of nature is a weapon of mass distraction.' Discuss.
b. Do we need nature?
c. 'What counts as nature cannot pre-exist its construction' (Braun, 2002: 17). Explore the implications of this contention.
d. 'Nature is dead! Long live nature!' Discuss.
e. Is nature a necessary illusion?
f. 'Much of the moral authority that has made environmentalism so compelling as a popular movement flows from its appeal to nature as a stable external source of non-human values against which human actions can be judged without much ambiguity' (Cronon, 1996: 26). Is Cronon's assessment a fair one?
g. 'There is no such thing as nature.' Critically assess this statement.
h. 'Whoever utters the word "nature" deserves to be needled by the question: "which nature?"' (Beck, 1992: 342). Explain and evaluate Beck's assertion.
i. 'Nature is a chaotic concept.' To what extent do you agree with this statement?
j. 'Sticks and stones may break my bones but words will never hurt me.' Assess the applicability of this schoolyard rhyme to the word 'nature'.
k. 'Social constructionism has helped destabilise the longstanding notion that bodies are "simply natural" or biological' (Longhurst, 2000: 23).
l. Critically evaluate social constructionist approaches to the human body.

m. 'A person's sexual identity is given not so much by their genital anatomy as by their sexual preferences' (Wade, 2002: 42). Do you agree?
n. 'Like all . . . powerful ideas, the idea of nature as wilderness – as something separate, pristine, eternal, and harmonious – has in many ways become more important than the reality it purports to describe' (Budiansky, 1996: 21). Assess this contention.
o. 'The naturalness of nature is, in one sense, inherently self-evident' (Adams, 1997: 82). Is this true?
p. Explain some of the problems of the idea of biological essentialism using examples from either the human or the animal world.
q. Is 'nature' or 'nurture' the most important factor in explaining *either* sexual preference *or* obesity?
r. 'Naming something gives it a reality; a name literally gives meaning to an object' (Unwin, 1996: 20). Discuss this claim in relation to either 'race' or gender.
s. 'Meanings can mould physical responses but they are constrained by them to' (Eagleton, 2000: 87). Explore how far ideas of nature can give rise to the material realities they purport merely to describe.
t. 'Nature knows best.' Discuss.
u. Are the boundaries between nature and society natural ones, and does the answer to this question matter?
v. To what extent are we subject to other people's representations of 'nature' and its collateral concepts?
w. How can the downsides of our epistemic dependence on either the mainstream news media or science be mitigated? Use real cases and examples to inform your answer.
x. Should we be worried that many seemingly innocent references to 'nature' are power by other means?

ENDNOTE: WHY WE (STILL) NEED TO TALK ABOUT NATURE

In this book, I've tried to speak to readers from a wide range of intellectual backgrounds. Mine has been a cross-disciplinary endeavour that transgresses intellectual borders. As the Preface makes clear, *Making sense of nature* is premised on a proposition that will strike neophytes as controversial, but is very old news to the likes of those whose work I've drawn upon in the previous chapters. In this endnote, I'd like to explain and defend that proposition. It once animated a large volume of research in the Anglophone social sciences and humanities; today, however, some consider it rather passé. It's the proposition that there's no such thing as nature.

This is not the same as saying there's no such thing as nature *anymore*. The sentiments expressed by Alan Weisman in his bestselling book *The world without us* (2008) are not ones that I share. Weisman represents a style of thinking and a genre of writing that many people find resonant: he rails against the destruction and disappearance of the natural world at the hands of a human population whose mounting ecological debt will soon become unpayable. Scientist Peter Ward does much the same in his recently published *The flooded earth: our future in a world without ice caps* (2010), as did environmentalist Bill McKibben in his bestsellers *Enough: genetic engineering and the end of human nature* (1993) and *The end of nature: humanity, climate change and the natural world* (1989). But, contra Weisman, Ward, McKibben and like-minded commentators, my point is that 'nature' has *never* existed – at least in the conventional senses of the term.

If anything has come to an 'end' in recent years, it is not 'nature' but a world in which we can use this familiar category with the same ease and regularity as we used to. We might say that nature exists only so long as we collectively believe it to exist (reversing the old adage that 'seeing is believing'). Our conceptions of nature are just that: *conceptions*, not the world itself (though very much a part of that world). Accordingly, when we alter or destroy those things we consider to be 'natural', it's not the world's intrinsic naturalness that has diminished but rather our capacity to describe it thus (and vice versa). Despite appearances, this is not a tautological statement. As geographer Kay Anderson so nicely puts it, 'There's no "natural" conception of nature, no stable inventory of the products that count as nature, and no universal register of questions timelessly posed by it' (Anderson, 2001: 70). She's not, of course, denying that those various things we regard as natural are real. Along with many others in the social sciences and humanities these past 30 years, she's simply insisting that we take our ideas about those things seriously. They have a life of their own, a certain degree of independence (or difference) from the world they ostensibly describe, explain, predict or evaluate. These ideas deserve to be scrutinised in their own right, not least because some of them are of world-changing importance by virtue of what they permit and proscribe.

In this light, then, what exists is not 'nature' but the myriad of things we, by convention, attach this label to.[1] By saying this, I'm directing our attention towards the process that has me using the scare-quotes – the process of naming, classifying, characterising, signifying, symbolising, delimiting and labelling. In this book, I've also been interested in the various communicative media that help to make this process possible – what I've called, for more than simplicity's sake, 'representations'. As explained in Chapter 2, these representations may be spoken or written, visual or verbal, interactive or unidirectional. As a corollary, I've been interested too in the various specialists and spokespeople who make nature intelligible to us via various media in a range of communicative genres.

I'm not what's sometimes pejoratively called an 'anti-realist'. Instead, I am very much a materialist, but my materialism regards ideas, images, inscriptions, sounds, symbols and other representations as both palpable and performative. They're as real as rocks, racoons or rivers; they direct our actions in various ways and they are, therefore, consequential. Indeed, these representations are among the tools we manufacture and use in order to forge *relations* with what we call nature (though certainly not the only tools). They connect us to life, rather than distancing us from it, and as such play a formative role in our daily existence. To presume anything less would be to adhere to a peculiarly lop-sided view of what matters in the world.[2]

Of course, some might happily agree that our representations are important, but dissent from my claim that there's no such thing as nature. There is, they might say, nature on the one hand and then a set of different representations we use to get a handle on it – some of which cut nature 'at the joints' and offer us a truthful perspective on it. But my point is that 'nature' is not natural by nature, but by ascription (and without remainder). As philosopher Hilary Lawson once put it, reflecting on epistemological variations among humans past and present, 'We don't have different accounts of the same "thing" but different understandings of "the thing" period' (Lawson, 1985: 128). Or, as another famous American philosophical pragmatist put it, what we call 'nature' 'has no preferred way of being represented' (Rorty, 1979: 300). In other words, our representations are at some level *constitutive*. They are not, to use some very well-worn metaphors, just templates, frames, filters, grids or sieves that allow us to impose some order on what we see, and to act accordingly. More actively, they're means of engagement and material forces – among our many 'prosthetics', if you will. We can never 'step outside' or 'around' them, even though they refer us to (and allow us to intervene practically in) a world that's ontologically irreducible to their content (and which, indeed, exceeds it). They condition what we perceive and how we act towards all manner of earthly phenomena, including our own flesh and blood.

Yet, in the case of 'nature', we routinely *conflate* the representations with the represented, as if the former holds up a mirror to the latter. Equally, we normally presume nature to exist outside or beyond any representations of it – be they factual or fictional, scientific or artistic. There's a good reason why. As William Cronon once observed, 'the very word... encourages us to ignore the context that defines it' (1996a: 35). For philosopher Kate Soper (1995: 34) nature is 'a kind of self-denying concept through which what is culturally ordained is presented as a pre-discursive external determination...' Cultural critic Jennifer Price made much the same point when she suggested that 'the definition of nature is . . . powerfully self-authenticating... [and] has built-in defences against questions' (1999: 250). It thus takes some degree of effort to 'avoid falling into the trap that this term has laid for us' (to paraphrase Cronon, 1996a).

That trap is set daily, and we become easily ensnared without realising it. The term 'nature' and its various meanings, I've argued, remain a very important part of our conceptual, linguistic and symbolic repertoire in the West. The fact that these meanings sometimes work in and through various 'collateral concepts' only serves to emphasise the point – concepts like 'gene', 'biodiversity', 'intelligence', 'wilderness', 'animal', 'rainforest', 'species', 'life' or 'race', for example. We need to pay close attention to these meanings, their contexts of use and their worldly effects. There's a lot of complexity involved, but one thing remains constant: when we talk about 'nature' we typically believe we're talking about something 'outside' ourselves (or 'within' us, if we're discussing human physiology or neurobiology). Yet in some measure, we're doing precisely the *opposite*, and it's as well to be aware of the fact. To misquote Einstein, nature is an illusion, albeit a very persistent and powerful one. I tried to explain how and why in Chapters 1 and 4.

In summary, I'd argue that there's no such thing as nature, and yet most people in the West proceed as if there *is*. It's part of our collective 'common sense', a seemingly timeless piece of intellectual furniture. Indeed, it almost seems natural to believe in nature's reality and existence. Unlike Bruno Latour, I do not regard this as a problem – or least not necessarily. In *We have never been modern* (1993) and several subsequent publications, such as *Politics of nature* (2004), Latour argues that Western societies have divided the world mentally according to a set of antonyms, and then wrongly supposed that this historically contingent act is simply a reflection of how the world really is. He makes the normative claim that we should abandon our binary ways of thinking so as to embrace what's been under our noses all along, namely an 'impure' world of 'actor-networks', 'hybrids' and 'imbroglios', which defies the supposed ontological division of the world into two principal components.

In my view, Latour is right to acknowledge the power of what he calls (metaphorically) 'the modern constitution', in which a distinction between 'the social' and 'the natural' is one of several key differences assumed to inhere in reality. But I think he's far too willing to believe that this distinction can be dislodged through sheer force of (his) argument. What's more, I find his 'double reality' presumption question begging. Latour concedes that the modern constitution governs our thought and actions, yet suggests that he (and we) can choose to step outside it in order to perceive a world somehow hidden in plain sight.[3] To my mind, this drastically underplays the inertia possessed by our existing mindsets and habits of thought. They're not dislodged easily or overnight. We must therefore interrogate them intelligently, rather than suggest that they can (or should) be abandoned willy-nilly through a process of mental revolution.

The paradox that Latour identifies thus lives on, in my view. Yes, the world is not 'really' divided in two. In Donna Haraway's (2003: 20) neat formulation, we should regard '"the relation" ... [a]s the smallest possible unit of analysis' rather than the *relata*. *And yet* we routinely employ categories that presume to draw lines – thick, thin, dashed and dotted – around what we take to be coherent pieces of reality whose identity at some level precedes their interactions with all the other pieces. As Sabine Weiland (2007: 68) noted, 'Manifest and stable divisions between the two realms [of "society" and "nature"] are a precondition of the functioning of many institutions and routines in modern society. To the extent that the divide has become blurred, we can discern efforts to counter this . . . by re-erecting or newly defining boundaries.'

Regardless of these arguments, it may be thought by some that a book like this is both untimely and unnecessary. So, with these readers in mind, let me offer a two-part apologia for having written it. First, *Making sense of nature* may seem 'behind the curve' intellectually because work by Isabel Stengers, John Law, Ann-Marie Mol, Michel Serres, Jane Bennett, Michel Callon, Gilles Deleuze, Tim Ingold, Keith Ansell-Pearson, Brian Massumi, Timothy Morton, Cary Wolfe, the already mentioned Latour, and various fellow travellers has, apparently, taken us 'beyond' a concern with 'the social', with 'discourse', with 'representation' and – indeed – with 'nature' too. However, as my comments about Latour indicate, such ideas cut little ice in the wider world or in much of academia outside the social sciences and humanities. 'Nature', and those myriad things we discuss using its collateral concepts, remain real for us to the extent that we continue to rely on 'the modern constitution', in large measure, to apprehend the world. The arguments of Weisman and others remind us of this fact. This is why I believe that it's not passé to (still) want to 'denaturalise' that which we routinely consider to be 'natural'.

I realise that I may be deemed guilty of a 'performative contradiction' here in at least two respects. First, I appear to be saying that things we call natural are in fact social, thus denying the former an ontological importance I'm according the latter. This evidently implicates me in an asymmetrical realism (or inconsistent materialism) that afflicts many social constructionist claims: in effect, 'the social' becomes synonymous with reality as such. In philosopher John Gray's words, 'By making human beliefs [and practices] the final arbiter of reality', I am apparently 'claiming that *nothing exists unless it appears in human consciousness*' (Gray, 2002: 55, emphasis added). Second, my argument in this book appears to rest on the very assumptions it's calling into question. I seem to be saying, like Latour, that the society–nature dualism is constructed. Yet it seems that I'm using one side of this self-same dualism to suggest that it somehow escapes the process of construction! I have no clever way of dodging the sharp horns of these related dilemmas. Am I so 'trapped' within 'the modern constitution' that I'm using its own categories in order to unsettle them?

Perhaps. However, as I said previously, I don't advance a 'muscular' conception of 'the social' in this book because it's neither my intention nor desire to disregard the importance of those things we have come to regard as natural. They undoubtedly matter (even though I pay limited attention to their multifaceted materiality in this book). As I've also said, the various 'socially constructed' representations of 'nature' that have concerned me in the previous chapters aren't presumed to be *sui generis*. While these representations are hardly 'determined' in content or style by the 'natural' things they depict, they are most certainly the result of practical engagements with the stuff of the world. How could it be otherwise? We can scarcely 'construct' something from nothing, and we're clearly made by the world as much as we make it. Haraway long ago expressed it well, when she said that 'If the world exists for us as "nature", this designates a kind of relationship, an achievement among many actors, not all of them human, not all of them organic' (Haraway, 1992: 297).

Even if this defence is deemed persuasive, some readers may wonder whether there's anything *new* to say about 'the social construction of nature' that hasn't already been said. After all, the literature on the subject is now colossal, and over the years has yielded some 'classic' publications that appear pretty hard to surpass, such as Alexander Wilson's (1992) *The culture of nature*, Kate Soper's (1995) *What is nature?* and Cronon's (1996a) edited book *Uncommon ground: rethinking the human place in nature*. Don't we now know enough – in fact, *more* than enough – about the social constitution of the natural? Well, we certainly know a lot, that much is true; however (and this is my second line of defence), I believe that *Making sense of nature* offers even those *au fait* with the literature upon which it draws some new insights and angles. This is not because it advances startlingly original ideas, resolves old arguments or presents novel evidence; instead, it seeks to combine existing theoretical and empirical research into the 'social construction of nature' in a way that's not really been done before. I 'add value' to the work this book is otherwise reliant upon by virtue of how I assemble, interpret and present it. Let me explain what I mean – lest it's not been clear to those who've waded through the text.

There are now several works in the history of ideas about 'nature' going back to Clarence Glacken's *Traces on the Rhodian shore* (1967), Carolyn Merchant's *The death of nature* (1980), Keith Thomas's *Man and the natural world* (1983), Robert Young's *Darwin's metaphor: nature's place in Victorian culture* (1985) and Joyce Salisbury's (1993) *The medieval world of nature*. Peter Coates's (1998) *Nature: Western attitudes since ancient times* is a slightly more recent addition to this line of inquiry. But *Making sense of nature* is not a work of intellectual history. There are also philosophical investigations into what the term 'nature' means today, such as John Habgood's (2002) *The concept of nature*. But my philosophical skills are more apprentice than craftsman; they're certainly insufficient to sustain a purely conceptual study. Then there are inquiries into the different ways that nature is these days thought about in society at large, such as Bruce Hull's (2006) manifesto for intellectual plurality and tolerance, *Infinite nature*. But my interest in this book has not been the exploration of epistemic diversity for its own sake, or the advocacy of minority perspectives (like 'aboriginal' ones) on the nature and significance of nature.

There are works exploring various strands of 'green' thinking about nature, but this book hasn't been about environmentalism *per se*. Neither have I been interested in how nature is made sense of within one place or a single national culture, much as I admire works like Price's (1999) *Flight maps: adventures with nature in modern America* and my sometime co-editor Bruce Braun's (2002) *The intemperate rainforest: nature, culture and power on Canada's west coast*. Equally, I've not focussed exclusively on the practices of one or other group whose job it is to shape social understandings of nature, such as the documentary makers Derek Bouse investigates in his engaging book *Wildlife films* (2000) or the news journalists discussed in Libby Lester's (2010) *Media and environment*. Likewise, I've not fixated on one category of 'natural phenomena' or on one 'natural environment' at the expense of all others. My topical and geographical perspective has thus been broader than works like *Representing animals* (Rothfels, 2002) and *Picturing tropical nature* (Stepan, 2001), whose titles announce their specific focus. Finally, the nature that's concerned me in these pages has not just been 'external' (what we colloquially call 'the environment') but so-called 'human nature' too. I've considered in one space what works like *Regulating Eden: the nature of order in North American parks* (Hermer, 2002), *The meaning of the gene* (Condit, 1999), *The meaning of race* (Malik, 1996) and *Nature via nurture* (Ridley, 2003) examine separately. Indeed, I've considered 'nature' in the full spectrum of its meanings and referents – in part to remind readers of just how pervasive a subject it is in our mental, emotional and practical lives.

In summary, while the idea that nature is 'socially constructed' is hardly new, my way of operationalising this idea and showing why it is (still) important has, I believe, amounted to roasting what's now an ageing chestnut over some fresh, bright-burning coals. By neither remaining at the level of high theory nor fixating empirically on just one or two elements of what we call 'nature' (animals, trees, genes, glaciers, etc.), I've tried to provide readers with a novel framework of understanding that is comprehensive, convincing and, ultimately, relevant to them making greater sense of their own lives.

ENDNOTES

1 As the historical geographer and environmental historian William Cronon so nicely put, 'Asserting that "nature" is an idea is far from saying . . . that there is no concrete referent out there in the world for the many human meanings we attach to the world "nature"' (Cronon, 1996: 21). He goes on: 'The non-human world is not (just) our creation, but nature is' (ibid.: 458). Note, then, that my argument in this book differs from that advanced by the likes of Emma Marris (in her 2011 book *Rambunctious garden: saving nature in a post-wild world*) and Nigel Dudley (in his 2011 book *Authenticity in nature*). Marris and Dudley echo my critical interpretation of the arguments of Weisman, Ward, McKibben and like-minded commentators; however, the basic philosophical logic of their erstwhile alternative positions remains, in fact, the same. Where Weisman and others decry 'the end of nature', Marris and Dudley enjoin us to work with the 'artificial nature' we've created over the centuries. Both sets of authors take it as read that, once upon a time, 'nature' existed – complete and free from human influence. For similar reasons, my stance on 'nature' should also be distinguished from that of several influential environmental historians, like William Ruddiman, Paul Martin and Callum Roberts. Each of them shows how 'the human impact' on nature precedes the industrial era by hundreds of years. While I don't doubt the veracity of their empirical claims, my point is that their arguments rely on a conception of 'nature' that becomes a quasi-baseline against which anthropogenic change is measured. The same might be said of Tim Flannery's bestseller *Here on earth: a new beginning* (2011), in which he emphasises the role of co-operation and symbiosis in species evolution to argue for a less rapacious, more biophiliac attitude towards the non-human world. So my principal interest is in the *process of semantic naturalisation*, not in what 'nature' actually is or is not. Though the term naturalisation has been used to describe attempts to 'acclimatise' non-native species to new environments, my use of the word differs significantly. For me, it describes conceptual and practical acts of placing boundaries around portions of the world as if those boundaries somehow inhere in that world prior to any human mind perceiving it.

2 I borrow the metaphor of language – indeed all forms of social communication – as a 'tool' from the philosopher Richard Rorty (1999: xxvi). Bruno Latour, in his many influential writings, has on many occasions recommended the term 'articulation', in both a metaphorical and more literal sense, as a way of capturing how we Westerners engage the world through our symbols and deeds. The post-Marxist theorists Ernesto Laclau and Chantal Mouffe did much the same in their influential joint writings in the 1980s. To 'articulate' means both to express (make manifest) and to connect (or to join) together: the double meaning is apposite for all three authors because they argue that to articulate in the former sense *requires* connecting to other people and material things in order to get things done in the world. I agree with all three in one sense but, as I explain in Chapter 2, I also think a 'hang over' of the so-called 'modern constitution' that Latour criticises is that we still operate *as if* 'representation' is something we routinely do. If this were not the case, Latour – and Rorty too – would not have gone to such length in print and in lectures to argue that we do not, in fact,

represent reality after all (or, to be more precise, not in the way we *think* we do). The challenge, I suggest, is not to 'do away' with representation but to think about it in unconventional ways (which is what Latour, Rorty and several others do).

3 Latour argues that we 'moderns' both rely on and yet routinely deny the existence of myriad socio-natural networks and hybrids. This rather paradoxical move on Latour's part reminds me of Soper's judgement: 'Nor can there be any claim to the effect that [the nature–culture] . . . dichotomy is itself conventional that does not tacitly rely for its force on precisely that . . . distinction . . . which is being rhetorically denied' (Soper, 1995: 39). Latour's books *We have never been modern* (1993) and *Politics of nature* (2004) illustrate well the paradox. The first is 'diagnostic', calling upon readers to change spectacles and see what's 'really going on' in the here and now. Yet the second is normative, pulling out all the stops to imagine a future world we have not yet brought into being. When encountering these books for the first time, some readers may ask if Latour is an 'environmentalist' and, if so, of what stripe. The answer is: he is not, at least in most recognised definitions of 'environmentalism', though (1) his work shares the relational worldview of many 'deep green' philosophies and (2) his most recent writings have expressed real alarm about the environmental changes attendant upon future climate change.

REFERENCES*

Abbott, A., 1995. Things of boundaries. *Social Research*, 62 (4), 857–82.
Abramson, R., 1992. Ice cores may hold clues to weather 200,000 years ago. *LA Times*, 2 December: A1.
Adam, J. G., 1948. Les reliques boisées et les essences des savanes dans la zone preforestière en Guniée française. *Bulletin de la Sociétié Botanique Française*, 98, 22–6.
Agarwal, B., 2010. *Gender and green governance*. Oxford: Oxford University Press.
Akera, A., 2007. *Calculating a natural world*. Cambridge, MA: MIT Press.
Anderson, K., 2000. The beast within: race, humanity and animality. *Environment and Planning D*, 18 (3), 301–20.
Anderson, K., 2001. The nature of 'race'. *In:* N. Castree and B. Braun, eds. *Social nature*. Oxford: Blackwell, 64–83.
Anderson, K. and Bows, A., 2011. Beyond 'dangerous' climate change: emission scenarios for a new world. *Philosophical Transactions of the Royal Society* A, 369, 20–44.
Anderson, K. and Bows, A., 2012. A new paradigm for climate change. *Nature Climate Change*, 2, 639–40.
Anderson, L., 2001. *Autobiography*. London: Routledge.
Antilla, L., 2005. Climate of scepticism. *Global Environmental Change*, 15 (3), 338–52.
Antilla, L., 2010. Self-censorship and science. *Public Understanding of Science*, 19 (2), 240–56.
Arendt, H., 1958. *The human condition*. Chicago, IL: University of Chicago Press.
Ashman, K. and Baringer, P. eds., 2001. *After the science wars*. London and New York: Routledge.
Atkinson, D., 2008. *Renewing the face of the Earth*. London: Canterbury Press.
Atwood, M., 2003. *Oryx and Crake*. London: Bloomsbury.
Aubréville, A., 1949. *Climate, forests and desertification in tropical Africa*. Paris: Society of Maritime & Colonial Geography.
Auerbach, E., 1957. *Mimesis: the representation of reality in Western literature* [trans. W. Trask]. New York, NY: Doubleday.
Bacon, M., 2011. *Pragmatism: an introduction*. Cambridge, UK: Polity.
Baetu, T., 2011. Genes after the human genome project. *Studies in History and Philosophy of Science Part C*, 43 (1), 191–201.
Bagemihl, B., 1999. *Biological exuberance: animal homosexuality and natural diversity*. New York, NY: St Martin's Press.
Baker, S., 2001. *Picturing the beast*. Champaign, IL: University of Illinois Press.
Baldwin, A., 2009. Ethnoscaping Canada's boreal forest. *Canadian Geographer*, 53 (4), 427–33.
Ball, T., 1992. New faces of power. *In:* T. Wartenberg, ed. *Rethinking power*. Albany, NY: SUNY Press, 14–31.
Bamshad, M. and Olson, S., 2003. Does race exist? *Scientific American*, December, 78–85.
Barnard, N., 2001. *Turn off the fat genes*. New York, NY: Three Rivers Press.

---

*Books marked in **bold** font are ones whose main arguments and focus overlap strongly with those of this book.

Barnes, T. and Duncan, J., 1992. *Writing worlds*. London: Routledge.
Barnett, C., 2004. Media, democracy and representation. *In:* C. Barnett and M. Low, eds. *Spaces of democracy*. London: Sage, 185–206.
Barnett, C., 2008. Convening publics: the parasitical spaces of public action'. *In:* K. R. Cox, M. Low, and J. Robinson, eds. *The Sage handbook of political geography*. London: Sage, 403–17.
Barnett, C. and Low, M., 2004. Introduction. *In:* C. Barnett and M. Low, eds. *Spaces of democracy*. London: Sage, 1–22.
Barrett, J. ed., 2007. *Transsexual and other disorders of gender identity: a practical guide to management*. Oxford: Radcliffe Publishing.
Barry, A., 2001. *Political machines*. London: Athlone.
Bastian, M., 2012. Fatally confused: telling the time in the midst of ecological crisis. *Environmental Philosophy*, 9 (1), 23–48.
Bateson, G., 2000/1972. *Steps to an ecology of mind*. Chicago, IL: Chicago University Press.
Baudrillard, J., 1994. *Simulacra and simulation* [trans. S. F. Glaser]. Ann Arbor, MI: University of Michigan Press.
Bauman, Z., 1991. *Modernity and the Holocaust*. Cambridge, UK: Polity.
Bauman, Z., 2001. *The individualized society*. Cambridge, UK: Polity.
Becher, T. and Trowler, P., 2001. *Academic tribes and territories*, 2nd ed. Buckingham: Open University Press.
Beck, S., 2011. Moving beyond the linear model of expertise? *Regional Environmental Change*, 11, 297–306.
Beck, U., 1992. *Risk society*. London: Sage.
Becker, H., 1982. *Art worlds*. Berkeley, CA: University of California Press.
Benjamin, A. and McCallum, B., 2008. *A world without bees*. London: Guardian Books.
Bennett, J. and Chaloupka, W., 1993. Introduction: TV dinners and organic brunch. *In:* J. Bennett and W. Chaloupka, eds. *In the nature of things*. Minneapolis, MN: University of Minnesota Press, vii–xx.
Bennett, T. *et al.*, eds., 2005. *New keywords*. Oxford: Blackwell.
Berger, J., 1972. *Ways of seeing*. Harmondsworth, UK: Penguin.
Berger, P. L. and Luckmann, T., 1966. *The social construction of reality*. New York, NY: Anchor Books.
Berlin, I., 1969/2002. *Liberty*. New York, NY: Oxford University Press.
Bernays, E., 1928. *Propaganda*. New York, NY: H. Liveright.
Bernstein, R., 2010. *The pragmatic turn*. Cambridge, UK: Polity.
Besio, K., Johnston, L. and Longhurst, R., 2008. Sexy beasts and devoted mums. *Environment and Planning A*, 40 (5), 1219–35.
Bhaskar, R., 1993. Ontology. *In:* W. Outhwaite, ed. *The dictionary of 20th century social thought*. Oxford: Blackwell, 429–30.
Bhatti, M. *et al.*, 2009. 'I love being in the garden': enchanting encounters in everyday life. *Social and Cultural Geography*, 10 (1), 61–76.
Biagioli, M., 2006. Patent republic: representing inventions, constructing rights and authors. *Social Research*, 73 (4), 1129–72.
Billings, D. and Urban, T., 1982. The socio-medical construction of transsexualism: an interpretation and critique. *Social Problems*, 29, 266–82.
Blok, A., 2011. War of the whales. *Science, Technology and Human Values*, 36 (1), 55–81.
Bluhdorn, I., 2000. *Post-ecologist politics*. London: Routledge.
Blumberg, M., 2008. *Freaks of nature: what anomalies tell us about development and evolution*. Oxford: Oxford University Press.
Bodmer, W. and McKie, R., 1997. *The book of man*. Oxford: Oxford University Press.
Boggs, C., 2000. *The end of politics*. New York, NY: Guilford Press.
Borges, J. L., 1979. The Congress. *In:* J. L. Borges, ed. *The book of sand*. Harmondsworth, UK: Penguin, 20–2.
Borges, J. L., 1981. *A universal history of infamy*. Harmondsworth, UK: Penguin.
Bornstein, K., 1994. *Gender outlaw*. New York, NY: Routledge.
Bostanci, A., 2006. Two drafts, one genome? *Science as Culture*, 15 (3), 183–98.

Botkin, D., 1990. *Discordant harmonies*. Oxford: Oxford University Press.
Bourdieu, P., 1992. *The logic of practice*. Cambridge, UK: Polity.
Bourke, J., 2011. *What it means to be human*. London: Virago.
Bouse, D., 2000. *Wildlife films*. College Station, PA: Penn State University Press.
Boykoff, M., 2009. We speak for the trees. *Annual Review of Environment and Resources*, 34, 431–58.
Boykoff, M., 2011. *Who speaks for the climate?* Cambridge: Cambridge University Press.
Boykoff, M. and Boykoff, J., 2004. Bias as balance. *Global Environmental Change*, 14 (2), 125–36.
Boykoff, M. and Boykoff, J., 2007. Climate change and journalistic norms. *Geoforum*, 38 (9), 1190–204.
Boykoff, M. and Mansfield, M., 2008. 'Ye olde hot air': reporting on human contributions to climate change in the UK tabloid press. *Enviromental Research Letters*, 3, 1–8.
Bowker, S. and Star, S. L., 1999. *Sorting things out*. Cambridge, MA: MIT Press.
Bradley R. S. and Hughes M. K., 1999. Northern hemisphere temperatures during the past millennium: inferences, uncertainties, and limitations. *Geophysical Research Letters*, 26 (6), 759–62.
Braun, B., 2002. *The intemperate rainforest: nature, culture and power on Canada's west coast*. Minneapolis, MN: Minnesota University Press.
Braun, B., 2003. On the raggedy edge of risk. *In:* D. Moore *et al.*, eds. *Race, nature and the politics of difference*. Durham, NC: Duke University Press, 175–203.
Braun, B., 2004. Nature and culture: on the career of a false problem. *In:* J. Duncan *et al.*, eds. *A companion to cultural geography*. Oxford: Blackwell, 151–77.
Braun, B., 2009. Nature. *In:* N. Castree *et al.*, eds. *A companion to environmental geography*. Oxford: Wiley-Blackwell, 19–36.
Braun, B. and Wainwright, J., 2001. Nature, post-structuralism and politics. *In:* N. Castree and B. Braun, eds. *Social nature*. Oxford: Blackwell, 41–63.
Braverman, I., 2012. *Zooland: the institution of captivity*. Stanford, CA: Stanford University Press.
Brockington, D., 2008a. Celebrity conservation: interpreting the Irwins. *Media International Australia*, 127, 96–108.
Brockington, D., 2008b. Powerful environmentalisms: conservation, celebrity and capitalism. *Media, Culture and Society*, 30 (4), 551–68.
Brockington, D., 2009. *Celebrity and the environment*. London: Zed Books.
Brooker, C., 2011. BBC needs more bias on climate. *Sunday Telegraph*, 24 July 2011.
Brown, M. and Rasmussen, C., 2010. Bestiality and the queering of the human animal. *Environment and Planning D*, 28 (2), 158–77.
Brown, W., 2008. *Regulating aversion: tolerance in the age of identity and empire*. Princeton, NJ: Princeton University Press.
Bryant, J. and Oliver, M. eds., 2009. *Media effects*. New York, NY: Routledge.
Budiansky, S., 1996. *Nature's keepers: the new science of nature management*. London: Phoenix Books.
Butler, J., 1993. *Bodies that matter*. New York, NY: Routledge.
Carolan, M., 2006. Science, expertise and the democratization of decision-making. *Society and Natural Resources*, 19 (7), 661–8.
Carolan, M., 2008. From patent law to regulation. *Capitalism, Nature, Socialism*, 17 (5), 749–65.
Carolan, M., 2010. The mutability of biotechnology patents. *Theory, Culture and Society*, 27 (1), 110–29.
Carrington, B., 2010. *Race, sport and politics*. London: Sage.
Carter, J., 2010. Displacing indigenous cultural landscapes: the naturalistic gaze at Fraser Island. *Geographical Research*, 48 (4), 398–410.
Carter, R., 2010. *Climate change: the counter-consensus*. London: Stacey International Publishing.

Carvalho, A., 2007. Ideological cultures and media discourses on scientific knowledge. *Public Understanding of Science*, 16 (2), 223–43.

Castells, M., 1996. *The rise of the network society*. Oxford: Blackwell.

Castells, M., 2007. Communication, power and counter-power in the network society. *International Journal of Communication*, 1 (1), 238–66.

Castree, N., 2005. *Nature*. London: Routledge.

Castree, N., 2009. Review of 'Geographies of nature'. *Progress in Human Geography*, 33, 2.

Castree, N., 2011. Nature. *In:* J. Agnew and J. Duncan, eds. *The companion to human geography*. Oxford: Wiley-Blackwell, 222–40.

Castree, N. and Braun, B., 2001. *Social nature: theory, practice and politics*. Oxford: Blackwell.

Challenger, M., 2011. *On extinction: how we became estranged from nature*. London: Granta Books.

Chilvers, J., 2009. Deliberative and participatory approaches to environment geography. *In:* N. Castree *et al.*, eds. *A companion to environmental geography*. Oxford: Blackwell, 400–17.

Clapp, R., 2004. Wilderness ethics and political ecology: remapping the Great Bear Rainforest. *Political Geography*, 23 (7), 839–62.

Clark, N., 2011. *Inhuman nature*. London: Sage.

Clegg, S. and Haugaard, M. eds., 2009. *Handbook of power*. London: Sage.

Clements, F., 1928. *Plant succession and indicators*. New York, NY: H. W. Wilson.

Clifford, W. K. 1877/1879. The ethics of belief. *In:* L. Stephen and F. Pollock, eds. *WK Clifford: lectures and essays*. London: Macmillan & Co., 1–10.

Clifford, J. and Marcus, G. eds., 1986. *Writing culture*. Berkeley, CA: University of California Press.

Coates, P., 1998. *Nature: Western attitudes since ancient times*. Cambridge, UK: Polity.

Code, L., 2006. *Ecological thinking*. Oxford: Oxford University Press.

Cohen, B., 1963. *The press and foreign policy*. Princeton, NJ: Princeton University Press.

Collier, P. and Horowitz, D. eds. 2005. *The anti-Chomsky reader*. San Francisco, CA: Encounter Books.

Collins, F., 2010. *The language of life*. New York, NY: HarperCollins.

Collins, H. and Evans, R., 2002. The third wave of science studies. *Social Studies of Science*, 32 (2), 235–96.

Collins, H. and Evans, R., 2007. *Rethinking expertise*. Chicago, IL: Chicago University Press.

Commoner, B., 1971. *The closing circle*. New York, NY: Knopf.

Condit, C. M., 1999. *The meanings of the gene*. Madison, WI: University of Wisconsin Press.

Conservation International, 2009. Deforestation, logging and GHG emissions fact sheet. Available from: http://www.conservation.org/Documents/CI_Climate_Deforestation_Logging_Greenhouse_Gas_Emissions_Facts-12-2009.pdf [Accessed 10 October 2012].

Cook, N., 2000. *Music: a very short introduction*. Oxford: Oxford University Press.

Coombe, R. and Herman, A., 2004. Rhetorical virtues: property, speech, and the commons on the world-wide web. *Anthropological Quarterly*, 77 (4), 559–74.

Cooper, R., Kaufman, J. and Ward, R., 2003. Race and genomics. *New England Journal of Medicine*, 348 (12), 1166–70.

Corbett, J., 2006. *Communicating nature*. Washington, DC: Island Press.

Corner, A., 2010. And yet it works. *Times Higher Education*, 22 July, 35–7.

Corning, P., 2011. *The fair society: the science of human nature and the pursuit of social justice*. Chicago, IL: Chicago University Press.

Cowie, F., 2009. *What's within? Nativism reconsidered*. Oxford: Oxford University Press.

Cox, R., 2010. *Environmental communication and the public sphere*, 2nd ed. London: Sage.

Cresswell, T., 1997. *In place/out of place*. Minneapolis, MN: University of Minnesota Press.

Crick, F., 1970. Central dogma of molecular biology. *Nature*, 227 (5258), 561–3.

Cronon, W., 1996a. In search of nature. *In:* W. Cronon, ed. *Uncommon ground: rethinking the human place in nature*. New York, NY: W. W. Norton, 23–68.
Cronon, W. ed., 1996b. *Uncommon ground: rethinking the human place in nature*. New York, NY: W. W. Norton.
Cronon, W. *et al.*, 1996a. Towards a conclusion. *In:* W. Cronon, ed. *Uncommon ground: rethinking the human place in nature*. New York, NY: W. W. Norton, 447–59.
Cronon, W. *et al.*, 1996b. The trouble with wilderness. *Environmental History*, 1 (1), 7–55.
Crouch, C., 2004. *Post-democracy*. Cambridge, UK: Polity.
Cruikshank, J., 2005. *Do glaciers listen?* Vancouver: UBC Press.
Crump, T., 1990. *The anthropology of numbers*. Cambridge: Cambridge University Press.
Daley, P. and O'Neill, D., 1991. Sad is too mild a word. *Journal of Communication*, 41 (4), 42–57.
Darwin, C. 1859/1998. *The origin of species*. Oxford: Oxford University Press.
Davidson, A., 2002. *The emergence of sexuality: historical epistemology and the formation of concepts*. Cambridge, MA: Harvard University Press.
Davis, J. S., 2005. Representing place: 'Deserted isles' and the reproduction of Bikini Atoll. *Annals of the Association of American Geographers*, 95 (3), 607–25.
Davis, J. S., 2007. Scales of Eden: conservation and pristine devastation on Bikini Atoll. *Environment and Planning D*, 25 (2), 213–35.
Davis, S., 1997. *Spectacular nature: corporate culture and the Sea World experience*. Berkeley, CA: University of California Press.
Deakin, R., 2007. *Wildwood*. Harmondsworth, UK: Penguin.
Dean, M., 1999. *Governmentality*. Sage: London.
Delaney, D., 2001. Making nature/marking humans. *Annals of the Association of American Geographers*, 91 (3), 487–503.
Delaney, D., 2003. *Law and nature*. Cambridge: Cambridge University Press.
Demeritt, D., 1998. Science, social constructivism and nature. *In:* B. Braun and N. Castree, eds. *Remaking reality*. New York, NY: Routledge, 173–93.
Demeritt, D., 2001. Being constructive about nature. *In:* N. Castree and B. Braun, eds. *Social nature*. Oxford: Blackwell, 22–40.
Demeritt, D., 2002. What is the social construction of nature? *Progress in Human Geography*, 26 (6), 767–90.
Dempsey, J., 2010. Tracking grizzly bears in British Columbia's politics. *Environment and Planning A*, 42 (5), 1138–56.
Dennett, D., 1991. *Consciousness explained*. Boston and New York: Little, Brown & Co.
Dewey, J., 1973. *Lectures in China, 1919–20*, edited by R. W. Clopson and T.-C. Ou. Honolulu, HI: University of Honolulu Press.
Disch, L., 2008. Representation as 'spokespersonship'. *Parallax*, 14 (3), 88–100.
Dodge, J., 2000. Foreword. *In*: G. Snyder, ed. *The Gary Snyder reader*. New York, NY: Counterpoint Press, xv–xx.
Doel, M., 2010. Representation and difference. *In:* B. Anderson and P. Harrison, eds. *Taking place: non-representational theories and geography*. Aldershot, UK: Ashgate, 117–30.
Donaldson, S. and Kymlicka, W., 2011. *Zoopolis*. Oxford: Oxford University Press.
Dorling, D., 2011. *So you think you know about Britain?* London: Constable and Robinson.
Dorst, A. and Young, C., 1990. *Clayoquot: on the wild side*. Vancouver: Western Canada Wilderness Committee.
Dowd, G., Stevenson, L., and Strong, J., eds. 2006. *Genre matters*. Portland, OR: Intellect Books.
Downey, J., 2007. Participation and/or deliberation? *In:* L. Dahlberg and E. Siapera, eds. *Radical democracy and the internet*. Basingstoke: Palgrave, 108–29.
Douglas, R. M., 2009. The green backlash? *Social Epistemology*, 23 (2), 145–63.
Doyle, J., 2011. *Mediating climate change*. Farnham, UK: Ashgate.
Dudley, N., 2011. *Authenticity in nature*. London: Earthscan.
Dupré, J., 2003. *Darwin's legacy*. Oxford: Oxford University Press.
Earle, C. *et al.* eds., 1996. *Concepts in human geography*. Lanham, MD: Rowman and Littlefield.

du Sautoy, M., 2008. *Symmetry: a journey into the patterns of nature.* New York, NY: HarperCollins.
Duster, T., 2003. Buried alive: the concept of race in science. *In:* A. Goodman *et al.*, eds. *Genetic nature/culture.* Berkeley, CA: University of California Press, 258–77.
Eagleton, T., 1991. *Ideology.* London: Verso.
Eckersley, R., 1999. The discourse ethic and the problem of representing nature. *Environmental Politics*, 8 (2), 24–49.
Ede, S., 2005. *Art and science.* London: I. B. Tauris.
Edelman, M., 1964. *The symbolic uses of politics.* Urbana, IL: University of Illinois Press.
Edwards, A. W. F., 2003. Human genetic diversity: Lewontin's fallacy. *BioEssays*, 25 (8), 798–801.
Edwards, P., 2010. *A vast machine: computer models, climate data and the politics of global warming.* Cambridge, MA: MIT Press.
Elam, M. and Bertilsson, M., 2003. Consuming, engaging and confronting science: the emerging dimensions of scientific citizenship. *European Journal of Social Theory*, 6 (2) 233–51.
Elias, N., 1969. *The civilising process*, vol. 1. Oxford: Blackwell.
Eliot, L., 2009. *Pink brain, blue brain.* New York, NY: Houghton Mifflin Harcourt.
Ellis, R., Waterton, C. and Wynne, B., 2010. Taxonomy, biodiversity and their publics. *Public Understanding of Science*, 19 (4), 497–512.
Emerson, R. W. 1836/2005. *Nature.* Harmondsworth, UK: Penguin.
Entine, J., 2000. *Taboo: why black athletes dominate sports and why we're afraid to talk about it.* Washington, DC: Public Affairs Publications.
Entman, R., 1993. Framing: towards clarification of a fractured paradigm. *Journal of Communication*, 43 (4), 51–8.
Entman, R., 2003. Cascading activation. *Political Communication*, 20 (4), 415–32.
Evans, J. and Hall, S. eds., 1999. *Visual culture: the reader.* London: Sage.
Evans, K., 2007. *Funny weather.* Toronto: Groundwood Books.
Evernden, N., 1992. *The social creation of nature.* Baltimore, MD: Johns Hopkins University Press.
Evernden, N., 2006. *The natural alien.* Toronto: Toronto University Press.
Fairhead, J. and Leach, M., 1996. Rethinking the forest–savanna mosaic. *In:* M. Leach and R. Mearns, eds. *The lie of the land.* Oxford: James Currey, 105–21.
Fairhead, J. and Leach, M., 1998a. *Reframing deforestation.* London: Routledge.
Fairhead, J. and Leach, M., 1998b. 'Representatives of the past. *In:* L. Rival, ed. *The social life of trees.* Oxford: Berg, 253–72.
Falk, R., 2010. What is a gene? Revisited. *Studies in the History and Philosophy of Biological and Biomedical Science*, 41 (4), 396–406.
Farnham, T., 2007. *Saving nature's legacy: origins of the idea of biological diversity.* New Haven, CT: Yale University Press.
Fausto-Sterling, A., 2004. Refashioning race. *Differences*, 15 (3), 1–37.
Feinberg, L., 1992. *Transgender liberation: a movement whose time has come*, New York, NY: World View Forum.
Feinberg, L., 1993/2003. *Stone butch blues.* New York, NY: Firebrand Books, republished by Alyson Books.
Fernandez-Armesto, F., 2004. *So you think you're human?* Oxford: Oxford University Press.
Festenstein, M., 1997. *Pragmatism and political theory.* Chicago, IL: Chicago University Press.
Fish, S., 1980. *Is there a text in this class?* Cambridge, MA: Harvard University Press.
Flannery, T., 2011. *Here on Earth: a new beginning.* Harmondsworth, UK: Penguin.
Fleck, L., 1935/1979. *Genesis and development of a scientific fact* [trans. F. Bradley and T. J. Trenn]. Chicago, IL: Chicago University Press.
Fletcher, M., 2007. Birds do it, bees do it ... *The Times*, 3 January.
Forsman, B., 2010. Unintelligent design: a discussion of Steve Fuller's 'Dissent over descent'. *Philosophy of the Social Sciences*, 40 (4), 446–55.
Foucault, M., 1970. *The order of things.* New York, NY: Pantheon.

Foucault, M., 1973. *The birth of the clinic: an archaeology of medical perception*. Brighton, UK: Tavistock Books.
Foucault, M., 1979. *Discipline and punish*. New York, NY: Pantheon.
Foucault, M., 1980. *Power/knowledge* [trans. C. Gordon]. New York, NY: Pantheon Books.
Foucault, M., 1981. *The history of sexuality*. Harmondsworth, UK: Penguin.
Foucault, M., 1982. The subject and power. *In:* H. Dreyfus and P. Rabinow, eds. *Michel Foucault*. Chicago, IL: Chicago University Press, 145–62.
Foucault, M., 2008. *The birth of biopolitics*. London: Palgrave.
Fox Keller, E., 2000. *The century of the gene*. Cambridge, MA: Harvard University Press.
Franken, A., 2003. *Lies and the lying liars who tell them*. New York, NY: E. P. Dutton.
Fraser, N., 1989. *Unruly practices*. New York, NY: Routledge.
Fraser, N., 1997. *Justice interruptus*. New York, NY: Routledge.
Frayn, M., 2007. *The human touch*. London: Faber & Faber.
Freudenburg, W. and Muselli, V., 2012. Re-examining climate change debates. *American Behavioural Scientist*, DOI: 10.1177.000276421248274.
Frohlick, S. and Johnson, L., 2011. Naturalizing bodies and places. *Annals of Tourism Research*, 38 (3), 1090–109.
Frow, J., 2006. *Genre*. London and New York: Routledge.
Fukuyama, F., 2002. *Our post-human future*. New York, NY: Picador.
Fuller, M., 2005. *Media ecologies*. Cambridge, MA: MIT Press.
Fuller, S., 1988. *Social epistemology*. Bloomington, IN: Indiana University Press.
Fuller, S., 2000. *The governance of science*. Buckingham, UK: Open University Press.
Fuller, S., 2003. *Kuhn versus Popper*. London: Icon Books.
Fuller, S., 2004. The case of Fuller vs Kuhn. *Social Epistemology*, 18 (1), 3–49.
Fuller, S., 2007a. *The knowledge book*. London: Acumen.
Fuller, S., 2007b. *New frontiers in science and technology studies*. Cambridge, UK: Polity.
Fuller, S., 2008. *Dissent over descent*. London: Icon Books.
Fullwiley, D., 2007. The molecularization of race: institutionalizing racial difference in pharmacogenetics practice, *Science as Culture*, 16 (1), 1–30.
Funtowicz, S. and Ravetz, J., 1993. Science for the post-normal age. *Futures*, 25 (7), 739–55.
Galison, P., 2004. Removing knowledge. *Critical Inquiry*, 31 (3), 229–43.
Gallie, W. B., 1956. Essentially contested concepts. *Proceedings of the Aristotelean Society*, 56 (2), 167–98.
Galtung, J. and Ruge, M. H., 1965. The structure of foreign news. *Journal of International Peace Research*, 1 (1), 64–90.
Gamson, W. and Modigliani, A., 1989. Media discourse and public opinion on nuclear power. *American Journal of Sociology*, 95 (1), 1–37.
Gaudillière, J.-P. and Rheinberger, H.-J. eds., 2004. *From molecular genetics to genomics: the mapping cultures of twentieth-century genetics*. New York, NY: Routledge.
Gelbspan, R., 1998. *The heat is on*. Cambridge, MA: Perseus Press.
Gellner, E., 1988. *Plough, sword and book*. Chicago, IL: Chicago University Press.
Gibson-Graham, J.-K., 1996. *The end of capitalism (as we knew it)*. Oxford: Blackwell.
Giddens, A., 1992. *The consequences of modernity*. Cambridge, UK: Polity.
Gieryn, T., 1983. Boundary work and the demarcation of science from non-science. *American Sociological Review*, 48 (6), 781–95.
Gieryn, T., 1995. Boundaries of science. *In:* S. Jasanoff *et al.*, eds. *Handbook of Science and Technology Studies*. London: Sage, 393–444.
Gieryn, T., 1999. *Cultural boundaries of science*. Chicago, IL: Chicago University Press.
Gilbert, W., 1992. Vision of the grail. *In:* D. J. Kevles and L. Hood, eds. *The code of codes*. Cambridge, MA: Harvard University Press, 83–97.
Gilroy, P., 2000. *Against race*. Cambridge, MA: The Belknap Press of Harvard University Press.
Glacken, C., 1967. *Traces on the Rhodian shore*. Berkeley, CA: University of California Press.
Goldfarb, J., 1991. *The cynical society*. Chicago, IL: Chicago University Press.

Golinski, J., 2005. *Making natural knowledge*. Chicago, IL: Chicago University Press.
Gordon, C. ed., 1980. *Power/knowledge*. New York, NY: Pantheon.
Gornitz, V. and NASA, 1985. A survey of anthropogenic vegetation changes in West Africa during the last century. *Climatic Change*, 7, 285–325.
Gottweis, H., 1997. Genetic engineering, discourses of deficiency and the new politics of population. *In:* P. J. Taylor *et al.*, eds. *Changing life*. Minneapolis, MN: University of Minnesota Press, 56–84.
Graber, D., 2003. The media and democracy. *Annual Review of Political Science*, 6, 139–60.
Gray, J., 2002. *Straw dogs: thoughts on humans and other animals*. London: Granta.
Griffiths, P. E. and Stotz, K., 2006. Genes in the post-genomic era. *Theoretical Medicine and Bioethics*, 27 (6), 499–521.
Gross, P. R. and Levitt, N., 1994. *Higher superstition: the academic Left and its quarrels with science*. Baltimore, MD: Johns Hopkins University Press.
Gross, P. R., Levitt, N. and Lewis, M. eds., 1997. *The flight from science and reason*. New York, NY: New York Academy of Sciences.
Grundmann, R., 2012. 'Climate gate' and the scientific ethos. *Science, Technology and Human Values*, DOI: 10.1177/0162243911432318.
Grundmann, R. and Stehr, N., 2012. *The power of scientific knowledge*. Cambridge: Cambridge University Press.
Gutmann, A., 1987. *Democratic education*. Princeton, NJ: Princeton University Press.
Haack, S., 2003. Pragmatism. *In:* N. Bunnin and E. P. Tsui-James, eds. *The Blackwell companion to philosophy*. Oxford: Blackwell, 774–89.
Haas, P. M., 1992. Introduction: epistemic communities and international policy co-ordination. *International Organization*, 46 (1), 1–36.
Habermas, J., 1962/1989. *The structural transformation of the public sphere*. Cambridge, MA: MIT Press.
Habgood, J., 2002. *The concept of nature*. Totnes, UK: Darton, Longman & Todd.
Hacking, I., 1999. *The social construction of what?* Cambridge, MA: Harvard University Press.
Hacking, I., 2002. *Historical ontology*. Cambridge, MA: Harvard University Press.
Hacking, I., 2004. Between Michel Foucault and Erving Goffman. *Economy and Society*, 33 (3), 277–302.
Hagendijk, R. and Irwin, A., 2006. Public deliberation and governance: engaging with science and technology in contemporary Europe. *Minerva*, 44 (2), 167–84.
Halberstam, J., 1992. SkinFlick. *Camera Obscura*, 27 (1), 37–52.
Hall, E., 2003. Reading maps of the genes. *Health and Place*, 9 (2), 151–61.
Hall, S., 1973. *Encoding and decoding in the television discourse*. Birmingham, UK: Centre for Cultural Studies, University of Birmingham.
Hall, S. ed., 1997. *Representation*. London: Sage.
Ham, J. and Senior, M., 1997. *Animal acts: configuring the human in Western history*. New York, NY: Routledge.
Hannaford, I., 1996. *Race: the history of an idea in the West*. Baltimore, MD: Johns Hopkins University Press.
Hannay, A., 2006. *On the public*. London: Routledge.
Hansen, A., 2010. *Environment, media and communication*. London: Routledge.
Haraway, D., 1992. The promises of monsters. *In:* L. Grossberg *et al.*, eds. *Cultural studies*. New York, NY: Routledge, 295–337.
Haraway, D., 2003. *The companion species manifesto*. Chicago, IL: Prickly Paradigm Press.
Haraway, D., 2008. *When species meet*. Minneapolis, MN: Minnesota University Press.
Hardt, M. and Negri, A., 2005. *Multitude*. Harmondsworth, UK: Penguin.
Hardwig, J., 1985. Epistemic dependence. *Journal of Philosophy*, 82 (7), 335–49.
Harley, B., 1992. Reading the maps of the Columbia encounter. *Annals of the Association of American Geographers*, 82 (3), 522–42.
Harré, R., Brockmeier, J. and Mulhausler, P., 1999. *Greenspeak: a study of environmental discourse*. London: Sage.
Harris, C., 1997. *The resettlement of British Columbia*. Vancouver: UBC Press.

Harris, S., 2011. *The moral landscape: how science can determine human values*. London: Bantam Books.
Hartigan, J., 2008. Is race still socially constructed? *Science as Culture*, 17 (2), 163–93.
Hartley, J., 1992. *The politics of pictures*. New York, NY: Routledge.
Harvey, D., 1974. Population, resources and the ideology of science. *Economic Geography*, 50 (2), 256–77.
Harvey, D., 1996. *Justice, nature and the geography of difference*. Oxford: Blackwell.
Haugaard, M., 2012. Rethinking the four dimensions of power. Unpublished manuscript, available from the author, University of Galway, Ireland.
Hausman, B., 1995. *Changing sex*. Durham, NC: Duke University Press.
Hayles, N. K., 1991. Constrained constructivism: locating scientific inquiry in the theatre of representation. *New Orleans Review*, 18 (1), 76–85.
Healy, S., 2007. Deadly dingoes. *Social Studies of Science*, 37 (3), 443–71.
Hemmings, C., 2005. Invoking affect. *Cultural Studies*, 19 (5), 548–67.
Herman, E. and Chomsky, N., 1988. *Manufacturing consent*. New York, NY: Vintage Books.
Hermer, S., 2002. *Regulating Eden: the nature of order in North American parks*. Toronto: University of Toronto Press.
Herschbach, D., 1996. Imaginary gardens with real toads. *In:* P. Gross *et al.*, eds. *The flight from science and reason*. New York, NY: New York Academy of Sciences, 11–30.
Hinchliffe, S., 2007. *Geographies of nature*. London: Sage.
Hind, D., 2010. *The return of the public*. London: Verso.
Hindess, B., 2004. Power, government, politics. *In:* K. Nash, ed. *The Blackwell companion to political sociology*. Oxford: Blackwell, 40–8.
Hirschfeld, L., 1996. *Race in the making: cognition, culture and the child's construction of human kind*. Cambridge, MA: MIT Press.
Hoberman, J., 1997. *Darwin's athletes*. New York, NY: Houghton Mifflin.
Hobson, K. and Neimeyer, S., 2012. What sceptics believe. *Public Understanding of Science*, DOI: 10.1177/0963662511430459.
Hoffman, A., 2011. Talking past each other? Skeptical and convinced logics of climate change. *Organization and Environment*, 24 (1), 3–33.
Holliman, R., 2011. Advocacy in the tail. *Journalism*, 12 (7), 832–46.
Holmberg, T., 2005. Questioning the number of the beast: constructions of humanness in the Human Genome Project narrative. *Science as Culture*, 14 (1), 23–37.
Houllier, F., 2012. Bring more rigour to GM research. *Nature*, 491, 327.
Howard, D., 2011. Why study the history of political thought? *Philosophy and Social Criticism*, 1–3, DOI 10.1177/0914537113992590.
Hubbard, R. and Wald, E., 1993. *Exploding the gene myth*. Boston, MA: Beacon Books.
Hull, B., 2006. *Infinite nature*. Chicago, IL: Chicago University Press.
Hulme, M., 2009. *Why we disagree about climate change*. Cambridge: Cambridge University Press.
Hulme, M., 2010a. 'Four meanings of climate change. *In:* S. Skrimshire, ed. *Future ethics: climate change and political action*. London: Continuum, 37–58.
Hulme, M., 2010b. The IPCC on trial: experimentation continues. *Environmental Research Web Talking Point*, 21 July. Available from: http://environmentalresearchweb.org/cws/article/opinion/43250 [Accessed 20 November 2012].
Hulme, M., 2010c. IPCC: cherish it, tweak it or scrap it? *Nature*, 463, 730–2.
Hulme, M., 2011. Reducing the future to climate: a story of climate determinism and reductionism. *Osiris*, 26, 1.
Hulme, M., 2012. 'Telling a different tale': literary, historical and meteorological reading of a Norfolk heatwave. *Climatic Change*, 113 (10), 5–21.
Hulme, M. and Mahoney, M., 2012. Climate change: what do we know about the IPCC? *Progress in Physical Geography*, 34 (5), 705–18.
Hume, D. 1739/1978. *A treatise of human nature*. Oxford: Clarendon Press.
Ingold, T., 1986. Hunting and gathering as ways of perceiving the environment. *In:* R. Ellen and K. Fukui, eds. *Redefining nature*. Oxford: Berg, 117–55.

Ingold, T., 2008. Bindings against boundings. *Environment and Planning A*, 40 (7), 1796–810.
Ingold, T., 2011. *Being alive: essays on movement, knowledge and description.* London: Routledge.
IPCC, 1995. *Climate change: the science.* Cambridge: Cambridge University Press.
IPCC, 2001. *Climate change: the scientific basis.* Cambridge: Cambridge University Press.
IPCC, 2007. *Climate change: the physical science basis.* Cambridge: Cambridge University Press.
Jacobson, N., 2007. Social epistemology. *Science Communication*, 29 (1), 116–27.
James, W., 1956. The will to believe. *In:* W. James, *The will to believe and other essays in popular philosophy*. New York, NY: Dover Publications.
Jasanoff, S. ed., 2004. *States of knowledge: the co-production of science and social order.* London and New York: Routledge.
Jenkins, S., 2011. The street protests of Western capitals are no Tahrir Square but mere scenery. *The Guardian*, 21 October, 33.
Jennings, N. and Hulme, M., 2010. UK newspaper (mis)representations of the potential for a collapse of the Thermohaline Circulation. *Area*, 42 (4), 444–56.
Jones, R., 2009. Categories, borders and boundaries. *Progress in Human Geography* 33 (2), 174–89.
Jones, S., 1993. *The language of the genes.* London: HarperCollins.
Jones, S. and van Loon, B., 2000. *Introducing genetics.* London: Icon Books.
Jowett, G. and O'Donnell, V., 2005. *Propaganda and Persuasion*, 4th ed. London: Sage.
Kahan, D., 2012. Why we are poles apart on climate change. *Nature*, 488, 255.
Kampfner, J., 2008. *Freedom for sale.* New York, NY: Basic Books.
Kant, I., 1784/1992. An answer to the question: what is Enlightenment? *In:* P. Waugh, ed. *Postmodernism: a reader*. London: Arnold, 89–95.
Kay, L., 1996. *The molecular vision of life.* Oxford: Oxford University Press.
Kelly, D., 2011. *Yuck! The nature and moral significance of disgust.* Cambridge, MA: MIT Press.
Kevles, D., 1995. *In the name of eugenics*, 2nd ed. Cambridge, MA: Harvard University Press.
Kinsman, P., 1995. Landscape, race and national identity. *Area*, 27 (3), 300–10.
Kirby, D., 2008a. Cinematic science: the public communication of science and technology in popular film. *In:* M. Bucchi and B. Trench, eds. *Handbook of public communication of science and technology*. London and New York: Routledge, 41–56.
Kirby, D., 2008b. Hollywood knowledge: communication between scientific and entertainment culture. *In:* D. Cheng *et al.*, eds. *Communicating science in social contexts.* Dordrecht: Springer, 165–80.
Kitcher, P., 1996. *The lives to come: the genetic revolution and human possibilities.* Harmondsworth, UK: Penguin.
Kleinman, D. L. and Suryanarayanan, S., 2012. Dying bees and the social production of ignorance. *Science, Technology and Human Values*, DOI 10.1177/0162243912442575 [Accessed 10 May 2012].
Knorr Cetina, K., 1981. *The manufacture of knowledge.* Oxford: Pergamon Press.
Knorr Cetina, K., 1999. *Epistemic cultures.* Cambridge, MA: Harvard University Press.
Kobayashi, A., 2009. Representation. *In:* R. Kitchin *et al.*, eds. *International encyclopedia of human geography*. Oxford: Elsevier, 378–400.
Koenig, B., 2008. *Revisiting race in a genomic age.* New Brunswick: Rutgers University Press.
Krech III, S., 1999. *The ecological Indian: myth and history.* New York, NY: W. W. Norton.
Kricher, J., 2009. *The balance of nature: ecology's enduring myth.* Princeton, NJ: Princeton University Press.
Kuhn, T., 1962. *The structure of scientific revolutions.* Chicago, IL: Chicago University Press.
Lakoff, G. and Johnson, M., 1980. *Metaphors we live by.* Chicago, IL: Chicago University Press.
Landström, C. *et al.*, 2011. Coproducing flood risk knowledge: redistributing expertise in critical participatory modelling. *Environment and Planning A*, 43 (7), 1617–33.

Lane, S. N. *et al.*, 2011. Doing flood risk science differently: an experiment in radical scientific method. *Transactions of the Institute of British Geographers*, 36 (1), 15–36.

Larsen, B., 2011. *Metaphors for environmental sustainability: redefining our relationship with nature.* New Haven, CT: Yale University Press.

Latour, B., 1987. *Science in action*. Cambridge, MA: Harvard University Press.

Latour, B., 1993. *We have never been modern* [trans. C. Porter]. Cambridge, MA: Harvard University Press.

Latour, B., 1999. *Pandora's hope*. Cambridge, MA: Harvard University Press.

Latour, B., 2004. *Politics of nature*. Cambridge, MA: Harvard University Press.

Latour, B., 2005. *Reassembling the social*. Oxford: Oxford University Press.

Lave, J. and Wenger, E., 1991. *Situated learning*. Cambridge: Cambridge University Press.

Law, J., 1994. *Organizing modernity*. Oxford: Blackwell.

Lawson, H., 1985. *Reflexivity*. La Salle, IL: Open Court Books.

Lazarsfeld, P. and Katz, E., 1955. *Personal influence*. New York, NY: The Free Press.

Leach, M. and Fairhead, J., 2000. Fashioned forest pasts, occluded histories? *Development and Change*, 31 (1), 35–59.

Leach, M. and Mearns, R., 1996. Environmental change and policy. *In:* M. Leach and R. Mearns, eds. *The lie of the land*. Oxford: James Currey, 1–33.

Leach, M. *et al.*, eds., 2005. *Science and citizens: globalisation and the challenge of engagement*. London and New York: Zed Books.

Lee, S., Koenig, B. and Richardson, S. eds., 2008. *Revisiting race in a genomic age*. New Brunswick: Rutgers University Press.

Legates, D. R., Idso, S., Idso, C., Baliunas, S. and Soon, W., 2003. Reconstructing climatic and environmental changes of the past 1000 years: a reappraisal. *Energy and Environment*, 14 (2), 233–96.

Leiss, W., 1974. *The domination of nature*. Boston, MA: Beacon Books.

Leiserowitz, A. *et al.*, 2012. Climate-gate, public opinion and the loss of trust. *American Behavioural Scientist*, DOI: 10.1177/0002764212548272.

Leopold, A., 1949. *A Sand county almanac*. Oxford: Oxford University Press.

Lester, L. and Hutchins, B., 2009. Power games: environmental protest, news media, and the internet. *Media, Culture and Society*, 31 (4), 579–96.

Lester, L., 2010. *Media and environment*. Cambridge, UK: Polity.

Lewis, J. *et al.*, 2008. A compromised fourth estate? *Journalism Studies*, 9 (1), 1–20.

Lewontin, R., 1972. The apportionment of human diversity. *Evolutionary Biology*, 6 (4), 391–98.

Lewontin, R., 1991. *Biology as ideology*. Concord, Ontario: Anansi Books.

Lewontin, R., 2000. *The triple helix: gene, organism and environment*. Cambridge, MA: Harvard University Press.

Liebes, T. and Katz, E., 1990. *The export of meaning*. New York, NY: Oxford University Press.

Lippman, A., 1992. Led (astray) by genetic maps: the cartography of the human genome and health care. *Social Science and Medicine*, 35 (12), 1469–76.

Lippmann, W., 1922. *Public opinion*. New York, NY: Harcourt.

Lippmann, W., 1925. *The phantom public*. New York, NY: Harcourt.

Livingstone, D., 2003. *Putting science in its place*. Chicago, IL: Chicago University Press.

Lloyd, G., 1993. *The man of reason*. London: Routledge.

Lock, A. and Strong, T., 2010. *Social constructionism: sources and stirrings*. Cambridge: Cambridge University Press.

Lomborg, B., 2010. *Smart solutions to climate change*. Cambridge: Cambridge University Press.

Lorenz, K., 1952. *King Solomon's ring*. London: Methuen.

Louv, C., 2005. *Last child in the woods: saving our children from nature deficit disorder*. Chapel Hill, NC: Algonquin Books.

Lovelock, J., 2007. *The revenge of Gaia*. Harmondsworth, UK: Penguin.

Luke, T., 1999. Environmentality as green governmentality. *In:* E. Darier, eds. *Discourses of the environment*. Oxford: Blackwell, 121–51.

Lynas, M., 2011. *The God species: how the planet can survive the age of humans.* London: Fourth Estate.
Mabey, R., 2005. *Nature cure.* London: Chatto & Windus.
Mabey, R., 2008. *Beechcombings: the narrahves of trees.* New York, NY: Vintage.
McAllister, I., McAllister, K. and Young, C., 2007. *The Great Bear Rainforest: Canada's forgotten coast.* Madeira Park: Harbour Publishing.
MacCannell, D., 1976. *The tourist.* Berkeley, CA: University of California Press.
McChesney, R. and Nichols, J., 2002. *Our media not theirs: the democratic struggle against corporate media.* New York, NY: Seven Stories Press.
McCombs, M. *et al.*, 2011. *The news and public opinion: media effects on civic life.* Cambridge, UK: Polity.
MacDonald, F., 2009. Visuality. *In:* N. Thrift *et al.*, eds. *International encyclopedia of human geography.* Amsterdam: Elsevier, 1–7.
McElheny, V., 2010. *Drawing the map of life.* Boston, MA: Basic Books.
McIntyre, S. and McKitrick, R., 2003. Corrections to the Mann *et al.* (1998) proxy data base and northern hemisphere average temperature series, *Energy and Environment,* 14 (6), 751–71.
McKee, R., 2008. Is this the end of hives of activity? *The Observer,* 20 July, Review Section, 12.
McKibben, B., 1989. *The end of nature: humanity, climate change and the natural world.* New York, NY: Random House.
McKibben, B., 1993. *Enough: genetic engineering and the end of human nature.* London: Bloomsbury Books.
McKibben, B., 1992/2006. *The age of missing information.* New York, NY: Penguin.
Maclaurin, J. and Sterelny, K., 2008. *What is biodiversity?* Chicago, IL: Chicago University Press.
McNair, B., 2006. *Cultural chaos: news, journalism and power in a globalised world.* London: Routledge.
Magee, J., 2009. *Art of nature.* London: Natural History Museum.
Magnusson, W. and Shaw, K. eds., 2003. *A political space: reading the global through Clayoquot.* Minneapolis, MN: Minnesota University Press.
Mahony, N., Newman, J. and Barnett, C., 2010. Rethinking the public. *In:* N. Mahony *et al.*, eds. *Rethinking the public.* Bristol: Policy Press, 1–14.
Malik, K., 1996. *The meaning of race.* London: Macmillan.
Malik, K., 2008. *Strange fruit.* Oxford: OneWorld.
Mann, M., 1986. *The sources of social power,* vol. 1. Cambridge: Cambridge University Press.
Mann, M., 1993. *The sources of social power,* vol. 2. Cambridge: Cambridge University Press.
Mann, M., 2012. *The hockey stick and the climate wars.* New York, NY: Columbia University Press.
Mann, M., Bradley, R. S. and Hughes, M. K., 1998. Global-scale temperature patterns and climate forcing over the past six centuries. *Nature,* 392, 779–87.
Mann, M., Bradley, R. S. and Hughes, M. K., 1999. Northern hemisphere temperatures during the past millennium: inferences, uncertainties, and limitations. *Geophysical Research Letters,* 26, 759–62.
Mann, M. *et al.*, 2003. On past temperatures and anomalous late-20th century warmth. *Eos,* 84 (27), 256–7.
Marks, J., 2002. *What it means to be 98% chimpanzee.* Berkeley, CA: University of California Press.
Markwell, K., 2001. An intimate rendezvous with nature? *Tourist Studies,* 1 (1), 39–57.
Marris, E., 2011. *Rambunctious garden: saving nature in a post-wild world.* London: Bloomsbury Books.
Marx, K. and Engels, F., 1846/1976. *The German ideology.* London: Lawrence & Wishart.
Mathews, M., 1994. *The horseman.* New York, NY: Prometheus Books.

M'charek, A., 2005. The mitochondrial Eve of modern genetics. *Science as Culture*, 14 (2), 161–83.
M'charek, A., 2009. Extravagance, or the good and the bad of genetic diversity. *In:* P. Atkinson *et al.*, eds. *The handbook of genetics and society*. London: Routledge, 422–36.
Meadows, D. *et al.*, 1972. *The limits to growth*. London: Pan.
Mels, T., 2002. Nature, home, and scenery: the official spatialities of Swedish national parks. *Environment and Planning D*, 20 (1), 135–54.
Merchant, C., 1980. *The death of nature: women, ecology and the scientific revolution*. New York, NY: Harper & Row.
Merton, R., 1942/1979. The normative structure of science. *In:* R. K. Merton, ed. *The sociology of science: theoretical and empirical investigations*. Chicago, IL: University of Chicago Press, 267–78.
Miller, J. B., 1992. Women and power. *In:* T. Wartenberg, ed. *Rethinking power*. Buffalo, NY: SUNY Press, 240–8.
Mills, S., 1997. *Discourse*. London: Routledge.
Mitchell, T., 1988. *Colonising Egypt*. Cambridge: Cambridge University Press.
Mitchell, T., 2002. *Rule of experts*. Berkeley, CA: University of California Press.
Mittermeier, R. A. *et al.*, 2008. *Climate for life: meeting the global challenge*. Arlington, VA: Conservation International.
Mol, A.-M., 2002. *The body multiple*. Durham, NC: Duke University Press.
Morley, D., 2005. Communication. *In:* T. Bennett *et al.*, eds. *New keywords*. Oxford, Blackwell.
Morris, D., 2005/1967. *The naked ape*. London: Vintage Books.
Morriss, P., 1997. Blurred boundaries. *Inquiry*, 40 (3), 259–90.
Mortimer-Sandiland, C. and Erikson, B. eds., 2010. *Queer ecologies*. Bloomington, IN: Indiana University Press.
Morton, T., 2010. *The ecological thought*. Cambridge, MA: MIT Press.
Moser, S., 2010. Communicating climate change: history, challenges, process and future directions. *WIREs Clim Change*, 2010 (1), 31–53. DOI: 10.1002/wcc.11.
Moss, L., 2003. *What genes can't do*. Cambridge, MA: MIT Press.
Mouffe, C., 2005. *On the political*. London: Routledge.
Mukuza, V., 2011. Linkages, contests and overlaps in the global intellectual property rights regime. *European Journal of International Relations*, 17 (4), 755–76.
Mumford, L., 1967. *Technics and human development*. New York, NY: Harcourt Brace Jovanovich.
Munton, R., 2003. Deliberative democracy and environmental decision-making. *In:* F. Berkhout *et al.*, eds. *Negotiating environmental change*. London: Edward Elgar, 109–36.
Murdoch, J., 1997a. Towards a geography of heterogeneous associations. *Progress in Human Geography*, 21 (3), 321–37.
Murdoch, J., 1997b. Inhuman/nonhuman/human. *Environment and Planning D*, 15 (4), 731–56.
Nabi, R. and M. B. Oliver eds., 2009. *The Sage handbook of media processes and effects*. Thousand Oaks, CA: Sage.
Nagel, T., 2008. Public education and intelligent design. *Philosophy and Public Affairs*, 36 (2), 187–205.
Nash, C., 2000. Performativity in practice. *Progress in Human Geography*, 24 (4), 653–64.
Negt, O. and Kluge, A., 1972/1993. *Public sphere and experience*. Minneapolis, MN: University of Minnesota Press.
Nicholas, R., 1914. État de cultures indigènes, août 1914. Archives Nationales de Conakry, Guinée, 1R12.
Nightingale, V., 2011. *Handbook of media audiences*. Oxford: Wiley-Blackwell.
NIPCC, 2009. *Climate change reconsidered*. Chicago, IL: The Heartland Institute.
NIPCC, 2011. *Climate change reconsidered: interim report*. Chicago, IL: The Heartland Institute.

Norton, B., 1999. Ecology and opportunity. *In:* A. Dobson, ed. *Fairness and futurity*. Oxford: Oxford University Press, 118–50.
Nussbaum, M., 2001. The secret sewers of vice. *In:* S. Bandes, ed. *The passions of law*. New York, NY: New York University Press, 19–62.
Nye, J., 2011. *The future of power*. New York, NY: Perseus Books.
Odysseos, L., 2011. Governing dissent in the Central Kalahari Game Reserve. *Globalizations*, 8 (4), 439–55.
Olson, R., 2008. *Sizzle: a global warming comedy*. Los Angeles, CA: Prairie Starfish Productions.
Olsson, G., 1980. *Birds in egg/eggs in bird*. London: Pion.
Olwig, K., 1996a. Nature: mapping the ghostly traces of a concept. *In:* C. Earle *et al.*, eds. *Concepts in human geography*. Lanham, MD: Rowman & Littlefield, 63–96.
Olwig, K., 1996b. 'Reinventing common nature. *In:* W. Cronon, ed., *Uncommon ground*. New York, NY: W. W. Norton, 379–408.
O'Neill, J., 2001. Representing people, representing nature, representing the world. *Environment and Planning C*, 19 (4), 483–500.
Oreskes, N., 2004. Science and public policy: what's proof got to do with it? *Environmental Sciences and Policy*, 7 (3), 369–83.
Oreskes, N. and Conway, E., 2010. *Merchants of doubt*. New York, NY: Bloomsbury Press.
Organisation for Economic Cooperation and Development (OECD), 1996. *The knowledge-based economy*. Paris: OECD.
Ortner, S., 1974. Is female to male as nature is to culture? *In:* M. Rosaldo and L. Lampere, eds. *Women, culture and society*. Stanford, CA: Stanford University Press, 68–87.
Osborne, P., 1991. Radicalism without limit? *In:* P. Osborne, ed. *Socialism and the limits of liberalism*. London: Verso, 201–21.
Outka, P., 2011. Posthuman/postnatural: ecocriticism and the sublime in Mary Shelley's *Frankenstein*. *In:* S. Lemenager *et al.*, eds. *Environmental criticism for the 21st century*. New York, NY: Routledge, 31–48.
Palmer, C. and Thornhill, R., 2000. *A natural history of rape: biological bases of sexual coercion*. Cambridge, MA: MIT Press.
Parry, S. and Dupré, J. eds., 2010. *Nature after the genome*. Oxford: Wiley-Blackwell.
Pasternak, C., 2007. *What makes us human?* Oxford: One World Books.
Paul, D., 1998. *Controlling human heredity*. Amherst, MA: Humanity Books.
Pauly, D., 2009. Aquacalypse now: the end of fish. *The New Republic*, 28 September.
Pauly, D. *et al.*, 2002. Towards sustainability in world fisheries. *Nature*, 418, 689–95.
Peace, A., 2001. Dingo discourse: constructions of nature and contradictions of capital in an Australian eco-tourist destination. *Anthropological Forum*, 11 (2), 175–94.
Peace, A., 2002. The cull of the wild: dingoes, development and death in an Australian eco-tourist location. *Anthropology Today*, 18 (5), 175–94.
Pearce, F., 2010. *The climate files*. London: Guardian Books.
Pellizzoni, L., 2001. The myth of the best argument. *British Journal of Sociology*, 52 (1), 59–86.
Penrose, J., 2003. When all the cowboys are Indians. *Annals of the Association of American Geographers*, 93 (3), 687–705.
Peters, J. D., 2004. Media and communications. *In:* J. Blau, ed. *The Blackwell companion to sociology*. Oxford: Blackwell, 16–29.
Peterson, A., 2008. Is the new genetics eugenic? *New Formations*, 60, 79–88.
Phillips, D., 2003. *The truth of ecology*. Oxford: Oxford University Press.
Pielke Jr., R., 2007. *The honest broker*. Cambridge: Cambridge University Press.
Pilgrim, S. and Pretty, J., 2010. *Nature and culture: rebuilding the lost connections*. London: Earthscan.
Pinch, S., 2009. Knowledge communities. *In:* R. Kitchin *et al.*, eds. *International encylopedia of human geography*. Oxford: Elsevier, 25–30.
Pinker, S., 2002. *The blank slate*. Harmondsworth, UK: Penguin.
Pinker, S., 2007. *The stuff of thought: language as a window into human nature*. Harmondsworth, UK: Penguin.

Plows, A., 2011. *Debating human genetics*. London: Routledge.
Poole, S., 2007. *Unspeak*. London: Abacus.
Poovey, M., 1998. *A history of the modern fact*. Chicago, IL: Chicago University Press.
Popper, K., 1945. *The open society and its enemies*. New York, NY: Harper & Row.
Porter, T., 1995. *Trust in numbers*. Princeton, NJ: Princeton University Press.
Powell, J., 2011. *The inquisition of climate science*. New York, NY: Columbia University Press.
Prendergast, C., 2000. *The triangle of representation*. New York, NY: Columbia University Press.
Price, J., 1999. *Flight maps: adventures with nature in modern America*. New York, NY: Basic Books.
Prinz, J., 2012. *Beyond human nature: how culture and experience shape our lives*. Harmondsworth, UK: Penguin.
Proctor, J., 1998. The social construction of nature. *Annals of the Association of American Geographers*, 88 (3), 352–76.
Proctor, J. and Pincetl, S., 1996. Nature and the reproduction of endangered space: the spotted owl in the Pacific Northwest and Southern California. *Environment and Planning D*, 14 (4), 683–708.
Prosser, J., 1998. *Second skins*. New York, NY: Columbia University Press.
Prudham, S., 2007. The fictions of autonomous invention. *Antipode*, 39 (3), 406–29.
Public Broadcasting Service, 2006. *African American Lives* [four part documentary first shown on PBS in February 2006]. Arlington, VA: PBS.
Rabinow, P., 1992. Artificiality and enlightenment. *In:* J. Crary and S. Kwinter, eds. *Zone 6: Incorporations*. New York, NY: Zone Books, 234–52.
Ranciére, J., 2005. *On the shores of politics*. London: Verso.
Rapley, C., 2012. Time to raft up. *Nature*, 488, 583–5.
Ray, L., 2004. Civil society and the public sphere. *In:* K. Nash and A. Scott, eds. *The Blackwell companion to political sociology*. Oxford: Blackwell, 219–29.
Raymond, J., 1979, reissued in 1994. *The transsexual empire: the making of the she-male*. Boston, MA: Beacon Books.
Reaka-Kudla, M. L., Wilson, D. E. and Wilson, E. O. eds., 1997. *Biodiversity II: understanding and protecting our biological resources*. National Academies Press: Washington.
Reardon, J., 2004. Decoding race and human differences in a genomic era. *Differences*, 15 (3), 38–65.
Riddell, C., 1980. *Divided sisterhood*. Liverpool: Self-published.
Ridley, M., 2003. *Nature via nurture*. New York, NY: HarperCollins.
Risch, N. *et al.*, 2002. Categorization of humans in biomedical research: genes, race and disease. *Genome Biology*, 3 (7), 2007.1–2007.7.
Ritvo, H., 1995. Border trouble: shifting the line between people and other animals. *Social Research*, 62 (3), 481–500.
Roach, C., 2003. *Mother/Nature*. Bloomington, IN: Indiana University Press.
Robbins, B. ed., 1993. *The phantom public sphere*. Minneapolis, MN: Minnesota University Press.
Robbins, P., 2001a. Fixed categories in a portable landscape: the causes and consequences of land-cover categorization. *Environment and Planning A*, 33 (2), 161–79.
Robbins, P., 2001b. Tracking invasive land covers in India, or why landscapes have never been modern. *Annals of the Association of American Geographers*, 91 (4), 637–59.
Robbins, T. and Maddock, T., 2000. Interrogating land cover categories: towards participatory change analysis in GIS. *Cartography and Geographic Information Science*, 27 (4), 371–84.
Roof, J., 2007. *The poetics of DNA*. Minneapolis, MN: Minnesota University Press.
Rorty, A. O., 1992. Power and powers. *In:* T. Wartenberg, ed. *Rethinking power*. Albany, NY: SUNY Press, 1–13.
Rorty, R., 1979. *Philosophy and the mirror of nature*. Princeton, NJ: Princeton University Press.
Rorty, R., 1999. *Philosophy and social hope*. Harmondsworth, UK: Penguin.

Rosaldo, M. and Lamphere, L. eds., 1974. *Woman, culture and society*. Stanford: Stanford University Press, 67–87.
Rose, G., 2007. *Visual methodologies*, 2nd ed. London: Sage.
Rose, H., 1998. Moving on from both state and consumer eugenics? *In:* B. Braun and N. Castree, eds. *Remaking reality*. London: Routledge, 84–99.
Rose, N., 1998. *Inventing ourselves*. Cambridge: Cambridge University Press.
Rose, N., 1999. *Powers of freedom*. Cambridge: Cambridge University Press.
Rose, N., 2001. The politics of life itself. *Theory, Culture and Society*, 18 (6), 1–30.
Rose, S., Kamin, L., and Lewontin, R., 1984. *Not in our genes*. Harmondsworth, UK: Penguin.
Rosenberg, N. *et al.*, 2002. Genetic structure of human populations. *Science*, 298, 2381–4.
Rosoff, P., 2010. In search of the mommy gene. *Science, Technology and Human Values*, 35 (2), 200–43.
Rossiter, D., 2004. The nature of protest. *Cultural Geographies*, 11 (2), 139–64.
Rothfels, N. ed., 2002. *Representing animals*. Bloomington, IN: Indiana University Press.
Rothman, B. K., 1998. *Genetic maps and the human imagination*. New York, NY: W. W. Norton.
Rousseau, J.-J., 1762. *The social contract*. Amsterdam: Mar Michel Rey.
Runciman, D. and Vieira, M., 2008. *Representation*. Cambridge, UK: Polity Press.
Rutherford, S., 2007. Green governmentality. *Progress in Human Geography*, 31 (3), 291–307.
Rutherford, S., 2011. *Governing the wild*. Minneapolis, MN: University of Minnesota Press.
Rutter, M., 2005. *Genes and behaviour: the nurture–nature interplay explained*. Oxford: Blackwell.
Sahlins, M., 1976. *The use and abuse of biology*. Ann Arbor, MI: Michigan University Press.
Said, E., 1978. *Orientalism*. New York, NY: Random House.
Salisbury, J., 1993. *The medieval world of nature*. New York: Garland Press.
Salter, M., 2011. Places everyone! *Political Geography*, 30 (1), 66.
Sarewitz, D., 2004. How science makes environmental controversies worse. *Environmental Science and Policy*, 7 (3), 385–403.
Sarewitz, D., 2011. The voice of science: let's agree to disagree. *Nature*, 478, 7.
Sarich, V. and Miele, F., 2004. *Race: the reality of human differences*. Boulder, CO: Westview Press.
Sayer, J. *et al.*, 1992. *Conservation atlas of tropical forests: Africa*. London: Macmillan.
Schafer, M., 2010. Taking stock: a meta-analysis of studies of the media's coverage of science. *Public Understanding of Science*, 1–14, DOI: 10.1177/0963662510387559.
Schnattscheider, E. E., 1975. *Semi-sovereign people*. Hinsdale, IL: Dryden Press.
Schneider, S. and Mesirow, L., 1976. *The Genesis strategy*. New York, NY: Plenum Press.
Scholes, R., 1982. *Semiotics and interpretation*. New Haven, CT: Yale University Press.
Scott, J., 1985. *Weapons of the weak: everyday forms of peasant resistance*. New Haven, CT: Yale University Press.
Scott, J., 1991. The evidence of experience. *Critical Inquiry*, 17, 773–97.
Scott, J., 2001. *Power*. Cambridge, UK: Polity Press.
Selinger, E. and Crease, R. eds., 2006. *The philosophy of expertise*. New York, NY: Columbia University Press.
Sen, A., 1999. *Development as freedom*. Oxford: Oxford University Press.
Sesardic, N., 2010. Race: a social destruction of a biological concept. *Biology and Philosophy*, 25 (2), 143.
Shapiro, M., 1988. *The politics of representation: writing practices in biography, photography, and policy analysis*. Madison, WI: University of Wisconsin Press.
Shearmur, J., 2010. Steve Fuller and intelligent design. *Philosophy of the Social Sciences*, 40, 433–45.
Shepherd, R., 2002. Commodification, culture and tourism. *Tourist Studies*, 2 (2), 183–201.
Sibley, D., 1995. *Geographies of exclusion*. London: Routledge.

Sibley, D., 2003. Psycho-geographies or rural space and practices of exclusion. *In:* P. Cloke, ed. *Country visions.* Harlow, UK: Prentice Hall, 218–31.
Skinner, D., 2006. Racialized futures: biologism and the changing politics of identity. *Social Studies of Science*, 36 (3), 459–88.
Slater, C., 2002. *Entangled Edens: visions of the Amazon.* Berkeley, CA: University of California Press.
Slater, C. ed., 2003. *In search of the rainforest.* Durham, NC: Duke University Press.
Sletto, B., 2002. Boundary making and regional identities in a globalized environment: rebordering the Nariva Swamp, Trinidad. *Environment and Planning D*, 20 (2), 183–208.
Smedley, A. and Smedley, B., 2005. Race as biology is fiction. *American Psychologist*, 60 (1), 16–26.
Smith, B. and Mark, D. M., 2003. Do mountains exist? Towards an ontology of landforms. *Environment and Planning B: Planning and Design*, 30 (3), 411–27.
Smith, N. and Katz, C., 1993. Grounding metaphor. *In:* M. Keith and S. Pile, eds. *Place and the politics of identity*. London: Routledge, 67–83.
Snyder, G., 1992. *No nature.* New York, NY: Pantheon Books.
Sokal, A., 1996a. Transgressing the boundaries: towards a transformative hermeneutics of quantum gravity. *Social Text*, 46/47, 217–52.
Sokal, A., 1996b. A physicist experiments with Cultural Studies. *Lingua Franca*, May/June, 32–36.
Sokal, A. and Bricment, J., 1998. *Fashionable nonsense: postmodern intellectuals' abuse of science.* New York, NY: Picador.
Solomon, R., 1993. *The passions.* Indianapolis, IN: Hackett Publishing.
Soon, W. and Baliunas, S., 2003. Proxy climatic and environmental changes of the past 1000 years. *Climatic Research*, 23 (1), 89–110.
Soper, K., 1995. *What is nature?* Oxford: Blackwell.
Soule, M. and Lease, G. eds., 1995. *Reinventing nature? Responses to postmodern deconstruction.* Washington, DC: Island Press.
Spence, M., 1999. *Dispossessing the wilderness.* Oxford: Oxford University Press.
Stafford, S., 2001. Epistemology for sale. *Social Epistemology*, 15 (3), 215–30.
Steeves, H. P. ed., 1999. *Animal others.* New York, NY: SUNY Press.
Stepan, N., 2001. *Picturing tropical nature.* Ithaca, NY: Cornell University Press.
Stoller, S., 2009. Phenomenology and the post-structural critique of experience. *International Journal of Philosophical Studies*, 17 (5), 707–37.
Stone, S., 1991. The empire strikes back. *In:* J. Epstein and K. Straub, eds. *Body Guards.* New York, NY: Routledge, 280–304.
Stotz, K. *et al.*, 2004. How scientists conceptualize genes. *Studies in History and Philosophy of Biological and Biomedical Sciences*, 35 (5), 647–73.
Stryker, S., 2006. (De)subjugated knowledges. *In:* S. Stryker and S. Whittle, eds. *The transgender studies reader.* New York, NY: Routledge, 1–18.
Stryker, S. and Whittle, S. eds., 2006. *The transgender studies reader.* New York, NY: Routledge.
Sturken, M. and Cartwright, L., 2001. *Practices of looking.* New York, NY: Oxford University Press.
Supreme Court of Canada. 2011. Available from: http://scc.lexum.org/en/2004/2004scc34/2004scc34.html [Accessed 10 December 2011].
Swyngedouw, E., 2007. Impossible/undesirable sustainability and the post-political condition. *In:* J. R. Krueger and D. Gibbs, eds. *The sustainable development paradox.* New York, NY: Guilford Press, 13–40.
Swyngedouw, E., 2010. Apocalypse forever? Post-political populism and the spectre of climate change. *Theory, Culture and Society*, 27 (2–3), 213–32.
Szersynski, B., Heim, W. and Waterton, C. eds., 2003. *Nature performed.* Oxford: Blackwell.
Takacs, J., 1996. *The idea of biodiversity.* Baltimore, MD: Johns Hopkins University Press.
Talisse, R. and Aikin, S., 2008. *Pragmatism: a guide for the perplexed.* London and New York: Continuum.

Taussig, K.-S. *et al.*, 2003. Flexible eugenics. *In:* A. Goodman *et al.*, eds. *Genetic nature/culture*. Berkeley, CA: University of California Press, 58–76.
Taverne, D., 2005. *The march of unreason*. Oxford: Oxford University Press.
Tegelberg, M., 2010. Hidden sights. *International Journal of Cultural Studies*, 13 (5), 491–509.
Templeton, A., 2003. Human races in the context of recent human evolution. *In:* A. Goodman *et al.*, eds. *Genetic nature/culture*. Berkeley, CA: University of California Press, 234–57.
Thomas, E., 1909. *The south country*. London: J. M. Dent & Co.
Thomas, K., 1983. *Man and the natural world*. Oxford: Oxford University Press.
Thompson, J., 1995. *The media and modernity*. Stanford, CA: Stanford University Press.
Thompson, J., 2004. The media and politics. *In:* K. Nash and A. Scott, eds. *The Blackwell companion to political sociology*. Oxford: Blackwell, 173–82.
Thompson, J., 2005. *The media and modernity*. Cambridge, UK: Polity Press.
Thompson, J., 2011. Shifting boundaries of public and private life. *Theory, Culture and Society*, 28 (4), 49–70.
Thompson, M., Ellis, R. and Wildavsky, A., 1990. *Cultural theory*. Boulder, CO: Westview Press.
Thrift, N., 2007. *Non-representational theory*. London and New York: Routledge.
Traweek, S., 1988. *Beamtimes and lifetimes*. Cambridge, MA: Harvard University Press.
Tudge, C., 2006. *The secret life of trees*. London: Allen Lane.
Turner, G., 2004. *Understanding celebrity*. London: Sage.
United Nations Environment Programme (UNEP), 2009. Vital forest graphics. Available from: http://www.unep.org/forests/AboutForests/tabid/29845/Default.aspx [Accessed 20 September 2011].
Urbinati, N., 2009. Unpolitical democracy. *Political Theory*, 38 (1), 65–92.
Urry, J., 2002. *The tourist gaze*, 2nd ed. London: Sage.
Vaidhyanathan, S., 2006. Critical information studies. *Cultural Studies*, 20 (2–3), 292–315.
Valentin, G. 1893. Rapport sur la Residence du Kissi. Archives Nationales de Senegal, Dakar, 1G188.
van Dooren, T., 2007. Terminated seed. *Science as Culture*, 16 (1), 71–94.
van Dooren, T., 2008. Inventing seed: the nature(s) of intellectual property in plants. *Society and Space*, 26 (4), 676–97.
van Dooren, T., 2010. Biopatents and the problem/promise of genetic leaks. *Capitalism, Nature, Socialism*, 21 (2), 43–63.
van Fraasen, B., 2010. Scientific representations: paradoxes of perspective. *Analysis Reviews*, 70, (3), 511–14.
van Hilvoorde, J. *et al.*, 2007. Flopping, klapping and gene doping: dichotomies between 'natural' and 'artificial' in elite sport. *Social Studies of Science*, 37 (1), 173–200.
Vines, G., 1999. Queer creatures. *New Scientist*, 2198, 7 August, 32–41.
Wade, N., 2002. Gene study identifies 5 main human populations linking them to geography. *New York Times*, 20 December, A37.
Wade, P., 2002. *Race, nature and culture*. London: Pluto.
Waldby, C., 2001. Code unknown: histories of the gene. *Social Studies of Science*, 31 (5), 779–91.
Ward, B. and Dubos, R., 1972. *Only one Earth*. Harmondsworth, UK: Penguin.
Ward, P., 2010. *The flooded earth: our future in a world without ice caps*. Boston, MA: Basic Books.
Warner, M., 1991. Introduction: fear of a queer planet. *Social Text*, 9 (4), 3–17.
Washington, H. and Cook, J., 2011. *Climate change denial: heads in the sand*. London: Earthscan.
Weiland, S., 2007. The power of nature and the nature of power. *Nature and Culture*, 2 (1), 67–86.
Weinberg, S., 1996. Sokal's hoax. *New York Review of Books*, 8 August, 15.
Weisman, A., 2008. *The world without us*. New York, NY: Picador.
Westbrook, R., 2005. *Democratic hope*. Ithaca, NY: Cornell University Press.

Wetherell, M. and Potter, J., 1992. *Mapping the language of racism*. New York, NY: Columbia University Press.
Whatmore, S., 1999. Hybrid geographies. *In:* D. Massey *et al.*, eds. *Human geography today*. Cambridge, UK: Polity, 3–21.
Whatmore, S. J. and Landström, C., 2011. Flood apprentices: an exercise in making things public. *Economy and Society*, 40 (4), 582–610.
Wheeler, M., 2012. The democratic worth of celebrity politics in an era of late modernity. *British Journal of Politics and International Relations*, 14 (3), 407–22.
White, E., 1988. *The beautiful room is empty*. London: Pan.
White, F., 1983. *The vegetation of Africa*. Paris: UNESCO.
Whitehead, A. N., 1934. *Nature and life*. Chicago, IL: Chicago University Press.
Whitmarsh, I. and Jones D. eds., 2010. *What's the use of race? Modern governance and the biology of difference*. Cambridge, MA: MIT Press.
Willems-Braun, B., 1997. Buried epistemologies. *Annals of the Association of American Geographers*, 87 (1), 3–31.
Williams, R., 1973. *The country and the city*. New York, NY: Oxford University Press.
Williams, R., 1976. *Keywords*. London: Fontana.
Williams, R., 1977. *Marxism and literature*. Oxford: Oxford University Press.
Williams, R., 1980. *Problems of materialism and culture*. London: Verso.
Williamson, J., 1978. *Decoding advertisements*. London: Boyars.
Wilson, A., 1992. *The culture of nature*. Oxford: Blackwell.
Wilson, E. O., 1992. *The diversity of life*. Cambridge, MA: Harvard University Press.
Wilson, E. O. ed., 1988. *Biodiversity*. Washington, DC: National Academy Press.
Winterson, J., 2011. *Why be happy when you could become normal?* London: Jonathan Cape.
Wise, S., 2000. *Rattling the cage*. New York, NY: Perseus Books.
Wolfe, A., 1993. *The human difference*. Berkeley, CA: University of California Press.
Wood, D., 2010. *Rethinking the power of maps*. New York, NY: Guilford Press.
Wood, D. and Fels, J., 2008. *The nature of maps*. Chicago, IL: Chicago University Press.
World Bank, 1998. *Knowledge for development*. Washington, DC: World Bank.
World Wildlife Fund (WWF), 2012a. *Earth book*. Available from: http://assets.wwf.org.uk/downloads/lpr2012_online_single_pages_11may2012.pdf [Accessed 21 August 2012].
World Wildlife Fund (WWF), 2012b. *Living Planet Report*. Available from: http://awsassets.panda.org/downloads/1_lpr_2012_online_full_size_single_pages_final_120516.pdf [Accessed 21 August 2012].
Wuthnow, R., 1989. *Communities of discourse: ideology and social structure in the Reformation, the Enlightenment and European socialism*. Cambridge, MA: Harvard University Press.
Wynne, B., 1992. Misunderstood misunderstanding. *Public Understanding of Science*, 1 (3), 281–304.
Wynne, B., 1996. May the sheep safely graze? *In:* S. Lash *et al.*, eds. *Risk, environment and modernity*. London: Sage, 44–83.
Yearley, S., 2012. Citizen engagement with the politics of air quality: lessons for social theory, science studies and environmental sociology. *In:* R. Lidskog and G. Sundqvist, eds. *Governing the air*. Cambridge, MA: MIT Press.
Young, R., 1985. *Darwin's metaphor: nature's place in Victorian culture*. Cambridge: Cambridge University Press.
Zizek, S., 2002. A plea for Leninist intolerance. *Critical Inquiry*, 28 (2), 542–66.

INDEX